CARBON MONOXIDE

Edited by

David G. Penney, Ph.D.

School of Medicine
Wayne State University
Detroit, Michigan

CRC Press
Boca Raton New York London Tokyo

Library of Congress Cataloging-in-Publication Data

Carbon monoxide/edited by David G. Penney.
p. cm.
Includes bibliographical references and index.
ISBN 0-8493-4796-3
1. Carbon monoxide—Toxicology. 2. Carbon monoxide—Metabolism.
3. Carbon monoxide—Physiological effect. I. Penney, David G.
RA1242.C2C37 1996
615.9′1—dc20 95-53854
CIP

Direct all inquiries to CRC Press, Inc., 2000 Corporate Blvd., N.W., Boca Raton, Florida 33431.

International Standard Book Number 0-8493-4796-3
Library of Congress Card Number 95-53854
Printed in the United States of America 1 2 3 4 5 6 7 8 9 0
Printed on acid-free paper

FOREWORD

This volume addresses many aspects of current issues involving carbon monoxide (CO) measurement, physiology, toxicology, behavior, and treatment. Outside atmospheric and domestic CO levels are examined in light of both regulations designed to lower emissions and the recent introduction of simple, inexpensive CO detection devices for home and business. The effects of CO on developing and adult animals are examined, beginning at the organ level and progressing to the cellular level, where new exciting information is emerging. The neural and behavioral effects of CO, primarily during acute exposure, are discussed. The management of CO poisoning using both conventional modalities and hyperbaric oxygen therapy is discussed, as well as experimental strategies that might protect cerebral function.

EDITOR

David G. Penney, Ph.D., is Professor of Physiology and Adjunct Professor of Occupational and Environmental Health in the School of Medicine, Wayne State University, Detroit Michigan. He is also Director of General Surgical Research at Providence Hospital, Southfield, Michigan, where he directs the scholarly activities of surgical residents and attending staff physicians.

Dr. Penney obtained his B.Sc. from Wayne State University in 1963, and his M.Sc. and Ph.D. from UCLA in 1966 and 1969, respectively. Before coming to Wayne State University in 1977, he was a faculty member at the University of Illinois, Chicago.

Dr. Penney's current research interests include carbon monoxide and cyanide poisoning in toxicology; normal and abnormal heart growth and hypertension in cardiology; and strategies for lessening the neural damage of acute cerebral ischemia and hypoxia in stroke research. He has a special interest in the medical/legal aspects of carbon monoxide toxicology.

During the past year, Dr. Penney has turned his energies increasingly to information technology in medical education. He is a member of the school's Associate Dean's Core Curricular Team and also heads the Dean's Office of Computer-Assisted Instruction. Major projects in this area have included the development of new computer teaching facilities in the Medical Library, and electronic faculty notes, Virtual Classroom, and other World-Wide Web-based teaching aids. He is also Director of the Medical Physiology course, a requirement for all first year medical students.

Dr. Penney's published work includes over 100 research papers, ten or more review articles, various book chapters, and several books in print or in press.

CONTRIBUTORS

Vernon A. Benignus, Ph.D.
Health Effects Research Laboratory
Human Studies Division
U.S. Environmental Protection Agency
Research Triangle Park, North Carolina

Darien W. Bradford, M.D.
Department of Surgery
The University of Texas Medical Branch
Division of Physiology
Galveston, Texas

Robert S. Fitzgerald, Ph.D.
Department of Environmental Health Sciences
The Johns Hopkins Medical Institutions
Baltimore, Maryland

Mark A. Helfaer, M.D.
Department of Anesthesiology and Critical Care Medicine
The Johns Hopkins University School of Medicine
Baltimore, Maryland

Masayuki Hiramatsu, Ph.D.
Department of Chemical Pharmacology
Faculty of Pharmaceutical Sciences
Meijo University
Nagoya, Japan

Tsutomu Kameyama, Ph.D.
Department of Chemical Pharmacology
Faculty of Pharmaceutical Sciences
Meijo University
Nagoya, Japan

Eric P. Kindwall, M.D.
Departments of Toxicology and Pharmacology
and
Plastic and Reconstructive Surgery
Director, Hyperbaric Unit
Medical College of Wisconsin
Milwaukee, Wisconsin

Toshitaka Nabeshima, Ph.D.
Department of Neuropsychopharmacology and Hospital Pharmacy
Nagoya University School of Medicine
Nagoya, Japan

Dieter Pankow, M.D.
Institute of Environmental Toxicology
Faculty of Medicine
Martin Luther University
Halle, Germany

David G. Penney, Ph.D.
Departments of Physiology
and
Occupational and Environmental Health
Wayne State University
School of Medicine
Detroit, Michigan

Claude A. Piantadosi, M.D.
Department of Medicine
Division of Pulmonary and Critical Care Medicine
Duke University Medical Center
Durham, North Carolina

Nanduri R. Prabhakar, Ph.D., D.Sc.
Department of Physiology and Biophysics
Case Western Reserve University School of Medicine
Cleveland, Ohio

Stephen R. Thom, M.D., Ph.D.
Department of Environmental Medicine
University of Pennsylvania School of Medicine
Philadephia, Pennsylvania

Peter Tikuisis, Ph.D.
Defence and Civil Institute of Environmental Medicine
North York, Ontario, Canada

Daniel L. Traber, Ph.D.
Department of Anesthesiology
Investigational Intensive Care Unit
The University of Texas
Medical Branch
Galveston, Texas

Richard J. Traystman, Ph.D.
Department of Anesthesiology and Critical Care Medicine
The Johns Hopkins University
School of Medicine
Baltimore, Maryland

Roger L. Wabeke, M.Sc., M.Sc.Ch.E., C.I.H., CH.M.M.
Department of Family Medicine
Division of Occupational and Environmental Medicine
Wayne State University School of Medicine
Detroit, Michigan

Suzanne R. White, M.D.
Children's Hospital Regional Poison Control Center
and Fellowship Director
Medical Toxicology
Wayne State University
School of Medicine
Detroit, Michigan

Contents

CHAPTER 1

CARBON MONOXIDE ANALYSIS

Roger L. Wabeke

CONTENTS

1 HISTORY OF CARBON MONOXIDE (CO) TESTING AND ANALYSIS

Clearly, the first "instruments" for "testing" the atmosphere for CO gas must have been human beings. With the advent of fire, it is not hard to imagine that as primitive mankind extended its social boundaries into colder and colder climates, or perhaps in retreating from advancing glaciers, people

0-8493-4796-3/96/$0.00+$.50

brought their fires into caves, dens, and other confined spaces. When some did not survive the night due to a smoldering cooking or heating fire generating lethal levels of CO gas, those who did survive could reason, after trial and fatal error, that failure to extinguish hot coals in poorly ventilated caves was extinguishing lives as well. If one snuffs out the fire before retiring, the lives of one's co-dwellers will be spared. This was a tragic yet serendipitous finding in early science and inhalation toxicology. Chemical asphyxiations from cooking fire gas inhalation became, perhaps, the first generation of air sampling indicators for CO gas.

A variant of this type of air "sampling" was skin color: if a cave dweller's skin became cherry red, maybe it was time to seek fresh air. If the complexion later became cyanotic, nobody should enter this cave. Red and blue skin hues were early qualitative colorimetric parameters of CO intoxication.

The body count most certainly must have been very high. It remains high today with many needless, often multiple, fatalities reported worldwide from poison control centers, hospital morgues, and workplace, home, and community accident investigations. In the author's experience, each of the 30 or more CO asphyxiation fatalities he has investigated over the years was easily preventable by simple industrial hygiene engineering controls that have been practiced for decades.

The second generation of air sampling instruments, after humans, was a biological surrogate: the bird. Caged canaries were brought into the European mines of the Industrial Revolution and placed near the miners. The logic was simple: If the bird died, it was time to get out. The biological theory was as simple: The higher metabolic rate of the tiny bird should cause it to succumb to CO chemical asphyxiation well before a worker should. One could argue that by the time the bird died, a miner could be at sublethal carboxyhemoglobin levels and virtually incapable of self-rescue. Japanese waltzing mice were also used with the same premise as the little yellow birds: Dead or convulsing mouse = time to leave for fresh air (and take living mice with you)!

The third generation of air sampling instruments for CO gas was exemplified by a relatively crude device: the miner's flame safety lamp. Detonations and combustion within the mines, especially coal mines, released CO gas among other products of incomplete combustion. The flame safety lamp, fueled by kerosene, had a flashback arrestor and a glass chimney calibrated with three etchings on the glass window. Normally, each miner carried his own nonsparking brass lamp.

This device was calibrated in outside air with the tip of the flame at the center etching. If the flame tip dropped below the lower etching while in the mine, this was an indication that there was insufficient oxygen to support life for much longer — that is, it was time to evacuate the mine. If the flame tip reached the upper etching on the glass chimney, combustible gases were accumulating and, sooner or later, with a source of ignition there will be an explosion — time to seek sunlight. As primitive as the device was, it undoubtedly saved many lives in poorly ventilated mines and shafts where there were gas accumulations of methane, CO, carbon dioxide, hydrogen sulfide, nitrogen (i.e., decreased oxygen concentrations), and other noxious and flammable gases. However, because the lower limit of explosibility of CO gas is 12.5% (125,000 ppm, or parts of gas per million parts of air), flame safety lamps arguably did little to protect miners from this gas alone (NIOSH, 1990). Significantly, this little lamp, as used in European and American mines, was the first semiquantitative device used to regularly monitor toxic, explosive gases with industrial hygiene significance in workplace atmospheres.

Until about 1910, little was done to accurately quantify CO in gas mixtures, whether in work environments, automobile exhaust gases, the home, industrial stacks, or other smoky places. The need became more pressing. Glass bottles, rubber bladders, evacuated flasks, and other rigid containers were used to transport environmental gas samples to laboratories for quantitative analysis; hence, the fourth generation of sampling and analysis for this chemical asphyxiant gas. Air and other gas samples containing CO were analyzed by several new techniques, including volumetric and gravimetric methods, as will be discussed later.

However, as more and more adverse health and toxic effects of CO were being reported in the industrial hygiene and medical literature, the need to reliably measure this gas in air and blood gases at lower and lower levels became apparent. In part, this was driven by the legitimate scientific inquiries into relating various inhaled doses of CO to corresponding levels of COHb and clinical signs and symptoms. Toxicologists and forensic pathologists questioned what the inhaled doses of CO might have been prior to testing postmortem blood specimens.

The fifth generation of instruments and devices for measuring CO developed in the 1950s. These mainly were direct-reading instruments that gave "real-time" measurements with varying degrees of accuracy and precision. Most were either electronic devices or colorimetric instruments that relied upon the chemical properties of CO known to gas chemists for years. Major instruments used in this generation are described later. Refinements of the principles applied for some of these early tools are with us today in the sixth generation of CO monitoring.

The sixth generation of instruments for measuring CO in air and other gas mixtures relies greatly on refinements of the basic principles established earlier by other industrial hygiene and atmospheric science chemists. The most significant advancements in this generation of analysis are the development of passive air sampling devices, considerably smaller instruments, "real-time" measurements, and lower limits of detection. Many of these newer instruments do not require air sampling pumps in the traditional sense, but rely instead on passive diffusion of CO molecules through a permeable membrane or solution to effect an electronic instrument response. This breakthrough permitted much more representative breathing zone air samples to be taken, for example, of many workers throughout an entire workshift. Moreover, this principle allows devices to be installed in potentially occupied spaces to serve as an alert to increasing amounts of CO in the atmosphere. Such a device, as with permanent wall-mounted instruments and network consoles of years past, permits incorporation of an electronic circuit with warning alarms such as a loud buzzer or bell to alert people to dangerous levels of CO.

Because there continued to be a need to measure CO at wide ranges in community air, industrial atmospheres, homes, exhaust stacks, automotive emission sources, and other sites, numerous measurement techniques and instruments were developed. In the future, investigators can expect many more improved refinements of existing methods as well as innovative techniques to qualify and quantify CO gas in the presence of interfering gases. For the moment, there is an armada of measurement devices and methods with sufficient accuracy and precision to characterize CO in most foreseeable situations and concentrations. Is there a seventh generation of CO monitoring instruments around the corner?

Finally, it should go without saying that any toxic gas without sensory properties (e.g., lack of odor, color, irritation, and taste at any concentration) such as CO requires measurement; one cannot reliably guess at the concentration, and estimates of CO gas levels are foolhardy. Twitching little yellow birds or cherry-red mice no longer suffice, either.

2 OCCUPATIONAL AND ENVIRONMENTAL CO EXPOSURE LIMITS

The first standards of CO exposure were established for workers in occupational settings. The U.S. Public Health Department recommended an 8-h time-weighted average exposure (TWAE) limit not to exceed 200 ppm (0.02%) in 1915 (LaDou, 1990). By today's standards, this was clearly too high. Certainly, workers experienced significant asphyxiation when exposed to this gas concentration for brief periods, say 2 to 3 h. While reported records of measurements of CO are rare during the 1910–1950 period, workplace exposures, at least of a chronic average exposure nature, most likely did not routinely exceed 100–150 ppm. Regardless, numerous cases of CO asphyxiation fatalities continued to be reported. Most of these were of the acute variety — that is, massive, albeit brief, exposures to CO gas. To this day, sequelae from chronic CO poisoning is disputed by some investigators in spite of increasing physiological evidence of subclinical findings.

The first meeting of the Threshold Limits Value (TLV) Committee of the American Conference of Governmental Industrial Hygienists (ACGIH) in 1946 resulted in recommending a 100 ppm CO maximum allowable concentration (MAC) for workers. This limit was to be used (measured and calculated) as an 8-h time-weighted average (ACGIH, 1992).

The MAC remained at this level until 1967, when the TLV Committee reduced it to 50 ppm because of increasing evidence of inadequate protection of workers at a 100 ppm concentration (ACGIH, 1992). The TLV still incorporated the time-weighted average concept; that is, the integrated dose of a worker to CO throughout an 8-h workday should not exceed the threshold dose. Implicit in the TLV concept is the notion that the TLVs are the worst acceptable concentrations — not a goal only above which workers could be ill and below which they are protected. In this regard, anyone applying TLVs should very carefully study the TLV booklet and heed the admonitions presented in this book's preface (TLVC, 1994). Questions on TLVs and their application should be referred to a professional industrial hygienist with experience with the chemical or physical agent in question.

In 1972, the National Institute for Occupational Safety and Health (NIOSH) recommended an 8-h TWAE limit of 35 ppm with a ceiling concentration of 200 ppm (NIOSH, 1972). These limits were based on NIOSH's review of the CO inhalation toxicity literature, which indicated that persons with mild coronary ischemia were at risk of myocardial infarction with an 8-h TWAE exceeding 50 ppm. The OSHA permissible exposure limit (PEL) is currently 35 ppm as an 8-h TWAE (U.S. Deptartment of Labor, 1993). One must consider the distinctions between regulatory limits and advisory recommendations and realize the delays that often accompany changes in limits and the lag between new scientific findings and promulgating new enforceable standards. Environmental health professionals are ethically bound to consider these findings and weigh them in their recommendations and interpretations.

In 1976, the TLV Committee recommended a short-term exposure limit (STEL) of 400 ppm as a companion to the 8-h TWAE of 50 ppm (TLVC, 1994). The STEL was not to be exceeded more than 15 min at a time in workplaces and only then with sufficient detoxification intervals so the 50 ppm TWAE limit was not exceeded.

In 1991, the TLV Committee proposed a 25 ppm 8-h TWAE, which was adopted in the 1994–1995 listing of TLVs (TLVC, 1994). No ceiling or STEL values are included at present, but one can see that a worker exposed, for example, to 200 ppm for 15 min and to 25 ppm for 465 min in an 8-h workday would have a TWAE exceeding 30 ppm. Careful control of peak excursions, therefore, assumes greater importance when workers' exposures approach high concentrations of airborne chemical contaminants having TLVs with low limits.

Environmental exposures to CO gas outside workplaces require limits more conservative than those used for industrial hygiene purposes. The reasons for this are several: there is no full period in which to detoxify; that is, with continuous exposure to CO, equilibration of carboxyhemoglobin (COHb) with polluted atmospheric air becomes problematic. Environmental CO limits must protect not only healthy workers when off the job but also the very young, the ill, the embryo/fetus, the neonate as well as the mother, hypersusceptibles, and the elderly. The U.S. Environmental Protection Agency (EPA), under the provisions of the Clean Air Act, has promulgated national primary and secondary air quality standards for CO (Department of Health, Education, and Welfare, 1985). The EPA's primary standard for protection of community health is 9 ppm CO as an 8-h average concentration not to be exceeded more than once per year in any community, rural or urban. The secondary EPA air quality standard for CO is 35 ppm. This is a maximum 1-h concentration not to be exceeded more than once per year in any community.

NIOSH has recommended 1500 ppm as the concentration of CO gas that is considered immediately dangerous to life and health (NIOSH, 1990). The IDLH level represents the maximum concentration from which, in the event of respirator failure, one could escape within 30 min without a respirator and without experiencing any escape-impairing or irreversible adverse health effects.

The ACGIH recommends a Biological Exposure Index of 20 ppm CO in end-exhaled air obtained from workers at the end of their workshift (TLVC, 1994). This corresponds to an 8-h TWAE of 25 ppm and a corresponding COHb level of 3.5%. Of course, subjects who regularly inhale tobacco smoke typically can have higher COHb levels.

3 RELIABILITY AND STATISTICS OF CO ANALYSIS

Several analytical and statistical concepts are important when expressing air concentrations of CO gas. All testing and measurements of CO should have the highest accuracy and precision possible in the context of the desired objectives.

What is meant by this? For example, it matters little if measured concentrations are 3000 ppm or 5000 ppm when the true (accurate) concentration is 4000 ppm (0.4%) in a confined space that pipefitters are about to enter. Either measurement, in a relative sense, is inaccurate, but completely adequate to bar entry into an atmosphere above the IDLH concentration of 1500 ppm. On the other hand, for example, research into diurnal variations of CO in a rural ambient environment will require instrumentation with greater accuracy and precision.

Accuracy refers to the closeness of the measured concentration to the true, or "real" concentration. *Precision* is the degree of repeatability of a single measurement. It has been said, somewhat facetiously, that an analyst can be very precise in his error; that is, he consistently errs in the same direction. The *true value* is the number that is completely consistent with the known value and the definition of the primary standard used. With respect to some common terms used to describe instrument performance, the following definitions are often used: *response time* is the time taken by instruments from start of a change in input before instrument output reaches a specified percentage of the ultimate range. Normally, this is set at either 90 or 95%. Response time is the measure of the minimum averaging time required to achieve valid, reliable output data. *Lag time* is the interval between start of the input signal and the observed start of the output response corresponding to that signal. Lag time is a measure of delay between input sample signal and output data and is always less than or equal to response time. *Minimum detectable concentration* is sensitivity as the CO concentration approaches zero. *Sensitivity* is the smallest change in pollutant concentration that can be reliably detected.

An extremely important concept in the evaluation of workers' exposures by the measurement of air contaminants for industrial hygiene purposes is the Action Level, a concentration defined as 50% of the PEL at the time. The Action Level for CO, then, using the 25 ppm TLV proposed by the ACGIH, is 12.5 ppm. Or, given the degree of accuracy and precision of most contemporary CO measuring instruments and devices, the Action Level rounds to 13 ppm.

The statistical utility of the Action Level to the practicing industrial hygienist is this: if the 8-h TWAEs of a worker on 2 separate days are below 13 ppm, it can be predicted, with 95% confidence, that all exposures of that worker will be below the PEL *provided that things do not change* (e.g., work practices, ventilation, CO generation rate, etc.) (Leidel et al., 1977). The beauty of this statistic is that once compliance with the Action Level has been achieved, the industrial hygienist then only has to ensure that environmental conditions have not changed since the initial evaluations. This is normally much easier to do than to proceed through more costly and time-consuming sets of air sampling.

Consider Jack, a forklift truck operator whose 8-h TWAE is 9 ppm on Tuesday, with no peak exposures exceeding 200 ppm. If Jack's TWAE on, say, next Monday is 5 ppm, again with all peak exposures below 200 ppm, we can state, with 95% confidence, that all of Jack's exposures in the future should not exceed 25 ppm if both test days were essentially identical and future exposures are identical in terms of ventilation, work practices, proximity to CO sources, engine operating characteristics, driving patterns, occupancy times in truck trailers and rail cars, etc.

In many cases, several short-term or so-called "grab" samples are collected in a worker's breathing zone during random times of the day. These "snapshots" of CO air quality are discontinuous, but they still may be used if a sufficient number is taken. OSHA uses the following procedure to help to determine compliance with the PEL based on a small number (n) of instantaneous (grab) samples collected at random intervals throughout the workday (NIOSH, 1972). Given: the results of n samples with a mean of m and a range r (difference between lowest and highest n). If, for from three to ten samples, m is greater than the total of three parameters: the standard plus the percentage of systematic instrument error multiplied by the standard plus $(t + r)/n$, then the true average concentration exceeds the standard ($p < 0.05$). The values of t and corresponding number of samples, n, are $2.35 = t$ and $n = 3$; $2.13 = t$ and $n = 4$; $2.01 = t$ and $n = 5$; $1.94 = t$ and $n = 6$;

1.89 = t and n = 7; 1.86 = t and n = 8; 1.83 = t and n = 9; and 1.81 = t and n = 10. OSHA recommends that for larger number of samples, the procedure given in Section 3-2.2.1 of the National Bureau of Standards Handbook 91 be followed (Leidel et al., 1977).

4 WHY, WHERE, WHEN, AND WHO TO SAMPLE FOR CO EXPOSURE ASSESSMENT

This section is not intended to be a comprehensive treatise on the approaches to air sampling. For such, the reader is referred to basic references of industrial hygiene and atmospheric pollution monitoring and standardized testing protocol (Leidel et al., 1977).

Why we obtain air samples for CO gas should be obvious but must also be clearly stated: to help ensure protection and conservation of human health and life and to determine compliance with regulatory and emission standards (OSHA, EPA, and local DNR); also, to help ensure protection of those animals with oxygen-conveying hemoglobin circulatory systems.

Where air samples for CO gas are obtained include personal breathing zone samples; general area air samples whether in the workplace, community, or home; and source zone air and gas samples such as automotive emissions, industrial stacks, air supplied respirator breathing air, and combustion processes. For industrial hygiene purposes, the best approach to estimate a worker's exposure to CO are the personal breathing zone air samples. In many cases, however, general area air samples may suffice to characterize the quality of the atmosphere in a work setting. Of course, such air samples must approximate (in both time and space) that to which workers could be exposed.

For very large workspaces, a grid pattern of air sampling may be established so that spot or grab samples are taken on either a random or regular basis throughout a workshift. From such data, a report containing a three-variable table (location, time, and concentration) listing the various CO gas concentrations at precise locations and different times can be prepared. This approach to characterization of a workspace, such as a large manufacturing assembly plant, a warehouse, or an automobile service garage, can be invaluable in defining and establishing an industrial hygiene program. Careful evaluation of the atmospheric monitoring data presented in such a table and grid can help in constructing a plot of equal CO gas concentration contours in a plant, an isopleth prepared on a plant floor plan.

It is not uncommon to directly measure CO in exhaust gases emitted from forklift trucks much as in automotive emissions testing (Michigan Department of Public Health, 1971). This detects vehicles that require service and engine maintenance. In general, vehicles without catalytic converters and with higher horsepower and engine rpm are the greatest generators of CO. Electric battery trucks are vehicles-of-choice for CO control and industrial hygiene purposes, and every attempt should be made to incorporate these into a plant's material handling program. All other factors being equal, and horsepower per horsepower, the gasoline engine generates more CO, in general, than propane-fueled engines, which, in turn, are generally higher contributors than diesel engines. Conversely, diesel engines produce, on average, more nitrogen oxides and exhaust particulates than a gasoline engine. However, a poorly tuned diesel engine can release far more CO than a finely tuned gasoline engine. The key is engine tuning, emissions testing, and proper, regular servicing and maintenance (Michigan Department of Public Health, 1971).

Confined spaces deserve special mention. Unfortunately, workers often die in holes, pits, tunnels, bins, silos, pipelines, tanks, ovens, and furnaces. Many of these fatalities would be prevented if full atmospheric testing was done before workers were permitted to enter, and air testing continued throughout occupancy of the space. At a minimum, these air samples must include testing for oxygen deficiency ($<21\%$ at 760 mm Hg), flammable vapors and gases, combustible airborne dusts, and toxicants (e.g., CO, H_2S, CO_2, Cl_2, C_6H_6, $HCCl_x$, etc.). Given the plethora of combustion processes and the ubiquitous nature of CO, plus the fact that gaseous molecules tend to move about, all confined space entry work must, in the author's view, include testing for CO regardless of the history of the confined space.

The author has discovered CO in the most unlikely confined spaces, at some of the worst possible times, and sometimes during forensic reconstruction of gas inhalation fatalities. The lack of warning properties for CO, its reputation as the "silent killer," and the relative ease with which CO air samples can be taken strongly argue for mandatory testing for CO (and other possible toxicants) during all confined space entry work. Every one of "Murphy's laws" can come into play during such highly hazardous work. Air quality testing is required per OSHA and state confined space occupational health standards (U.S. Department of Labor, 1993).

Location of air samples with respect to ceiling height can also be significant. The author investigated a construction site where concrete paving of a new large floor in a warehouse was being completed. Cement trucks came, delivered, and left. CO air samples at ground level were unremarkable in spite of several other pieces of construction combustion equipment (e.g., power screeds, compressors, direct-fired air heaters, natural gas salamanders, sand tampers). CO concentrations 20 ft above the floor at the ceiling truss level ranged from 190 to 320 ppm. Electricians installing conduit at this level soon became ill, with substantiating COHb levels. Engine exhaust emissions from the several cement trucks at the site were rising to ceiling level, where ventilation was poor. Because the hot exhaust gases from the trucks vented about 10 ft above the floor, their contribution to CO at ground level was negligible. Cold air stratified at ground level, interfering with ventilation and air turbulence where electricians worked.

Stack emission testing for CO from combustion processes can be helpful in determining reentry of contaminants into building air intakes (Industrial

Ventilation Committee, 1993). Knowledge of amount of CO in a stack gas, the volumetric gas flow rate from the stack, stack height, distance to nearest building air inlets, and various meteorological parameters permits calculation of theoretical concentrations of CO being reentrained into a building's air supply. While this is useful in an empirical sense, actual CO air sampling under the most adverse atmospheric conditions is normally superior to the projected "calculated" amounts.

Industrial hygienists often judge which employees in a group would likely have the highest exposures, whether such workers are production employees, maintenance workers, or service workers. These people are evaluated on the premise that if their exposures are well controlled, others working in the area will likely also be protected. This approach requires considerable professional judgment.

When to obtain air samples for CO also requires careful consideration. Entry into any hazardous area, such as confined spaces, requires testing before the entry is permitted. However, for most day-in and day-out types of routine industrial hygiene testing for CO, the choice of when to sample becomes more questionable. Obviously, air samples should be taken while people are present or, in the case of stationary CO sensors and alarms, before people enter areas with potential exposures to CO.

Air samples taken on the day shift may not be representative of air quality on the afternoon or "graveyard" shift when production conditions are different. Ventilation in industrial plants and commercial buildings is often greatly reduced during the night, so that dilution ventilation that proved acceptable during the day may be deficient at night. Testing should be done on all shifts to determine if diurnal variations of CO are an issue.

Air samples obtained from Bob, Jane, and Alan may suggest a nonproblem, but exposures of Sally, Curt, Sue, and Nick may be an order of magnitude higher because of different work practices, anthropomorphic differences, production rates, work shift differences, materials used, etc. Further, so-called "average" exposures can be misleading. For an understanding of log-normal distributions of concentrations of air samples, the reader is referred to a standard publication on this issue (Leidel et al., 1977). The author was once told by a plant manager that, "on average," all of his workers were under the TLV for CO (mean = 24 ppm when the TLV was 50 ppm, $n = 19$ workers, range of TWAEs = 3 to 145 ppm, two workers had TWAEs exceeding 50 ppm). This manager, perhaps well meaning but very misguided, did not understand three very significant points: (1) at least two of his employees were at risk of adverse health effects from CO inhalation, (2) there was noncompliance with the OSHA PEL, and (3) the 50 ppm TLV for CO is the worst acceptable concentration, and every feasible attempt should be made to control exposures to the lowest practicable level.

Consider, for example, Ted's 8-h work shift from 8:00 a.m. to 4:00 p.m. For simplicity, assume Ted eats his lunch on the job and does not leave his work station for any significant period during his workday. Further assume

Ted has potential exposures to CO, and we wish to characterize his exposure and to determine compliance with the OSHA PEL. How might we go about doing this?

Depending on the instrument and method selected, we might take one full-shift continuous air sample from 8:00 a.m. to 4:00 p.m. The integrated result of this single test would give us Ted's actual 8-h TWAE that day, but little else. We would know nothing about Ted's peak excursion exposures that day. It is highly unlikely that exposures to most industrial air contaminants are steady-state. The typical scenario is one of substantial fluctuations throughout the day, sometimes of two or more orders of magnitude. Dependent variables include ventilation, contaminant generation rates, Ted's proximity to air contaminant sources, atmospheric conditions, production rates, presence of other workers, among others.

We might consider two 4-h air samples back to back. This would be useful because we could start to get an idea of differences in exposure pre- and postlunch. Regardless, we would have to calculate Ted's 8-h TWAE based on the separate individual results. Still, we would not have much of a better idea of peak exposures. A hidden advantage of this approach is to gain a notion if Ted deliberately "spiked" one or both samples. It is difficult to constantly observe a worker. In the author's experience, deliberate contamination of air samples by workers when no one was looking is very rare, yet it sometimes happens. However, it is rare for a worker to be consistent; that is, if Ted chooses to "spike" both air sampling devices, it is unlikely he would be uniform in doing so, and, moreover, deliberately contaminated samples are often "off the wall" and do not make sense to seasoned investigators. In this case, we simply resample Ted (with counseling) and discard all data that does not fit within reason. More careful observation of this worker is in order.

Another approach to CO air sampling, particularly with passive dosimeters and noncumbersome personal air sampling pumps, is to attach two devices on a worker's collar at the start of the workshift. Remove one after, say, 4 h and the other at the end of the work shift. The full-shift air sample provides the TWAE. The difference between the two results provides a notion of what the air quality was like during each of the two halves of the shift. Furthermore, if a worker was to "spike" a sample, it is unlikely that he or she would be consistent with both devices.

Going even further, we might collect four 2-h samples or, for example, one 2-h and two 3-h air samples. These and the previous samples are called full-period consecutive air samples. While the analytical costs increase only slightly with more air samples, the gains in exposure and dose information increase exponentially. Costs of testing devices and samples are normally puny when compared to the salaries of industrial hygienists and others who collect air samples.

We might, for varying reasons, collect a sample from 9:00 a.m. to noon and one from 1:30 p.m. to 3:30 p.m. This type of sampling, referred to as consecutive partial period sampling, may be less desirous than full-shift continuous sampling. High (or low) exposures can be missed during intervals. Still, we have little idea of significant peak exposures to CO with this type of sampling.

Instantaneous (grab) samples can collected throughout the workday either on a regular schedule (e.g., one air sample every 30 min: 16 samples in an 8-h workshift) or randomly (e.g., one air sample taken at nonpredetermined times in each of 7 h of work: seven air samples in an 8-h workshift). Statistics of such air sampling were previously presented. This type of air sampling might "catch" some peaks, but with large time intervals between samples, many could, and probably would, be missed.

Several short-term air samples (e.g., 3-min duration) may be randomly or regularly taken throughout the workday much like the peak or grab samples. These have the same advantages and disadvantages. However, in all cases, no matter which approach is used, the objective is to ensure that a sufficient number and types of air samples are taken to characterize all of the workers' 8-h TWAEs and to ensure there is compliance with the PEL and STEL. Stationary CO monitors must accurately and precisely quantify CO in the atmosphere to activate an alarm, or other sensor, to alert people to dangerous atmospheres at the IDLH or some fraction of the IDLH, STEL, or TLV/PEL (e.g., 200 ppm for 10 min, 25 ppm for 30 min, or 400 ppm-min). The alert level depends, of course, on the setting, such as confined work space in an industrial setting or air supply to hospital neonatology incubators.

For industrial hygiene purposes, the best approach is to select a combination of methods that provides an integrated exposure concentration over time, such as 9 h out of a 10-h workshift. Incidentally, for example, for a 10-h workshift, the TLV becomes 20 ppm (25 ppm × 8 h = y ppm × 10 h; y = 20 ppm). Whether numerous grab samples are obtained throughout the workday or continuous sampling devices or other methods are used is a judgment call best based on the conditions at the time. In many cases, a direct-reading "real-time" strip chart recorder can be used that both integrates the exposure concentrations during the sampling intervals and gives a graphic printout of the peak CO occurrences at specific points in time.

Calculation of a worker's 8-h TWAE is a simple procedure as the following example shows: Consider a worker exposed to 50 ppm CO for 4 h, 250 ppm for 1 h, and 100 ppm for 3 h out of an 8-h workshift. The TWAE for the 8-h workday is: $[(50 \times 4) + (250 \times 1) + (100 \times 3)]/8 = 750$ ppm-h/8 h = 94 ppm, an overexposure requiring prompt controls (e.g., sharing test results with employee and others with similar exposures, worker education and training, perhaps better ventilation and changes in work practices, recording test results, developing follow-up action plan to close the loop, etc.).

Test records of air sampling for CO exposure should contain, at a minimum, the facility, employee name, employee identification number, date, actual test results, time interval of sampling, TWAE calculations, exposure conditions, person conducting test, report distribution, follow-up "close-the-loop" information — that is, actions taken, type of air sample (personal, general area, source zone, etc.), description of instruments, air quality standards, temperature, altitude, instrument calibration procedures, remarks and possible interferences, names of other employees whose exposure is represented by these test results, sample identification number (if any), and employee comments.

5 PROBLEMATIC ISSUES IN CO HAZARD EVALUATION AND RISK ASSESSMENT

Numerous variables enter into CO exposure assessments that can profoundly affect the reliability, accurate interpretation, and correct application of the test results. Among these are the failures to:

- Read, understand, and follow instructions supplied with CO monitoring instruments.
- Calibrate instruments with known, verified concentration CO span gas standards.
- Regularly recalibrate CO monitoring instruments. Record calibration parameters.
- Record all data and conditions that could affect reliability of test results.
- Record the established accuracy and precision of the test method for all results.
- Sample sufficient number of employees, workstations, and CO exposure areas.
- Obtain sufficient number of air samples from representative employees in order to fully characterize all employees' daily time-weighted average CO gas exposures.
- Evaluate employees on all workshifts, or reasonably project what the CO exposures could be under other working or environmental conditions and at different times.
- Account for interfering gases and vapors (positive or negative).
- Account for exposures to other air toxicants that could exacerbate exposures to CO (e.g., other chemical asphyxiants such as H_2S, HCN, aromatic amines, nitriles, nitrites, etc., and/or a reduced partial pressure of oxygen — that is, altitude effects).
- Follow through on abatement and engineering, administrative work practices, and personal protective equipment control recommendations.
- Carefully explain purposes of your testing to employees and solicit their cooperation, understanding, and ideas.
- Personally inform affected employees of test results and their health significance.

- Obtain a sufficient number of air samples.
- Determine compliance with TWAE, STEL, peak, and IDLH exposure limits or, in community and environmental air sampling, EPA CO exposure standards.
- Evaluate the most "unlikely" places where CO gas could accumulate.
- Record all relevant exposure data (e.g., ventilation, production rates, CO sources).
- Evaluate exposures on at least two representative days to ensure compliance with the Action Level.
- Cite the current and any proposed changes to CO exposure standards.
- Maintain instruments and alarms (including regular calibration with verifiable span gases).
- Test all confined spaces for CO concentrations (at a minimum) before entry and while working inside.
- Perform basal statistical calculations on test results. Explain statistical results in the report.
- Provide sufficient explanations in reports (toxicology of CO, OSHA and EPA limits, ventilation standards, test methods, health relevance, engineering controls, etc.).
- Account for extensive overtime (i.e., >10%) in TWAE calculations and determination of CO inhalation dose.
- Revisit problematic sites on a regular basis. Establish a time-bounded compliance program and accountability plan.
- Establish comprehensive industrial hygiene programs with minimal compliance to all NIOSH recommendations (NIOSH, 1972).
- Provide comprehensive hazard analyses and risk assessments in clear reports to management and others. Clearly state consequences of failure to abate excessive exposures (i.e., adverse health effects, citations and fines, work stoppages, adverse publicity, etc.)
- Exercise a fair degree of humility in CO exposure assessment while entertaining the notion that all, or most, of "Murphy's laws" lurk nearby and are ready to haunt you.
- Establish an end-of-shift biological monitoring program to test alveolar air for CO. Such a program can be invaluable when coupled with regular atmospheric testing.
- Break the tips off colorimetric detector tubes.
- Use the correct type of detector tube.
- Insert the detector tube in the correct direction.
- Turn an electronic monitoring device "on" (or plug into "live" electric circuit).
- Correlate end-of-shift alveolar air CO or carboxyhemoglobin test results with air CO sample test results and covering correlation in reports to management and others.

- Account for possible erroneous "readings" when obtaining CO gas samples in "non-standard" atmospheres, e.g., measuring CO in nitrogen or CO in steam.
- Correct for altitude effects and pressure differences and for temperature extremes when sampling hot or cold gases.

6 ANALYTICAL METHODS AND INSTRUMENTS

There are numerous techniques to measure CO in air and other gas mixtures. These include volumetric, gravimetric, chemical, and electrochemical methods. The methods available to continuously measure CO include nondispersive infrared analysis (NDIR), mercury vapor analysis, catalytic and electrolytic analysis, electrochemical analysis, and gas chromatography. The methods available to provide "spot" or integrated CO measurements include infrared spectrophotometric analysis, colorimetric analysis, nondispersive infrared analysis, and gas chromatography.

In any testing for CO, however, it is necessary to reliably and regularly calibrate the testing device with known concentrations of CO in air or some other diluent gas. Accurately prepared CO gas standards must be readily available in the field for the calibration of methods and measuring devices before CO can be quantified with any reliable degree of precision and accuracy. The best procedure, in most cases, is to use a volumetric gas dilution for the preparation of gas standards. For example, if 100 ml of pure CO gas are diluted to 1 m^3 (1,000,000 ml) with air devoid of trace CO, the final gas concentration becomes 100 ppm CO (volume/volume, or 0.01%). Pure CO can be obtained from the commercial suppliers of reagent grade gases. An analytical report of the pure gas should accompany the supply. If large quantities of standard CO calibration gases are needed, the dilution can be made into pressurized gas bottles and tanks. Care should be taken to ensure the accuracy of the standard because CO in pressurized gas bottles and tanks may be unstable and decay at CO concentrations of 1 ppm or less.

A known volume of pure CO is injected into an evacuated tank of known volume. The tank is subsequently filled to a standard pressure with CO-free air (or, sometimes, with CO-free helium or nitrogen). The concentration of the CO in the tank is calculated. Standard CO gas samples in 5- to 100-l volumes may be prepared in plastic bags. The diluent CO-free gas is metered into a vacuum-evacuated plastic bag with a wet test meter or rotameter. As it flows into the bag or bladder, a measured volume of pure CO is injected by gas syringe into the line (usually thick rubber or plastic pressure tubing) connecting the metering device to the bag. The final concentration is calculated from the total volumes of diluent gas and the CO in the plastic bag. Some investigators prepare a stock mixture of, say, 1% (10,000 ppm) and make serial dilutions from this mixture as needed for instrument measurements on any one day.

The diluent gas must be CO-free. Some gas chemists recommend that pure helium, which is reliably CO-free, is preferable to nitrogen and air as a diluent gas. A less expensive method is to periodically use helium as a true "zero" gas to estimate the CO concentration of the nitrogen diluent gas. A suitable correction factor is then used to account for the CO concentration in the nitrogen. Passing the diluent gas through a CO sieve, such as heated "Hopcalite," can further assist in achieving a CO-free diluent gas.

Gravimetric methods are not routinely used in air pollution and industrial hygiene studies to measure CO because relatively large samples are required for quantitative analyses. Gravimetric methods are useful, however, for checking CO standards (McCullough et al., 1947). The reaction: CO (gas) + HgO (solid) yields at 175° to 200°C Hg (vapor) + CO_2 (gas). During very slow passage of the gas sample containing small amounts of CO, the red mercuric oxide is reduced by the CO and loses weight. The loss in weight and the measured volume of the sample are used to calculate the concentration of CO in the standard gas mixture. Toxic mercury vapor must be collected to ensure nonrelease to the investigator's breathing zone and to the environment.

A somewhat similar method is used where the gas containing the CO is passed through "Hopcalite" at 195°C and the carbon dioxide gas formed is absorbed in tubes that contain "Ascarite" (Salsburg et al., 1947). The gas volumes are determined with a stopwatch and a rotameter or similar gas flowmeter. From the weight of the CO_2 absorbed in the tubes, the CO concentration in the feed gas is calculated. This method is accurate to about 2% for CO concentrations from 22 to 870 ppm.

Another gravimetric method relies on oxidation of CO to CO_2 by decomposed silver permanganate ($AgMnO_4$) as a catalyst (Lysyj et al., 1959). The released CO_2 is absorbed into an "Ascarite" tube and weighed by difference.

There is one reliable chemical assay method for CO gas standards that relies on reduction of iodine pentoxide by CO (Adams and Simmons, 1950). The reaction is: 5 CO + I_2O_5 yields 5 CO_2 + I_2. When the gas mixture containing CO is passed through the heated I_2O_5, iodine vapor is liberated, which is then collected in a gas washing bottle and then titrated with a standardized solution of sodium thiosulfate.

For general air pollution CO measurement purposes, NDIR analyzers are the most commonly used automated, continuous devices for the determination of atmospheric CO. These are generally accepted as being the most reliable reference method for calibration of other CO measuring instruments. A NDIR analyzer operates on the principle that CO gas has a sufficiently characteristic infrared absorption spectrum. Absorption of infrared radiation by the CO molecule can be used as a measure of CO gas concentration even in the presence of other gases. The size, shape, range, and sensitivity of these instruments varies from one manufacturer to the next, but the basic components are similar. Most commercial instruments include a hot filament wire source of IR radiation, a rotating "chopper," a sample cell, a reference gas cell, and an electronic photometric detector.

The IR detector senses pressure changes on each side of a diaphragm that separates portions of the gases being irradiated through the reference and sample cells of the instrument. These pressure changes are converted to electrical signals that correspond to the difference between radiation received from the reference and sample cells. The signal is amplified and rectified and read on a meter or a recorder calibrated to various CO span gas concentrations.

NDIR analyzer response times are determined by the physical dimensions of the system, flow rate of the air or gas sample, and the response time of the recorder and the meter. The response times may range from less than 1 min to as long as 5 min.

CO_2 and H_2O vapor interfere in the determination of CO by NDIR techniques. Filter cells are used to reduce these interferences. These are typically placed in front of sample cells or in front of both sample and reference cells. Filter cells that are filled with CO_2 and H_2O vapor absorb radiation at interfering wavelengths so that normal atmospheric concentrations of CO_2 and H_2O vapor in the sample have minimal effects on the IR radiation that reaches the detector.

Optical filters are also used successfully to limit the infrared wavelength and band width to a range in which the CO_2 and H_2O vapor are "transparent" and thus are "invisible" to the instrument.

Most commercial NDIR instruments operate on the double-beam principle at atmospheric pressure. They are able to detect minimum CO concentrations of about 0.5 to 1 ppm. The sensitivity of NDIR instruments and the minimum concentrations they are able to detect are proportional to cell path length, operating pressures, and electronic amplification. Measuring ranges typically cover 1 to 50 ppm CO or 1 to 100 ppm CO. The newer NDIR instruments operate from 1 to 25 ppm CO at atmospheric pressure and with expanded cell path lengths below 0.5 m. Since CO concentrations in many industrial hygiene situations often exceed these concentrations, NDIR may have limited usefulness in CO-problematic industries. However, the serial dilution of industrial plant air with CO-free air may extend sampling into the range of the instrument. Moreover, since NDIR instruments tend to be bulky, their cumbersome nature may preclude their use in many industrial hygiene situations.

Other techniques may be used to reduce or prevent CO_2 interference. Tubes filled with "Ascarite" remove CO_2 from the entering gas stream. Other systems add nominal quantities of CO_2 (300–400 ppm, i.e., typical ambient concentrations) to the zero and span gases that are used to calibrate the NDIR instrument.

Water vapor interference may be reduced by filter cells, but they cannot handle high humidities, which are frequently encountered in atmospheric monitoring. In a NDIR instrument that has no H_2O vapor filter, a H_2O

vapor pressure of 5 mm of Hg can account for as much as 7 ppm CO additive instrument response. At 30 mm Hg of H_2O vapor pressure, the additive response can be as high as 15 ppm CO, a significant potential error. Various drying agents can be used with varying degrees of success; some, like calcium sulfate, are ineffective. To overcome this, some commercial NDIR instruments saturate the incoming air at a constant temperature with water vapor. This eliminates water vapor as a variable.

The advantages of NDIR instruments are several. They are not sensitive to gas flow rate. No wet chemicals are required. They are reasonably independent of most ambient air temperature changes. They are sensitive over wide concentration ranges. They have short response times. NDIR systems can be operated by nontechnical personnel.

NDIR disadvantages include zero drift, high cost, nonlinearity, and span drift. Newer instruments have minimum drift because high-quality thermostats and solid state electronics are used in their manufacture. Such instruments also have automatic zeroing, spanning, and recalibrating capability. Some newer instruments also can be purchased with essentially linear outputs. These features should be considered when multiple-station CO monitoring networks are designed and installed.

Galvanic cells are used as continuously monitoring electrochemical CO sensor analyzers (Hersch, 1964). When an airstream containing CO is passed into a chamber packed with iodine pentoxide and heated to 150°C, the following reaction occurs: 5 CO + I_2O_5 yields 5 CO_2 + I_2. The liberated iodine vapor is absorbed by an electrolyte and transferred to the cathode of a galvanic cell. At the cathode, the iodine is reduced, and the resulting current is measured by a galvanometer. Such instruments have been used to measure CO in traffic near freeways. Mercaptans, hydrogen, H_2S, olefins, acetylene, and water vapor interfere, but with combinations of filters and absorbents, interferences can be reasonably controlled.

A coulometric method employing a modified Hersch-type cell has been used to measure CO in ambient air on a continuous basis (Dubois et al., 1966). The I_2O_5 reaction with CO liberates iodine vapor as given in a previous equation where 5 mol of CO release 1 mol of iodine vapor, which is then passed into a Ditte cell. The electrical current generated is measured by an electrometer-recorder combination. Interferences are the same as with the galvanic CO analyzer. This method can be used for minimum detectable concentration of 1 ppm. The reproducibility and accuracy are good if gas flow rates and temperatures are well controlled. This method requires careful column preparation and use of selective filters to remove interfering vapors and gases. It has a relatively slow response time, which may be an added disadvantage in some work.

A continuous CO analyzer using the reaction of CO with hot red mercuric oxide and the photometric determination of the mercury vapor released is based on the equation: CO + HgO (solid) yields at 210°C CO_2 and Hg (vapor). Olefins and oxygenated hydrocarbons react quantitatively and provide positive interferences. Because these are normally present in low concentrations in the atmosphere relative to CO, their contribution is negligible in practical work. Free hydrogen can also interfere with this method, but this is not a serious interference because very little free hydrogen is present in normal atmospheres. Ozone, theoretically, can interfere mole for mole, but ozone is normally present in much lower concentrations than CO. The mercury vapor device has a response time similar to that of the NDIR analyzer, but because of interferences and some electronic instability, it does not appear to be suitable for routine atmospheric monitoring. The instrument is, however, a useful, portable continuous analyzer with a capability of analyzing very low gas concentrations from 0.025 ppm to 10 ppm CO with detectable changes of only 0.002 ppm. For this reason, it has found application in the determination of geophysical "background" CO concentrations throughout the world.

An automated gas chromatographic system can measure both methane and CO (Stevens et al., 1968). The system comprises a gas sampling valve, a gas backflush valve, a precolumn, a molecular sieve column, a catalytic reactor, and a flame ionization detector. The precolumn prevents H_2O, CO_2, and nonmethane hydrocarbons from entering the molecular sieve. The catalytic reactor quantitatively converts CO to CH_4, which is then measured by the flame ionization detector. The system is designed for semicontinuous operation with a capability of performing one analysis approximately every 5 min. It has a linear output for both CO and methane. Its very wide operating range permits its use in both heavily polluted and clean atmospheres. Simultaneous CO and methane concentrations over three orders of magnitude from 0.1 to 1000 ppm can be measured with this instrument. Its semicontinuous characteristics suggest applications in special surveys rather than for routine air sampling and monitoring. A disadvantage is that the instrument must be operated by chemists and technically trained personnel.

Intermittent or spot samples can be collected in the field and later analyzed in the laboratory. A detailed outline of this method has been described by Mueller (1965). The sample containers may be rigid (glass bottles or stainless steel tanks) or nonrigid, such as relatively nonpermeable plastic gas sampling bags. Because of location or cost, intermittent tests may rarely be needed, but they may be the only practical method for air monitoring. However, with such recent advances in CO sampling and analysis, such situations are likely to be encountered less frequently. Gas samples can be taken over a few minutes or accumulated intermittently to obtain, after analysis, either spot or integrated results.

Field NDIR continuous monitoring analyzers previously discussed can be used for later laboratory analysis of intermittent samples. NDIR instruments designed by some manufacturers are primarily for laboratory use and must be modified for ambient air monitoring on a continuous basis in the field.

CO may be measured by its infrared absorption at 4.6 μm (Kemmner et al., 1966). Other gases may absorb at 4.6 μm, but they can be differentiated from CO by examining the complete infrared spectrum from 2 to 15 μm. The sensitivity and range of the technique depends on the sophistication of the instrument used. A minimum detectable concentration of 2.5 ppm CO has been obtained by instruments using scale expansion and a 1-m path length cell. It can be expected that a similar infrared wavelength spectrophotometer equipped with a folded 10-m cell and ordinate scale expansion would have a 0.3 ppm CO detection capability. H_2O vapor and CO_2 must be removed from the gas sample before the infrared scan can be made. As discussed earlier, beds of magnesium perchlorate and "Ascarite" are normally used to remove interfering gases and vapors. The technique is specific and accurate, but it requires very expensive and non-portable equipment.

A gas chromatograph with a flame ionization detector can be used to measure CO after the gas is passed over a nickel catalyst (Porter and Volman, 1962); 1 mol of CO produces 1 mol of methane. Because methane can be detected at about 0.01 ppm, theoretically CO could be detected at such a low concentration. This technique does not suffer from interferences because the measurement is made on a CH_4 peak derived from a separated CO peak. Operating temperatures and other parameters must be carefully controlled, however, before reliable results can be obtained.

CO reacts in an alkaline solution with the silver salt of p-sulfamoylbenzoate to form a colored silver solution (Ciuhandu, 1957). Concentrations of 10 to 20,000 ppm CO can be measured by this method. The procedure has been modified to determine CO concentrations in incinerator effluents (Levaggi and Feldstein, 1966). Samples are collected in an evacuated flask and reacted. The absorbance of the resulting colloidal solution is measured spectrophotometrically. Acetylene and formaldehyde interfere, but can be removed by passing the sample through mercuric sulfate on silica gel. CO concentrations from 5 to 18,000 ppm can be measured with an accuracy of 90 to 100% of the true value.

A National Bureau of Standards colorimetric indicating gel that incorporates palladium and molybdenum salts has been devised to measure CO in the field (Shepherd et al., 1955). The laboratory method involves colorimetric comparison with freshly prepared indicating gels exposed to known concentrations of CO. The method has an accuracy range of 5 to 10% of the amount of CO involved. A minimum detectable concentration of 0.1 ppm is achievable. This technique requires relatively simple and inexpensive equipment, but oxidizing and reducing gases interfere. Preparation of indicator tubes is a tedious and time-consuming task.

Length-of-stain indicator tubes employ a method using potassium pallodosulfite. This is a commonly used method particularly in industrial hygiene and in exhaust and stack emission testing (Silverman and Gardner, 1965). CO reacts with the chemicals in the tube to produce a discoloration (from yellow to brown-black). The potassium pallodosulfite is reduced by CO to metallic platinum, which produces a dark brown stain. The reaction is: $K_2Pd(SO_3)_2$ + CO yields K_2SO_3 + Pd + CO_2 + SO_2. The length of the discoloration is an exponential function of the CO concentration. This method and other indicator tube manual methods using bellows or piston pumps to suck air through the tube are estimated to be accurate to ±25% of the amount of CO present at concentrations of about 100 ppm. Such indicator tube manual methods have been used frequently in air pollution studies. Colorimetric length-of-stain indicator tubes can be used in passive sampling devices that permit collection of air samples over many hours.

Colorimetric techniques and length-of-stain discoloration methods are generally recommended for use only when the other physicochemical monitoring systems are not available. They are helpful in range-finding CO concentrations and to ascertain quickly and inexpensively if an acute exposure situation exists, something totally inappropriate for a laborious laboratory analysis with results reported, for example, 2 weeks later. As such, these methods may be used in the field for gross mapping, where the highest accuracy is not of paramount importance and might be of immediate and great value during emergencies. Since interpretation of stain length is somewhat subjective, one or two detector tubes from each batch lot should be calibrated against a known concentration of CO gas in a plastic bag — perhaps at two different concentrations, e.g., 20 and 150 ppm. CO gas detector tubes often come in different concentration ranges from the same manufacturer (e.g., 8 to 500 ppm, 300 to 30,000 ppm, and 20,000 to 400,000 ppm). They should be stored in a cool place and not used past their expiration date without careful calibration. The author stores detector tubes at <25°F and has found that prolonged storage (>4 years) past the manufacturer's expiration date has caused no perceptible changes in the tube response. Careful reading of the instruction package insert is necessary especially with respect to interfering gases, use at extremes of temperature and pressure, and storage conditions.

In the past 20 years or so, there has been a surge of electrochemical detectors for CO introduced into commercial markets primarily intended for industrial hygienists and safety engineers. There have been continuous refinements in these instruments. Most can be purchased with multiple sensing heads (i.e., for CO, CH_4, H_2S, and O_2). Other gases and vapors that can be measured with these instruments include HCl, Cl_2, hydrocarbons, NO_2, NH_3, amines, SO_2, and chlorinated solvents, among others. Many of these provide data logging, peak readings, and instantaneous, STEL, and TWA alarms and readings. For CO, as an example, one instrument can measure 0–999 ppm CO at ±1 ppm. The electrical output from the instrument's chemical cell can be used to activate an alarm for peak concentrations at 200 ppm and when a worker's TWAE reaches 35 ppm. The electronics are solid state. Units are highly portable and relatively rugged and, considering the amount of information that they provide, inexpensive. As with any instrument, careful reading of the instruction manual is necessary. In other words, read, understand, and follow everything before using the instrument. Failing to do this can result in grossly erroneous measurements and, possibly, loss of life.

Most of these instruments rely on passive gas diffusion cells to permit entry of the gases into the chemical cell. Many can actually be used as personal monitoring devices worn directly by individual workers. Variations on the electrochemical cell principle have resulted in instruments that can be used as stack emission analyzers (measuring CO, O_2, SO_2, NO_x, stack gas temperature, draft, and soot) and exhaust emission analyzers (measuring CO, O_2, NO_x, CH_x, exhaust temperature, and soot).

Recently, passive CO detectors have been developed for home use. These devices constantly monitor the air for CO and activate a loud, piercing alarm to alert occupants of dangerous concentrations of CO gas. The U.S. Consumer Product Safety Commission recommends installation of these detectors in every home where CO gas could accumulate due to faulty fossil fuel furnaces, gas-fired hot water and kerosene or gas space heaters, automobiles idling in attached garages, etc. The units are mounted on ceilings or walls much like smoke detectors. Installation is easy. Some do not require batteries and operate on electric line current. The sensor life for some, according to the manufacturer's literature, is as long as 10 years.

In summary, there are numerous methods and instruments today to accurately and precisely measure CO in industry, the home, emission sources, and ambient air of our communities. Detection limits for some methods are in the parts per billion level, while most devices are designed to operate in the usual ppm ranges of industrial hygiene significance. With full understanding of the operating characteristics, interferences, and limitations of the instruments, highly reliable measurements are attainable. Frequent calibration of all instruments with verifiable CO span gas is of paramount importance.

REFERENCES

Adams, E.G. and Simmons, N.T., The determination of carbon monoxide by means of iodine pentoxide, *Journal of Applied Chemistry,* 1, 20, 1950.

American Conference of Governmental Industrial Hygienists, *Documentation of the Threshold Limit Values: Carbon Monoxide,* ACGIH, Cincinnati, OH, 1992.

Ciuhandu, G., Photometric determination of carbon monoxide in air, *Analytical Chemistry,* 155, 321, 1957.

Department of Health, Education, and Welfare, *Air Quality Criteria for Carbon Monoxide,* U.S. Government Printing Office, Washington, D.C., 1985.

Dubois, L., Zdrojewski, A., and Monkman, J.R., The analysis of carbon monoxide in urban air at the ppm level, and the normal carbon monoxide value, *Journal of the Air Pollution Control Association,* 16, 135, 1966.

Hersch, P. *Advances in Analytical Chemistry and Instrumentation,* Vol. 3, Reilly, C. A., Ed., Interscience Publishers, New York, 1964, 183.

Industrial Ventilation Committee, *Industrial Ventilation*, 20th ed., American Conference of Governmental Industrial Hygienists, Lansing, MI, 1993.
Kemmner, G., Nonnenmacher, K., and Wehling, W., Analytical application of infrared spectroscopy with gratings: Quantitative analysis of gas tracings, *Analytical Chemistry*, 222, 149, 1966.
LaDou, J., Ed., *Occupational Medicine*, Appleton and Lange, Norwalk, CT, 1990, 32.
Leidel, N.A., Busch, K.A., and Lynch, J.R., *Occupational Exposure Sampling Strategy Manual*, National Institute for Occupational Safety and Health, Washington, D.C., 1977.
Levaggi, D.A. and Feldstein, M., The colorimetric determination of low concentrations of carbon monoxide, *American Industrial Hygiene Association Journal*, 25, 64, 1966.
Lysyj, I., Zarembo, J.E., and Hanley, A., Rapid method for determination of small amounts of carbon monoxide in gas mixtures, *Analytical Chemistry*, 31, 902, 1959.
McCullough, J.D., Crane, R.A., and Beckman, A.O., Determination of CO in air by use of red mercuric oxide, *Analytical Chemistry*, 19, 999, 1947.
Michigan Department of Public Health, *Michigan's Occupational Health: Industrial Lift Trucks — Reducing Carbon Monoxide and Operational Costs*, Lansing, MI, 1971.
Mueller, P.K., Detection and analysis of atmospheric pollutants, paper presented at the Engineering Extension Course of Combustion Generated Air Pollution, University of California, Berkeley, 1965.
National Institute for Occupational Safety and Health, Criteria for a recommended standard: Occupational exposure to carbon monoxide, U.S. Department of Health, Education, and Welfare, Washington, D.C., 1972.
National Institute for Occupational Safety and Health, NIOSH Pocket Guide to Chemical Hazards, U.S. Department of Health and Human Services, Washington, D.C., 1990.
Porter, K. and Volman, D.H., Flame ionization detection of carbon monoxide for gas chromatographic analysis, *Analytical Chemistry*, 34, 748, 1962.
Salsburg, J.M., Cole, J.W., and Yoe, J.H., Determination of carbon monoxide — A microgravimetric method, *Analytical Chemistry*, 19, 1947.
Shepherd, M.S., Schuhmann, D., and Kilday, M.V., Determination of carbon monoxide in air pollution studies, *Analytical Chemistry*, 27, 380, 1955.
Silverman, L. and Gardner, G.R., Potassium pallado sulfite method for carbon monoxide detection, *American Industrial Hygiene Association Journal*, 26, 97, 1965.
Stevens, R.K., O'Keefe, A.E., and Ortman, G.C., A gas chromatographic approach to the semicontinuous monitoring of atmospheric carbon monoxide and methane, paper presented at 156th Natl. Meeting American Chemical Society, Atlantic City, NJ, 1968.
Threshold Limit Value Committee, *1994–1995 Threshold Limit Values for Chemical Substances and Physical Agents and Biological Exposure Indices*, American Conference of Governmental Industrial Hygienists, Cincinnati, OH, 1994.
U.S. Department of Labor, Code of Federal Regulations: Confined Space Entry, Vol. 29, Part 1910, U.S. Government Printing Office, Washington, D.C., 1993.

CHAPTER 2

Carbon Monoxide Formation Due To Metabolism of Xenobiotics

Dieter Pankow

CONTENTS

0-8493-4796-3/96/$0.00+$.50

1 INTRODUCTION

Carbon monoxide (CO) was first discovered as a metabolite of a xenobiotic in the 1970s. CO formation caused by exposure to dichloromethane (DCM) was suggested as the result of the chance observation that a physician had much higher than expected levels of carboxyhemoglobin (COHb) on each of two mornings following his use of varnish remover (80% DCM, 20% methanol) the evening before. An increase of COHb level from 0.4 to 2.4% was also produced when the physician was exposed 1 h under controlled conditions to 210 ±10 ppm DCM. Subsequent studies with volunteers showed that all of them produced CO as a metabolite of DCM. The physician and the volunteers were nonsmokers (Stewart et al., 1972; Stewart and Hake, 1976).

Also, animal experiments showed the metabolic CO formation: In order to estimate combined effects of CO and DCM, rats were exposed to 100 ppm CO and 1000 ppm DCM. The COHb level due to CO exposure was expected to be about 10%, but it was measured as 20% (Fodor et al., 1973). Later advances in the research dealing with metabolically formed CO included studies with radioactive DCM. The demonstration of ^{13}CO formation following intraperitoneal administration of ^{13}C-DCM, 3 mmol/kg, via infrared spectroscopy, of $C^{18}O$ formation in rat liver microsomes after application of both dibromomethane (DBM) and $^{18}O_2$, and of the reduction of the V_{max} of CO formation of 5.4 to 0.7 nmol CO/(min × mg protein) due to replacement of hydrogen in DCM by deuterium (Kubic et al., 1974; Kubic and Anders, 1978) are evidence for the origin of CO. Other dihalomethanes are oxidized to CO, too. The rate of conversion of the various dihalomethanes to CO was found to be diiodomethane (DIM) > DBM > bromochloromethane (BCM) > DCM (Gargas et al., 1986; Kubic and Anders, 1975; Pankow et al., 1992; Stevens et al., 1980) suggesting a relationship between decreased electronegativity of the halogen within the dihalomethane and increased biotransformation. Difluoromethane, fluorochloromethane, carbon tetrachloride, chloroform, chloromethane, iodomethane, trichloroethene, carbon disulfide, methanol, formaldehyde, dimethoxymethane, trichlorofluoromethane (Freon 11), and dichlorodifluoromethane (Freon 12) were also tested in rats and produced no detectable elevation in blood COHb levels (Fetz et al., 1978; Gargas et al., 1986; Kubic et al., 1974). Fodor and Roscovanu (1976) reported that the trihalogenated methane derivatives chloro-, bromo-, and iodoform lead to an increased COHb level in the blood of rats; however, data have not been published.

Other endogenous CO sources are the CO formation in connection with the hem degradation (Sjöstrand, 1970) or during the peroxidative degradation of unsaturated fatty acids (Wolff and Bidlack, 1976).

The purpose of this article is to review factors influencing the CO formation due to metabolism of xenobiotics, mainly of DCM, to show differences between CO hypoxia caused by CO or DCM exposure, and to characterize the possible CO risk due to DCM exposure.

2 XENOBIOTICS THAT ARE METABOLIZED TO CARBON MONOXIDE (CO)

2.1 DIHALOMETHANES

2.1.1 Dichloromethane

Dichloromethane (DCM, CH_2Cl_2, methylene chloride) is a volatile liquid that has found wide use as a paint remover, degreaser, and aerosol propellant, and as a solvent in photographic film production, in plastics manufacturing, and in laboratories. Two pathways have been associated with the metabolism of DCM.

One pathway that is localized in the endoplasmic reticulum (giving rise to the microsomal fraction), is catalyzed by cytochrome P-450 and involves the oxidative dechlorination of DCM resulting in the production of CO, chloride, and carbon dioxide (CO_2) via the proposed intermediates chloromethanol and formylchloride (Kubic and Anders, 1980; Gargas et al., 1986; Reitz et al., 1986).

The second pathway is localized in the cytosol, uses the system glutathione/glutathione S-transferase, and produces CO_2 and chloride (Ahmed and Anders, 1976, 1978). The first pathway has high affinity and low capacity, the cytosolic pathway low affinity and high capacity.

Kubic et al. (1974) showed that pretreatment of rats with phenobarbital or 3-methylcholanthrene failed to alter COHb levels after a single i.p. administration of DCM, 3 mmol/kg. This indicated that neither cytochrome P-450s 2B and 2C nor cytochrome P-450s 1A1 and 1A2 plays an essential role as catalyst of the DCM oxidation. In studies on the oxidation of DCM to CO, Guengerich et al. (1991) observed variation in V_{max} among four human liver microsomal preparations, but when these data were compared to rates of chlorzoxazone 6-hydroxylation, a correlation coefficient of 0.94 was found. Cytochrome P-450 2E1 (CYP2E1) is the primary catalyst of chlorzoxazone 6-hydroxylation in human liver (Peter et al., 1990). CYP2E1 is induced by a variety of agents including isoniazid, ethanol, acetone, pyrazole, imidazole, benzene, xylenes, and trichloroethylene (Ingelman-Sundberg and Jørnvall, 1984; Koop and Coon, 1986; Ryan et al., 1985, 1986). Increased rates of the oxidation of DCM to CO as measured by COHb formation was observed by pretreatment with these and other inducers provided they are no longer present to compete with the DCM for oxidative metabolism (Table 1). After simultaneous uptake of DCM and a substrate of CYP2E1 such as ethanol, methanol, butanol, acetone, chloroform, tetrachloromethane, benzene, m-, p- or o-xylene, styrene, ethylbenzene, tetrahydrofuran, aniline, pyrazole, trichloroethylene, isoniazid, or acetaminophen, the DCM-derived COHb formation is decreased or completely blocked. Some examples are given in Figure 1. This inhibition is well known after simultaneous exposure to DCM and ethanol (Balmer et al., 1976;

TABLE 1

Dichloromethane (DCM)-Derived (6.2 mmol/kg p.o.) Carboxyhemoglobin (COHb) Formation in Male Wistar Rats (age: 60–80 d) Pretreated with Vehicle (Water, *Oleum pedum tauri* or Saline) or with an Inducer of Cytochrome P-450 2E1 (CYP2E1); COHb Determinations 6 h after DCM administration (n = 58); COHb Basis Level: 0.7 ± 0.3% (n = 26); Own Determinations.

CYP2E1 Inducer	Dose [mmol/kg]	Time[a] [h]	COHb [%] (Vehicle-Pretreated Rats)	COHb [%] (CYP2E1-Inducer Pretreated Rats)
m-Xylene	1 × 16.3 p.o.	24	11.1 ± 0.4	24.7 ± 0.6*
Methanol	1 × 198 p.o.	24	8.1 ± 0.5	23.0 ± 1.0*
Isoniazid	1 × 0.36 i.p.	24	8.9 ± 0.6	22.9 ± 1.2*
Acetylhydrazine	1 × 0.36 i.p.	24	8.9 ± 0.6	22.6 ± 1.7*
Trichloroethylene	1 × 24.8 p.o.	24	11.8 ± 1.3	22.8 ± 0.9*
o-Xylene	1 × 16.3 p.o.	24	11.6 ± 0.6	21.1 ± 0.9*
Benzene	1 × 16.9 p.o.	16	11.1 ± 1.1	20.7 ± 1.3*
Toluene	1 × 18.8 p.o.	20	11.3 ± 1.0	18.6 ± 1.1*
Acetone	1 × 30 i.p.	48	9.3 ± 0.5	17.2 ± 2.1*
Styrene	1 × 18 p.o.	24	10.7 ± 0.5	16.4 ± 1.3*
Ethanol	1 × 174 p.o.	18	9.3 ± 0.5	15.6 ± 0.7*
p-Xylene	1 × 16.3 p.o.	24	11.6 ± 0.6	15.2 ± 0.9*
Aspirin	1 × 2.8 p.o.	24	10.4 ± 0.6	15.0 ± 0.6*
Pyrazole	3 × 0.73 i.p.	24	9.5 ± 1.0	14.4 ± 0.6*
Imidazole	2 × 4.4 i.p.	24	8.8 ± 0.5	13.0 ± 0.6*

[a] Time interval between vehicle/CYP2E1 inducer and DCM administration.
* $p < 0.05$ (t-test).

Ciuchta et al., 1979; Glatzel et al., 1987). An inhibition was also demonstrated in rats pretreated with toluene, 5 mmol/kg i.p., 30 min prior to an exposure to 5000 ppm DCM (Ciuchta et al., 1979) and in rats after simultaneous oral administration of both toluene and DCM, and if DCM was administered 12 h after toluene uptake, but the mean COHb formation in blood was significantly enhanced by prior administration of toluene at 20 to 28 h (Figure 2). Similar biphasic effects in dependence of the time interval between the uptake of CYP2E1 inducer and DCM were observed in rats following the combined exposure to ethanol or other aromatic compounds with DCM (Pankow et al., 1990, 1991b).

Relatively low doses of the mechanism-based inhibitors of CYP2E1, diethyldithiocarbamate (DDTC) or disulfiram, are necessary to inhibit the carboxyhemoglobinemia 6 h after oral administration of DCM, 6.2 mmol/kg

FIGURE 1
Carboxyhemoglobin (COHb) level in the blood of male Wistar rats 6 h after simultaneous oral administration of dichloromethane, 6.2 mmol/kg (1:10 in *Oleum pedum tauri*) and different aromatic compounds (A), ethanol (B), or drugs (C); the data are the mean ± SEM, n = 5–6, in A for better visibility without SEM.

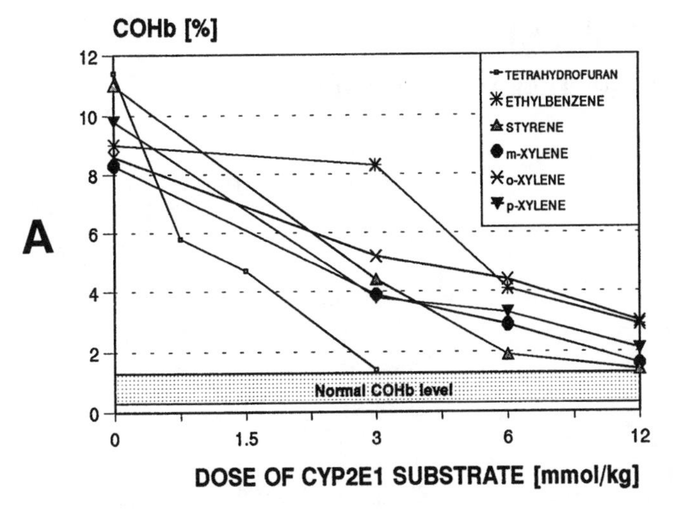
COHb [%]
12
10
8
6
4
2
0
TETRAHYDROFURAN
ETHYLBENZENE
STYRENE
m-XYLENE
o-XYLENE
p-XYLENE
A
Normal COHb level
0
1.5
3
6
12
DOSE OF CYP2E1 SUBSTRATE [mmol/kg]

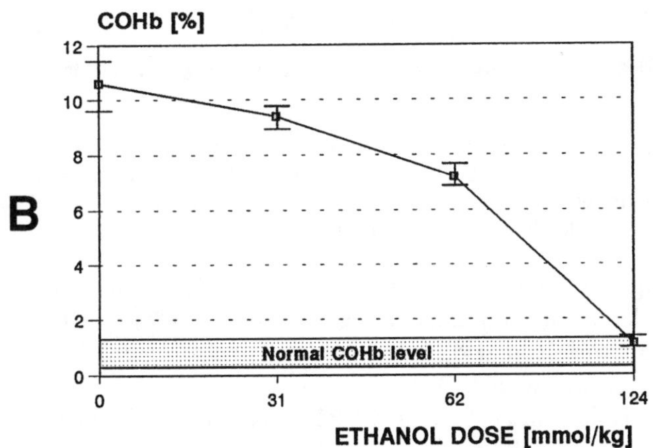
COHb [%]
12
10
8
6
4
2
0
B
Normal COHb level
0
31
62
124
ETHANOL DOSE [mmol/kg]

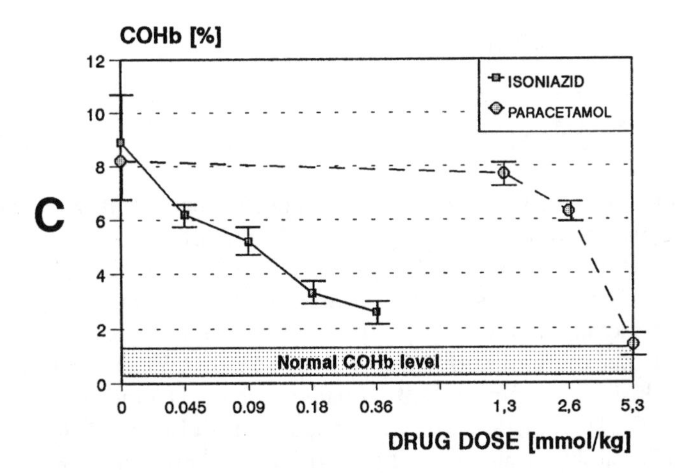
COHb [%]
12
10
8
6
4
2
0
ISONIAZID
PARACETAMOL
C
Normal COHb level
0
0.045
0.09
0.18
0.36
1,3
2,6
5,3
DRUG DOSE [mmol/kg]

FIGURE 1

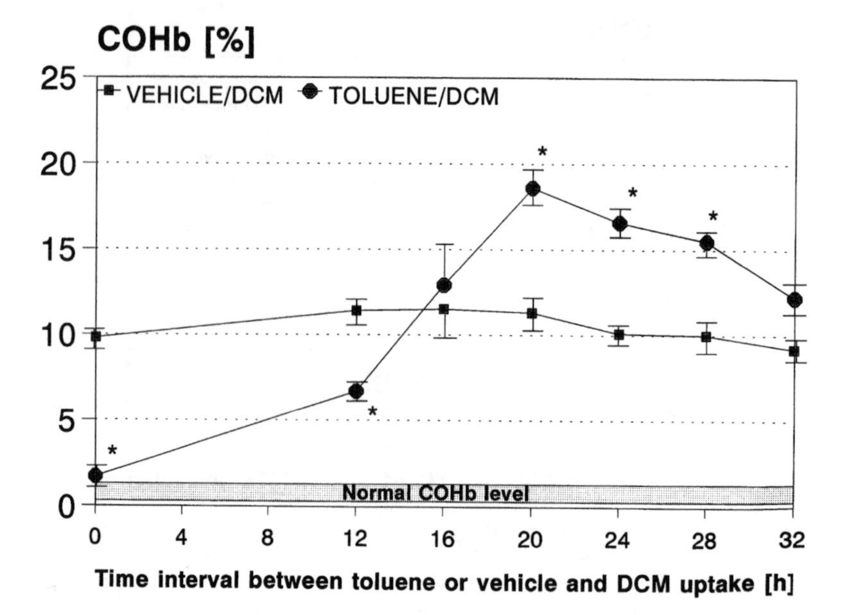

FIGURE 2
Carboxyhemoglobin (COHb) level in the blood 6 h after oral administration of dichloromethane, 6.2 mmol/kg (1:10 in *Oleum pedum tauri*) to male Wistar rats simultaneously treated or pretreated with a single dose of toluene, 18.8 mmol/kg p.o. (1:20 in *Oleum pedum tauri*) or with vehicle; the data are the mean ± SEM (n = 5), * $p < 0.05$.

(400 µl/kg) as a 10% (v/v) solution in *Oleum pedum tauri* to rats (Figure 3). The stimulation of the oxidative metabolism of DCM to CO due to treatment of rats with isoniazid, 360 µmol/kg i.p., 24 h before, is also inhibited after pretreatment with DDTC plus isoniazid.

These results indicate that CYP2E1 is a major, if not the principal, catalyst involved in the oxidation of DCM:

$$CH_2Cl_2 \xrightarrow[\text{NADPH, } O_2]{\text{CYP2E1}} [Cl_2CH(OH)] \xrightarrow[-HCl]{} ClCHO \xrightarrow[-HCl]{} CO$$

Studies using metabolic inhibitors suggest that significant amounts of CO_2 are also derived from this pathway (Gargas et al., 1986).

Volunteers were exposed for 6 h to 100 and 350 ppm DCM. DCM and COHb levels in blood, as well as DCM and CO concentrations in the expired air, were measured during exposure and for 24 h thereafter. The results indicated that the biotransformation of DCM via CYP2E1 is saturable and this conclusion was further substantiated in rats (DiVincenzo and Kaplan, 1981b; ECETOC, 1984; Kurppa et al., 1981; McKenna et al., 1982). The pathway is saturated at a concentration between 300 and 500 ppm DCM in the inhaled air or at a oral dose of about 6 mmol/kg as mentioned.

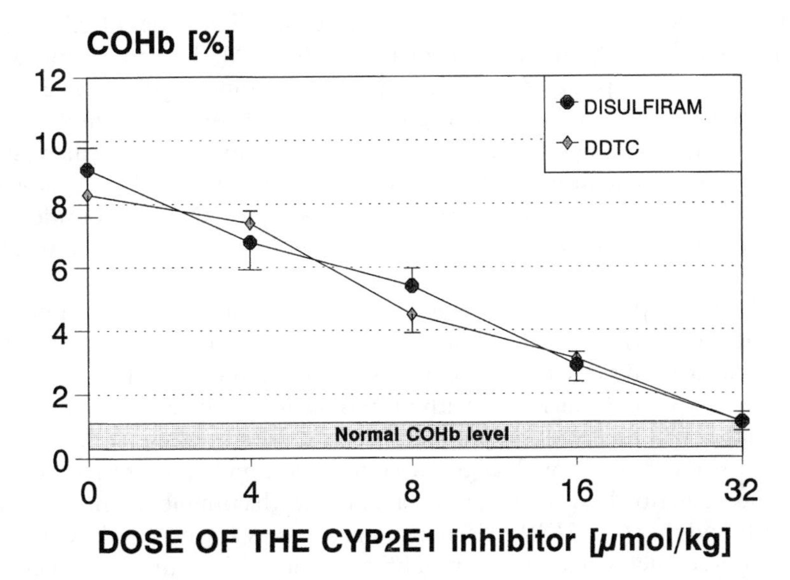

FIGURE 3
Carboxyhemoglobin (COHb) level in the blood of male Wistar rats 6 h after simultaneous oral administration of dichloromethane, 6.2 mmol/kg (1:10 in *Oleum pedum tauri*) and diethyldithiocarbamate (DDTC) or disulfiram; the data are the mean ± SEM, n = 6, * $p < 0.05$.

The resulting blood level is 10–12% COHb (Andersen et al., 1991; Fodor and Roscovanu, 1976; Kurppa and Vainio, 1981; McKenna et al., 1982; Pankow et al., 1991; Peterson, 1978; Schwetz et al., 1975). The COHb level drops more slowly after DCM than after CO exposure. This is interpreted as indicating that the metabolism of DCM to CO is saturated, leading to a deposition of DCM in a variety of body tissues, mainly in fat, from which it is slowly released. The dechlorination of DCM thus continues after termination of the exposure (Fodor and Roscovanu, 1976; McKenna et al., 1982; Ratney et al., 1974; Stewart and Hake, 1976). Relatively high concentrations of DCM in blood and fat tissue were measured after an accidental scenario in rats, exposure to 250,000 ppm DCM for 20 s, but a maximum of only 3–4% COHb resulted. The inhibition of CYP2E1 by CO and the hepatotoxicity of DCM (Balmer et al., 1976; Condie et al., 1983; Kitchin and Brown, 1989) were discussed to explain the low COHb formation (Wolna et al., 1994).

The perfused liver of rats was used as an experimental model to investigate the interaction of DCM-derived CO with cytochrome P-450: Monitoring was done of the spectral change following DCM in perfused liver. A hemoglobin-free perfusion medium saturated with carbogen was introduced via the portal vein, with the effluent flowing out from the inferior vena cava; DCM, 9 mmol/l, was administered in the perfusate. The spectrum after DCM infusion showed a peak at 450 nm and it was concluded that CO formed during cytochrome P-450 (CYP2E1)-dependent metabolism of DCM binds to cytochrome P-450 in the perfused liver (Takano and

Miyazaki, 1988). In another study hepatocytes at cell densities of 1.25×10^4 to 10^5 cells/0.2 ml/cm^2 were incubated with DCM, 10^{-4} to 10 mmol/l, for 24 h. The CO production increased with increasing exposure time, total cell number, and DCM concentration for values up to 4 mmol/l, but decreased at DCM concentrations greater than 4 mmol/l and with increasing cell density (Mizutani et al., 1988). Also, studies in rats and mice showed that the COHb level following exposure to 4000 ppm DCM is considerably reduced when compared with the COHb level following exposure to 500 ppm DCM (ECETOC, 1984).

The total uptake of DCM is higher for fat people than for thin people or people of normal weight (Lundberg, 1987). In rats, the DCM-derived COHb formation as well as the isoniazid-evoked stimulation of the oxidative metabolism of DCM increased with increasing age of the animals (Table 2). The higher CO formation in the older rats is caused by a higher fat content in the tissues. Animals with high fat content accumulate more DCM with the result described. Moreover, using a rat model that mimics human obesity, it was found that CYP2E1 levels in liver microsomes are increased: Rats were made obese (defined as body weight exceeding 125% that of pellet-fed control rats) by feeding with an energy-dense diet and the CYP2E1 activity was evaluated by determination of the catalytic activities associated with CYP2E1, including p-nitrophenol hydroxylation, N-nitrosodimethylamine demethylation and aniline hydroxylation (Raucy et al., 1991).

Interestingly, fasting also affects the expression of CYP2E1 (Hong et al., 1987) and the DCM-derived COHb was enhanced in rats following food deprivation for 48 h: 14.6 ±1.1% COHb vs. control value of 8.3 ±0.4% COHb were measured (Pankow and Hoffmann, 1989).

TABLE 2

Dichloromethane (DCM)-derived (6.2 mmol/kg p.o.) Carboxyhemoglobin (COHb) Formation in Rats Pretreated with Saline, 1 ml/kg i.p., or Isoniazid (INH), 360 μmol/kg i.p., in Dependence of the Age; COHb Determination 6 h after DCM Administration; the Data Are the Mean ± SEM (n = 6–8)

Group Age [days]	COHb [%] (DCM 24 h after NaCl)	COHb [%] (DCM 24 h after INH)
A 44	8.0 ± 0.8	15.1 ± 1.2*
B 77	8.5 ± 0.6	16.7 ± 0.7*
C 149	10.8 ± 0.6	20.5 ± 1.1*
D 303	11.8 ± 0.4	24.0 ± 1.7*
Probability of error $p < 0.05$	A vs. C A vs. D B vs. D	A vs. C A vs. D B vs. D

* $p < 0.05$ (t-test).

During repeated exposures to DCM, CO formation and COHb levels did not show a significant increase compared with levels following a single exposure. This was evident in volunteers exposed to 500 ppm DCM or less (DiVincenzo and Kaplan, 1981b; Fodor and Roscovanu, 1976), in rats and mice exposed to 1250 ppm DCM or less (Kim and Carlsson, 1986; Kurppa and Vainio, 1981; Schwetz et al., 1975), in hamsters exposed to 500, 1500 or 3500 ppm DCM (Burek et al., 1984) and in rats after repeated intravenous or oral administrations of DCM (Angelo et al., 1986, Pankow et al., 1985).

In controlled human studies, physical exercise increased the absorption of DCM, the metabolism of DCM to CO, the COHb formation and the pulmonary expiration of CO. There was no further increase of COHb in individuals exercising at moderate to heavy workloads, because the pulmonary excretion of CO increases while performing heavy workloads (Åstrand et al., 1975; DiVincenzo and Kaplan, 1981a; Hake, 1979).

The V_{max} values of the hepatic metabolism of DCM to CO are similar in men and rats (ECETOC, 1989). COHb levels were also similar in both rats and mice at 500 ppm DCM and did not increase at higher concentrations (ECETOC, 1989; Schwetz et al., 1975). The COHb levels in hamsters after DCM, 6.2 mmol/kg p.o., following pretreatment with vehicle or isoniazid, 360 μmol/kg, were 18.6 ±0.4 and 27.9 ±0.9%, respectively, compared to COHb levels in rats of 8.5 ±0.6 and 16.7 ±0.7%, respectively, using the same doses. In the blood of hamsters exposed 6 h to 500, 1500, or 3500 ppm DCM, COHb values of 24.1 ±2.2, 33.5 ±4.6, and 28.5 ±5.6, respectively, were measured (Burek et al., 1984).

After oral administration of DCM there is a circadian rhythm of the COHb formation in both naive rats and, more pronounced, rats pretreated with isoniazid or m-xylene, which are known to be inducers of CYP2E1 (Pankow et al., 1993). These results correspond with the high aniline hydroxylase activity in the microsomes of rat liver between 05.00 h and 13.00 h as determined by Bélanger and Labrecque (1989). In addition the enhanced exhalation rate of CO in nocturnal active rats may play a role.

The metabolism of DCM to CO is the basis on which current occupational exposure limits are set. A threshold limit value of 50 ppm DCM (American Conference of Governmental Industrial Hygienists) or a maximum allowable concentration (MAK-value; German Science Foundation) at the workplace of 100 ppm DCM is recommended. These values have been applied as 8-h time-weighted averages. In general, COHb levels of ≈ 2% and ≈ 4%, respectively, result following these exposures.

As mentioned earlier, the metabolism of DCM to CO seems to be saturable. Moreover, after exposure to high concentration of DCM the formed CO may be bound by cytochrome P-450 (inclusively CYP2E1) producing an inhibition of the further oxidation of DCM. Data from a group of nonsmoking industrial workers indicate that the mean maximum COHb level does not exceed 10% (Ott et al., 1983), but there are some case reports with distinctly higher COHb levels. These data (Table 3) are consistent with

TABLE 3

Case Reports of Exposure to Dichloromethane (DCM) with Resulting Carboxyhemoglobin (COHb) Levels More than 10%

Condition of DCM Exposure	Time after Exposure [h]	COHb [%]	Ref.
30 min cleaning of an open vessel (Diameter: 1 m, 1–2 m deep)	1	12	Wells and Waldron, 1984
Nitromors (DCM, methanol, cellulose acetate, triethanolamine, paraffin wax, detergents), ingestion, 300 ml	36	12.1	Hughes and Tracey, 1993
Cleaning of watch components	?	13	Liniger and Sigrist, 1994
986 ppm, 3 h	0.5	15	Stewart and Hake, 1976
Cleaning computer equipment, 6–8 h	16	20	Rudge, 1990
500 ppm, 8 h	0	24	Fodor and Roscovanu, 1976
Using paint remover 6 h	0	26	Langehennig et al., 1976
Approx. 168000 ppm (*n* = 2), death	?	30	Manno et al., 1992
6 h using paint remover	0	40	Langehennig et al., 1976
> 1 h using paint remover	?	50	Fagin et al., 1980

the hypothesis of a high existing CYP2E1 activity before the DCM poisoning. It is a speculation that disturbances of the lung function and resulting decreases of the CO exhalation may play an additional role.

DCM and methanol are commonly employed organic solvents that are also used in formulations. The results of a study to investigate the effect of methanol on the COHb formation after DCM gavage in rats demonstrate that the metabolism of DCM to CO may be inhibited or stimulated by methanol. The effect depends on the interval between methanol and DCM administration; the dose of methanol or DCM; and the frequency in the case of simultaneous administration of both DCM and methanol (Pankow and Jagielki, 1993). Other investigations showed that the COHb formation in rats exposed to 5000 ppm DCM for 1 h was not significantly different from the COHb formation in rats exposed to DCM in combination with 2500 ppm or 11,000 ppm methanol, and rats pretreated with methanol, 30 or 90 mmol/kg i.p. 30 min prior to a DCM inhalation responded with a significant decrease of the COHb formation, and the same tendency was seen in monkeys (Ciuchta et al., 1979). Only the minor pathway of methanol oxidation in the rat is catalyzed by CYP2E1 and the *in vitro* activity of CYP2E1, with methanol as substrate, is low in comparison to other substrates (Coon et al., 1984).

Therefore, high doses of methanol (> 148 mmol/kg p.o.) are necessary to reduce the DCM metabolism to CO on account of competitive inhibition. It seems that methanol is a stronger inducer than a substrate of CYP2E1: The DCM-derived COHb formation was enhanced (1) already after pretreatment with methanol, 12.4 mmol/kg p.o., and (2) after repeated simultaneous administration of DCM and methanol. Methanol ingestion results in a transient decrease of the glutathione content of the liver. In rats treated with glutathione-depleting chemicals such as diethylmaleate, phorone, or buthionine sulphoximine, there were no enhancements of the carboxyhemoglobinemia caused by DCM. The COHb formation was not influenced by an increase of the hepatic glutathione concentration due to repeated administration of butylated hydroxyanisole. Therefore, the two pathways of DCM, the oxidative via CYP2E1 and the metabolism via glutathione/glutathione S-transferase, seem to be independent (Pankow and Jagielki, 1993).

2.1.2 *Dibromomethane*

The rate of metabolism of DBM to CO is higher than that of DCM to CO both *in vitro* (Kubic and Anders, 1975; Stevens et al., 1980) and *in vivo* (Table 4).

TABLE 4

Dibromomethane(DBM)-derived Carboxyhemoglobinemia in Rats

DBM Exposure	Time after Exposure (h)	COHb [%]	Ref.
310 ppm, 4 hr	0	16	Gargas et al., 1986
3 mmol/kg i.p. in sesame oil	8	16	Fozo and Penney, 1993
6 mmol/kg i.p. in sesame oil	12	18	
3 mmol/kg i.p. in corn oil	6	20	
6.2 mmol/kg p.o. in *Oleum pedum tauri*	16	25	Kubic et al., 1974
1000 ppm, 3 h	0	22	Pankow et al., 1992
	2.5	27	Fodor and Roscovanu, 1976

The Michaelis constant (Km) for the microsomal metabolism of DBM to CO was about 16 mM; the maximal velocity V_{max} = 8 nmol CO/mg microsomal protein per min. Cytochrome P-450 was found to bind DBM; it results in a type I binding spectrum (Kubic and Anders, 1975).

Pretreatment with isoniazid > phenobarbital > 3-methylcholanthrene increased the rate of conversion of DBM to CO (Kubic and Anders, 1975; Pankow et al., 1992; Stevens et al., 1980). A mean COHb level of nearly 40% was measured following the gavage of DBM, 6 mmol/kg p.o., in isoniazid-pretreated rats. 2-Diethylamino-ethyl-2,2-diphenylvalerate-HCl (SKF 525A), ethylmorphine, hexobarbital, diethyl maleate, ethanol, and pyrazole inhibited the oxidation of DBM to CO (Gargas et al., 1986; Kubic and Anders, 1975; Stevens et al., 1980). These results demonstrate that CYP2E1 is the main enzyme catalyzing the oxygen insertion in DBM, and other cytochrome P450s as CYP2B1/2B2 or as CYP1A1/1A2 play an additional role to activate the reaction:

$$CH_2Br_2 \xrightarrow[\text{NADPH, } O_2]{\text{Cytochrome P450}} [Br_2CH(OH)] \xrightarrow[-HBr]{} BrCHO \xrightarrow[-HBr]{} CO$$

DBM was given once daily in corn oil solution for 4 d at a dose of 3 mmol/kg i.p. These repeated administrations led to no appreciable changes in COHb levels in comparison with the carboxyhemoglobinemia after a single DBM gavage (Kubic et al., 1974).

2.1.3 Diiodomethane

In the blood of rats exposed 0.5 or 3 h to 1000 ppm DIM, the COHb levels were 9 and 23%, respectively (Fodor and Roscovanu, 1976). Rats received DIM orally, 3.1 mmol/kg in *Oleum pedum tauri*; 30 min later the COHb level was increased from 0.7 ±0.3 to 8.9 ±0.1% and this value doubled after 10 h (Figure 4). A poisoning with a peak level of 14.2% COHb reached at 11 h after ingestion of 25 ml pure DIM by a 20-month-old girl was described (Weimerskirch et al., 1990). The patient exhibited lethargy, incoordination, hepatic failure, and bone marrow suppression and died 9 d after ingestion.

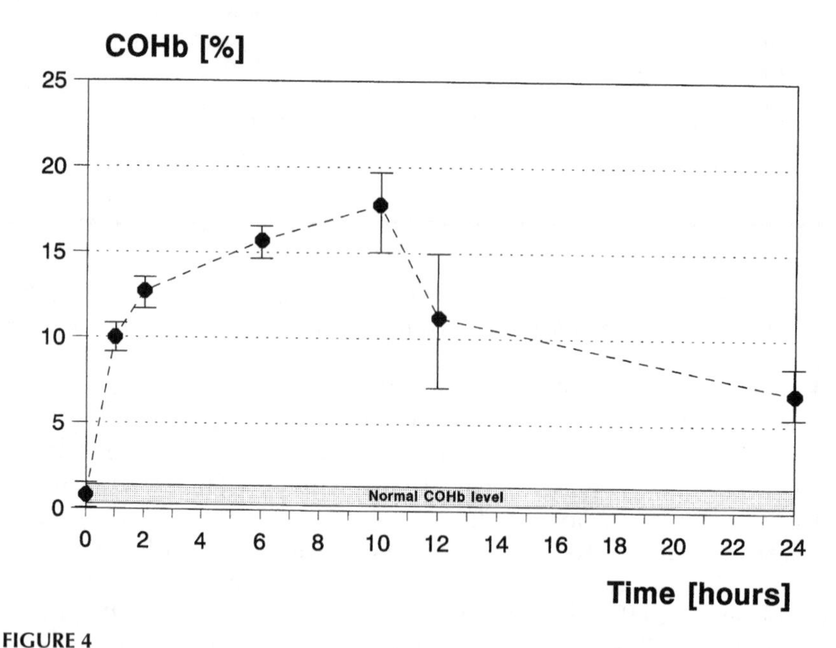

FIGURE 4
Course of the carboxyhemoglobin (COHb) level in the blood of male Wistar rats after oral administration of diiodomethane, 3.1 mmol/kg (1:10 in *Oleum pedum tauri*); the data are the mean ± SEM, n = 5.

2.1.4 *Bromochloromethane*

Blood COHb levels in rats associated with BCM exposure are similar as following DCM exposure (Table 5). Eight hours after administration of BCM, 6.2 mmol/kg p.o., in the blood of rats pretreated once daily for 4 d with saline, 1 ml/kg i.p., isoniazid, 360 μmol/kg i.p., or phenobarbital, 310 μmol/kg i.p., mean values of 10.7 ±0.6, 24.1 ±0.8 ($p < 0.05$), and 12.5 ±0.3% COHb ($p < 0.05$), respectively, were determined (Pankow et al., 1992). The BCM levels in the blood of rats pretreated with isoniazid were significantly lower than in rats pretreated with saline. Therefore, CYP2E1 is the major catalyst involved in the oxidative metabolism of BCM and other cytochrome P-450s such as CYP2B1/2B2 play a minor role in the dehalogenation of BCM.

TABLE 5

Bromochloromethane (BCM)-derived Carboxyhemoglobinemia in Rats

BCM exposure	Time after Exposure (h)	COHb [%]	Ref.
3 mmol/kg in corn oil	4	5	Kubic et al., 1976
1000 ppm, 4 h	0	9	Gargas et al., 1986
5000 ppm, 30 min	3	11	Andersen et al., 1991
6.2 mmol/kg p.o. in *Oleum pedum tauri*	8	11	Pankow et al., 1992

2.2 Trihalomethanes

Fodor and Roscovanu (1976) described that "trihalogenated methane derivatives were found to lead to an increased blood-COHb-level" but data were not given. Chloroform is oxidized mainly via CYP2E1 to trichloromethanol, which spontaneously dehydrochlorinates to the phosgene. The end products of the phosgene reaction with cellular water are carbon dioxide and hydrochloric acid. CO has been identified as a minor metabolite of anaerobic chloroform pathway, both from *in vitro* studies (Ahmed et al., 1977; Wolf et al., 1977) and *in vivo* animal studies (Anders et al., 1978). The *in vitro* rate of metabolism follows the sequence $CHCl_3 = CHBrCl_2 < CHBr_2Cl < CHBr_3 < CHI_3$ (Ahmed et al., 1980). Our own determinations showed that the COHb level in rats 6 h after administration of chloroform, 6.2 mmol/kg p.o. in *Oleum pedum tauri*, was 0.4 ± 0.1% ($n = 6$) and consistent with that of controls (0.4 ± 0.1% COHb; $n = 6$). Rats receiving dibromochloromethane, 3.1 mmol/kg p.o. in *Oleum pedum tauri*, responded with an increase to 2.8 ± 0.2% COHb ($n = 6$) after 6 h in comparison with 0.7 ± 0.1% COHb ($n = 6$) in the blood of vehicle-treated rats.

2.3 Tetrahalomethanes

^{14}CO is a product of the anaerobic incubation of $^{14}CCl4$ with hepatic microsomes and NADPH. Spectral studies indicated that the precursor of the CO is probably the cytochrome P-450-Fe^{2+} dichlorocarbene complex (Ahr et al., 1980). Twenty-four hours after a single oral administration of tetrachloromethane, 5.2 mmol/kg in *Oleum pedum tauri*, the COHb levels in the blood of rats was higher (1.5 ±0.1% COHb) than in controls (0.7 ±0.2% COHb), but both values were in the normal range.

2.4 1,3-Benzodioxoles

Under anaerobic conditions the optical difference spectra of several 1,3-benzodioxoles (methylenedioxyphenyl compounds) in NADPH-reduced rat liver microsomes exhibited a peak at 451.5 nm. Subsequent addition of hemoglobin showed formation of a peak at 419 nm, suggesting the presence of CO. The metabolic CO formation, which was confirmed by gas chromatography assay, was increased following phenobarbital pretreatment, not changed following 3-methylcholanthrene pretreatment, and decreased following cobaltous chloride pretreatment of the rats. The finding that [$2^{13}C$]-1,3 benzodioxoles are metabolized to ^{13}CO with the same degree of enrichment as the substrate shows clearly that CO is derived from the 1,3-benzodioxole and does not arise from heme breakdown or lipid peroxidation (Anders et al., 1984; Yu et al., 1980).

3 EFFECTS OF CO PRODUCED BY METABOLISM OF XENOBIOTICS

As it has been proved that DCM is metabolized to CO, the question has been raised whether the impairment of functions after DCM exposure is caused by DCM itself or by CO. Intracellular effects of CO depend on the CO partial pressures in the tissues. These CO partial pressures may be higher following DCM than CO exposure: *DCM-derived CO* is produced by means of CYP2E1, bound to membranes of the endoplasmic reticulum. The formed CO may be distributed among various cell fractions. Primarily, CO reacts with cytochrome P-450, cytochrome c oxidase, and myoglobin. Secondarily, CO binds to hemoglobin.

The first step in the process of *exogenous CO inhalation* is an increase in the CO concentration in the alveolar gas with diffusion from the gas phase through the pulmonary membrane and into the blood combining with hemoglobin to form COHb. In a second step CO binds to cellular hemoproteins. Recently, the cytochrome c oxidase activity in different tissues of rats were investigated following oral administration of DCM, 12.4 mmol/kg (about 10% COHb 6 h after gavage) or following accidental scenario — that is, the rats were exposed nose-only to DCM, 250,000 ppm for 20 s (3–4% COHb after 2 h). The cytochrome c oxidase activity was reduced 6 h after DCM ingestion in the tissues of brain, lung, and muscle by 28–42% and 20 min after the inhalative uptake of DCM in the tissues of brain, liver, kidney, and muscle by 42–51%. The effects were reversible and not evident in rats pretreated with the mechanism-based inhibitor of CYP2E1, diethyldithiocarbamate. Therefore, it seems that the effect of DCM is caused by the DCM metabolite CO (Lehnebach et al., 1994). Müller (1994) showed a decrease of the venous oxygen partial pressure of more than 30% after exposure of rats to 250,000 ppm DCM for 20 s. This may be explained by inhibition of the oxygen transport (COHb formation), by impairing the release of oxygen from oxyhemoglobin in the presence of COHb, and by the observed reduction of the cytochrome c oxidase activity that produced a moderate inhibition of the cellular respiration. In conclusion, the degree of DCM-derived CO hypoxia is determined not only by the COHb level.

The nature of the symptoms found in poisoning indicate that the combined action of DCM and CO are responsible for the development of vascular and CNS disturbances. Acute oral intoxication is accompanied by gastrointestinal hemorrhagy, without doubt, an effect of DCM. By comparing the CNS-depressant effects such as vigilance decrement, visual-flicker-fusion-depression and visual-motor disturbances in humans exposed to 300–800 ppm DCM with the lack of effects due to exposure to CO producing comparable COHb, it was concluded that metabolically formed CO could not be considered responsible for the observed DCM effects (Winneke, 1981). The combined tracking-monitoring-performance and the vigilance performance was impaired both by DCM and CO exposure with about 5% COHb (Putz et al., 1976) and this finding supports the hypothesis that CO is involved in the effect evoked by DCM. The decrease of nerve conduction velocity in rats following DCM administration may be caused by CO. Gavage of ethanol preventing the COHb formation does not prevent the DCM-related retardation of the nerve conduction. Therefore, it seems that the retardation is mainly caused by direct effects of the solvent (Glatzel et al., 1987). Peak COHb levels in the blood of rats were 16% 8 h after i.p. administration of DBM, 3 mmol/kg, diluted 1:3 in sesame oil, and 17% after exposure to 225 ppm CO for 120 min. The rectal body temperature dropped by about 1.0°C in both the DBM- and CO-exposed rats. The blood pressure was significantly decreased in rats following CO exposure, but not after DBM injection. Neither DBM nor CO exposure altered the heart rate or the blood glucose or lactate concentration (Fozo and Penney, 1993).

4 CONCLUSIONS: IS THERE A CO RISK?

The conversion of dihalomethanes to CO constitutes an important component of its short-term health risks. COHb levels as low as 2% have been associated with decreased exercise tolerance in coronary artery disease patients exposed to CO alone. The acute health effects of CO are well known, with headaches and light-headedness typically reported in the 10 to 20% COHb range, and nausea, dyspnea on exertion, decreased perceptual motor performance and motoric nerve conduction velocity, obtundation, convulsions, coma, and death occurring at progressively higher levels. Under conditions of high CYP2E1 activity such as preuptake of an inducer of CYP2E1, diabetes, fasting, or energy-dense diet, the COHb formation after exposure to dihalomethanes is relatively high and there really is a risk of CO poisoning (Table 3). The CO risk is high in smokers: An additive effect on blood COHb level by simultaneous exposure to CO (component of tobacco smoke) and DCM was observed (Fodor et al., 1973; Kurppa et al., 1981; Skrabalak and Babish, 1983), and sulfur and nitrogen containing heterocycles including thiazole, pyrazine, pyridazine, pyrimidine, thiophene, and triazole, which are present in tobacco and tobacco smoke, induce CYP2E1 (Kim and Novak, 1993).

REFERENCES

Ahmed, A.E. and Anders, M.W., Metabolism of dihalomethane to formaldehyde and inorganic chloride. *Drug Metab. Dispos.,* 4, 357, 1976.

Ahmed, A.E. and Anders, M.W., Metabolism of dichloromethane to formaldehyde and inorganic halide. II. Studies on the mechanism of the reaction, *Biochem. Pharmacol.,* 27, 2021, 1978.

Ahmed, A.E., Kubic, V.L., and Anders, M.W., Metabolism of haloforms to carbon monoxide. I. *In vitro* studies, *Drug Metab. Dispos.,* 5, 198, 1977.

Ahmed, A.E., Kubic, V.L., Stevens, J.L., and Anders, M.W., Halogenated methanes: Metabolism and toxicity, *Fed. Proc.,* 39, 3150, 1980.

Ahr, H.J., King, L.J., Nastainczyk, W., and Ullrich, V., The mechanism of chloroform and carbon monoxide formation from carbon tetrachloride by microsomal cytochrome P-450, *Biochem. Pharmacol.,* 29, 2855, 1980.

Anders, M.W., Stevens, J.L., Sprague, R.W., Shaath, Z., and Ahmed, A.E., Metabolism of haloforms to carbon monoxide. II. *In vivo* studies, *Drug Metab. Dispos.,* 6, 556, 1978.

Anders, M.W., Sunram, J.M., and Wilkinson, C.F., Mechanism of the metabolism of 1,3-benzodioxoles to carbon monoxide, *Biochem. Pharmacol.,* 33, 577, 1984.

Andersen, M.E., Clewell, H.J., III, Gargas, M.L., MacNaughton, M.G., Reitz, R.H., Nolan, R.J., and McKenna, M.J., Physiologically based pharmacokinetic modeling with dichloromethane, its metabolite, carbon monoxide, and blood carboxyhemoglobin in rats and humans, Toxicol. Appl. Pharmacol., 108, 14, 1991.

Angelo, M.J., Pritchard, A.B., Hawkins, D.R., Waller, A.R., and Roberts, A., The pharmacokinetics of dichloromethane. I. Disposition in B6C3F1 mice following intravenous and oral administration, *Food Chem. Toxicol.,* 24, 965, 1986.

Åstrand, I., Övrum, P., and Carlsson, A., Exposure to methylene chloride. I. Its concentration in alveolar air and blood during rest and exercise and its metabolism, *Scand. J. Work Environ. Health,* 1, 78, 1975.

Balmer, M.F., Smith, F.A., Leach, L.J., and Yuile, C.L., Effects in the liver of methylene chloride alone and with ethyl alcohol, *Am. Ind. Hyg. Assoc. J.,* 36, 345, 1976.

Bélanger, P.M. and Labrecque, G., Temporal aspects of drug metabolism, in *Chronopharmacology. Cellular and Biochemical Interactions,* Lemmer, B., Ed., Marcel Dekker, New York, 1989, 15.

Burek, J.D., Nitschke, K.D., Bell, T.J., Wackerle, D.L., Childs, R.C., Beyer, J.E., Dittenber, D.A., Rampy, L.W., and McKenna, M.J., Methylene chloride: A two-year inhalation toxicity and oncogenicity study in rats and hamsters, *Fund. Appl. Toxicol.,* 4, 30, 1984.
Ciuchta, A.P., Savell, G.M., and Spiker, R.C., The effect of alcohols and toluene upon methylene chloride-induced carboxyhemoglobin in the rat and monkey, *Toxicol. Appl. Pharmacol.,* 49, 347, 1979.
Condie, L.W., Smallwood, C.L., and Laurie, R.D., Comparative renal and hepatotoxicity of halomethanes: Bromodichloromethane, bromoform, chloroform, dibromochloromethane and methylene chloride, *Drug Chem. Toxicol.,* 6, 563, 1983.
Coon, M.J., Koop, D.R., Reeve, L.E., and Crump, B.L., Alcohol metabolism and toxicity: role of cytochrome P-450, *Fund. Appl. Toxicol.,* 4, 134, 1984.
DiVincenzo, G.D. and Kaplan, C.J., Effect of exercise or smoking on the uptake, metabolism, and excretion of methylene chloride vapor, *Toxicol. Appl. Pharmacol.,* 59, 141, 1981a.
DiVincenzo, G.D. and Kaplan, C.J., Uptake, metabolism, and elimination of methylene chloride vapor by humans, *Toxicol. Appl. Pharmacol.,* 59, 130, 1981b.
ECETOC (European Chemical Industry Ecology and Toxicology Centre), Methylene chloride, No. 4, Brussels, 1984.
ECETOC (European Chemical Industry Ecology and Toxicology Centre), Methylene chloride (dichloromethane): an overview of experimental work investigating species, differences in carcinogenicity, and their relevance to man, No. 34, Brussels, 1989.
Fagin, J., Bradley, J., and Williams, D., Carbon monoxide poisoning secondary to inhaling methylene chloride, *Br. Med. J.,* 281, 1461, 1980.
Fetz, H., Hoos, W.R., and Henschler, D., On the metabolic formation of carbon monoxide from trichloroethylene, *Naunyn-Schmiedeberg's Arch. Pharmacol.,* Suppl. 302, R22, 1978.
Fodor, G., Prajsnar, D., and Schlipköter, H.W., Endogene CO-Bildung durch inkorporierte Halogenkohlenwasserstoffe der Methanreihe, *Staub-Reinhalt. Luft,* 33, 258, 1973.
Fodor, G. and Roscovanu, A., Erhöhter Blut-CO-Gehalt bei Mensch und Tier durch inkorporierte halogenierte Kohlenwasserstoffe, *Zbl. Bakt. Hyg. I. Abt. Orig. B,* 162, 34, 1976.
Fozo, M.S. and Penney, D.G., Dibromomethane and carbon monoxide in the rat: Comparison of the cardiovascular and metabolic effects, *J. Appl. Toxicol.,* 13, 147, 1993.
Gargas, M.L., Clewell, H.J., and Andersen, M.E., Metabolism of inhaled dihalomethanes *in vivo*: Differentiation of kinetic constants for two independent pathways, *Toxicol. Appl. Pharmacol.,* 82, 211, 1986.
Glatzel, W., Tietze, K., Gutewort, R., and Pankow, D., Interaction of dichloromethane and ethanol in rats: Toxicokinetics and nerve conduction velocity, *Alcoholism: Clin. Exp. Res.,* 11, 450, 1987.
Guengerich, F.P., Kim, D.-H., and Iwasaki, M., Role of human cytochrome P-450 IIE1 in the oxidation of many low molecular weight cancer species, *Chem. Res. Toxicol.,* 4, 168, 1991.
Hake, C.L., Simulation studies of blood carboxyhemoglobin levels associated with inhalation exposure to methylene chloride, *Toxicol. Appl. Pharmacol.,* 48, A56, 1979.
Hong, J., Pan, J., Gonzalez, F.J., Gelboin, H.V., and Yang, C.S., The induction of a specific form of cytochrome P450 (P450j) by fasting, *Biochem. Biophys. Res. Commun.,* 142, 1077, 1987.
Hughes, N.J. and Tracey, J.A., A case of methylene chloride (Nitromors) poisoning, effects on carboxyhaemoglobin levels, *Human Exper. Toxicol.,* 12, 159, 1993.
Ingelman-Sundberg, M. and Jörnvall, H., Induction of the ethanol-inducible form of rabbit liver microsomal cytochrome P-450 by inhibitors of alcohol dehydrogenase, *Biochem. Biophys. Res. Commun.,* 124, 375, 1984.
Kim, S.G. and Novak, R.F., The induction of cytochrome P4502E1 by nitrogen- and sulfur-containing heterocycles: Expression and molecular regulation, *Toxicol. Appl. Pharmacol.,* 120, 257, 1993.
Kim, Y.C. and Carlson, G.P., The effect of unusual workshift on chemical toxicity. I. Studies on the exposure of rats and mice to dichloromethane, *Fund. Appl. Toxicol.,* 6, 162, 1986.
Kitchin, K.T. and Brown, J.L., Biochemical effects of three carcinogenic chlorinated methanes in rat liver, *Teratol. Carcinog. Mutag.,* 9, 61, 1989.

Koop, D.R. and Coon, M.J., Ethanol oxidation and toxicity: role of alcohol P-450 oxygenase, *Alcoholism: Clin. Exp. Res.,* 10, 44S, 1986.

Kubic, V.L. and Anders, M.W., Metabolism of dihalomethanes to carbon monoxide. II. *In vitro* studies, *Drug Metabol. Disp.,* 3, 104, 1975.

Kubic, V.L. and Anders, M.W., Metabolism of dihalomethanes to carbon monoxide. III. Studies on the mechanism of the reaction, *Biochem. Pharmacol.,* 27, 2349, 1978.

Kubic, V.L., Anders, M.W., Engel, R.R., Barlow, C.H., and Caughey, W.S., Metabolism of dihalomethanes to carbon monoxide. I. *In vivo* studies, *Drug Metab. Dispos.,* 2, 53, 1974.

Kurppa, K., Kivistö, H., and Vainio, H., Dichloromethane and carbon monoxide inhalation: Carboxyhemoglobin addition, and drug metabolizing enzymes in rat, *Int. Arch. Occup. Environ. Health,* 49, 83, 1981.

Kurppa, K. and Vainio, H., Effects of intermittent dichloromethane inhalation on blood carboxyhemoglobin concentration and drug metabolizing enzymes in rat, *Res. Commun. Chem. Pathol. Pharmacol.,* 32, 535, 1981.

Langehennig, P.L., Seeler, R.A., and Berman, E., Paint removers and carboxyhemoglobin, *N. Engl. J. Med.,* 295, 1137, 1976.

Lehnebach, A., Kuhn, C., and Pankow, D. Dichloromethane as an inhibitor of cytochrome c oxidase, *ISSX Proc.,* 5, 43, 1994.

Liniger, B. and Sigrist, T., Kohlenmonoxid-Hämoglobinämie infolge Dichlormethan-Exposition mit dermatologischen Auswirkungen, *Hautarzt,* 45, 8, 1994.

Lundberg, I., Methylene chloride. Arbete och Hälsa, 40, 75, 1987.

Manno, M., Rugge, M., and Cocheo, V., Double fatal inhalation of dichloromethane, *Human Exp. Toxicol.,* 11, 540, 1992.

McKenna, M.J., Zempel, J.A., and Braun, W.H., The pharmacokinetics of inhaled methylene chloride in rats, *Toxicol. Appl. Pharmacol.,* 65, 1, 1982.

Mizutani, K., Shinomiya, K., and Shinomiya, T., Hepatotoxicity of dichloromethane, *Forensic Sci. Int.,* 38, 113, 1988.

Müller, S., Short time exposure of rats to high concentration of dichloromethane — effects on the cardiovascular system, *Naunyn-Schmiedeberg's Arch. Pharmacol.,* Suppl. 349, R107, 1994.

Ott, M.G., Skory, L.K., Holder, B.B., Bronson, J.M., and Williams, P.R., Health evaluation of employees occupationally exposed to methylene chloride. Metabolism data and oxygen half-saturation pressure, *Scand. J. Work Environ. Health,* 9 (Suppl. I), 31, 1983.

Pankow, D. and Hoffmann, P., Diochloromethane metabolism to carbon monoxide can be induced by isoniazid, acetone and fasting, *Arch. Toxicol.,* Suppl. 13, 302, 1989.

Pankow, D. and Jagielki, S., Effect of methanol or modifications of the hepatic glutathione concentration on the metabolism of dichloromethane to carbon monoxide in rats. *Human Exp. Toxicol.,* 12, 227, 1993.

Pankow, D., Kretschmer, S., and Hoffmann, P., Inhibition and stimulation of dichloromethane metabolism to carbon monoxide in the rat following administration of ethanol, *Eur. J. Pharmacol.,* 183, 566, 1990.

Pankow, D., Kretschmer, S., and Weise, M., Effect of pyrazole on dichloromethane metabolism to carbon monoxide, *Arch. Toxicol.,* Suppl. 14, 246, 1991a.

Pankow, D., Matschiner, F., and Weigmann, H.J., Influence of aromatic hydrocarbons on the metabolism of dichloromethane to carbon monoxide in rats, *Toxicology,* 68, 89, 1991b.

Pankow, D., Ponsold, W., and Moritz, R.P., Zur Sauerstofftransportkapazität des Blutes unter Einfluß von Ethanol und Dichlormethan, *Z. Gesamte Hyg. Ihre Grenzgeb.,* 31, 460, 1985.

Pankow, D., Wolna, P., and Hoffmann, P., Is there a circadian rhythm of cytochrome P450IIE1 activity?, in *Chronobiology and Chronomedicine. Basic Research and Application,* Gutenbrunner, C., Hildebrandt, G., Moog, R., Eds., Verlag Peter Lang, Frankfurt, 1993, 26.

Pankow, D., Weise, M., and Hoffmann, P., Effect of isoniazid or phenobarbital pretreatment on the metabolism of dihalomethanes to carbon monoxide, *Pol. J. Occup. Med. Environ. Health,* 5, 245, 1992.

Peter, R., Böcker, R., Beaune, P.H., Iwasaki, M., Guengerich, F.P., and Yang, C.S., Hydroxylation of chlorzoxazone as a specific probe for human liver cytochrome P450-E1, *Chem. Res. Toxicol.,* 3, 566, 1990.

Peterson, J.E., Modeling the uptake, metabolism and excretion of dichloromethane by man, *Am. Ind. Hyg. Assoc. J.*, 39, 41, 1978.

Putz, V.R., Johnson, B.L., and Setzer, J.V., A comparative study of the effect of carbon monoxide and methylene chloride on human performance, *J. Environ. Pathol. Toxicol.*, 2, 97, 1976.

Ratney, R.S., Wegman, D.H., and Elkins, H.B., *In vivo* conversion of methylene chloride to carbon monoxide, *Arch. Environ. Health*, 28, 223, 1974.

Raucy, J.L., Lasker, J.M., Kraner, J.C., Salazar, D.E., Lieber, C.S., and Corcoran, G.B., Induction of cytochrome P450-E1 in the obese overfed rat, *Mol. Pharmacol.*, 39, 275, 1991.

Reitz, R.H., Smith, F.A., and Andersen, M.E., *In-vivo* metabolism of ^{14}C-methylene chloride, *Toxicologist*, 6, 260, 1986.

Roth, R.P., Drew, R.T., Lo, R.J., and Fouts, J.R., Dichloromethane inhalation, carboxyhemoglobin concentrations, and drug metabolizing enzymes in rabbits, *Toxicol. Appl. Pharmacol.*, 33, 427, 1975.

Rudge, F.W., Treatment of methylene chloride induced carbon monoxide poisoning with hyperbaric oxygenation, *Mil. Med.*, 155, 570, 1990.

Ryan, D.E., Koop, D.R., Thomas, P.E., Coon, M.J., and Levin, W., Evidence that isoniazid and ethanol induce the same microsomal cytochrome P-450 in rat liver, an isozyme homologous to rabbit liver cytochrome P-450 isozyme 3a, *Arch. Biochem. Biophys.*, 246, 633, 1986.

Ryan, D.E., Ramanathan, L., Iida, S., Thomas, P.E., Haniu, M., Shively, J.E., Lieber, C.S., and Levin, W., Characterization of a major form of rat hepatic microsomal cytochrome P-450 induced by isoniazid, *J. Biol. Chem.*, 260, 6385, 1985.

Schwetz, B.A., Leong, K.J., and Gehring, P.J., The effect of maternally inhaled trichloroethylene, perchloroethylene, methyl chloroform, and methylene chloride on embryonal and fetal development in mice and rats, *Toxicol. Appl. Pharmacol.*, 32, 84, 1975.

Sjöstrand, T., Early studies of CO production, *Ann. N.Y. Acad. Sci.*, 174(I), 5, 1970.

Skrabalak, D.S. and Babish, J.G., Safety standards for occupational exposure to dichloromethane, *Regul. Toxicol. Pharmacol.*, 3, 139, 1983.

Stevens, J.L., Ratnayake, J.H., and Anders, M.W., Metabolism of dihalomethanes to carbon monoxide. IV. Studies in isolated rat hepatocytes, *Toxicol. Appl. Pharmacol.*, 55, 484, 1980.

Stewart, R.D., Fisher, T.N., Hosko, M.J., Peterson, J.E., Baretta, E.D., and Dodd, H.C., Experimental human exposure to methylene chloride, *Arch. Environ. Health*, 25, 342, 1972.

Stewart, R.D. and Hake, C.L., Paint remover hazard, *J. Am. Med. Assoc.*, 235, 398, 1976.

Takano, T. and Miyazaki, Y., Metabolism of dichloromethane and the subsequent binding of its product carbon monoxide, to cytochrome P-450 in perfused rat liver, *Toxicol. Lett.*, 40, 93, 1988.

Weimerskirch, P.J., Burkhart, K.K., Bono, M.J., Finch, A.B., and Montes, J.E., Methylene iodide poisoning, *Ann. Emerg. Med.*, 19, 1171, 1990.

Wells, G.G. and Waldron, H.A., Methylene chloride burns, *Br. J. Ind. Med.*, 41, 420, 1984.

Winneke, G., The neurotoxicity of dichloromethane, *Neurobehav. Toxicol. Teratol.*, 3, 391, 1981.

Wolf, C.R., Mansuy, D., Nastainczyk, W., Deutschmann, G., and Ullrich, V., The reduction of polyhalogenated methanes by liver microsomal cytochrome P-450, *Mol. Pharmacol.*, 13, 698, 1977.

Wolff, D.G. and Bidlack, W.R., The formation of carbon monoxide during peroxidation of microsomal lipids, *Biochem. Biophys. Res. Commun.*, 73, 850, 1976.

Wolna, P., Wünscher, U., and Damme, B., Toxicokinetics of accidental scenarios with dichloromethane, *Naunyn-Schmiedeberg's Arch. Pharmacol.*, Suppl. 349, R107, 1994.

Yu, L.-S., Wilkinson, C.F., and Anders, M.W., Generation of carbon monoxide during the microsomal metabolism of methylenedioxyphenyl compounds, *Biochem. Pharmacol.*, 29, 1113, 1980.

CHAPTER 3

Modeling the Uptake and Elimination of Carbon Monoxide

Peter Tikuisis

CONTENTS

0-8493-4796-3/96/$0.00+$.50

1 INTRODUCTION

Carbon monoxide (CO) has received a considerable amount of attention because of its potentially lethal consequences in relatively small doses, coupled with the fact that human senses cannot detect CO. The dangers are exacerbated by the occurrence of symptoms that are so nonspecific as to be ignored or to cause delay in treatment. It is desirable, therefore, to have the capability of predicting CO uptake and elimination for preparedness in the event of an exposure, whether planned or accidental. One means of achieving this capability is through the development of mathematical models. Ideally, the model should take into account not only the level of CO exposure, but also the subject's physiological characteristics and level of activity. This chapter will explore the various models that have emerged and will describe in detail the most successful model to date. This will be followed by guidelines for its use in certain applications.

The uptake of CO and its effects were closely examined by Haldane (1895) a century ago. Attempts to quantify CO uptake followed about 50 years later through the efforts of Forbes et al. (1945) and Pace et al. (1946). With increased data on human exposure to CO and improved understanding of the physical and physiological processes underlying CO uptake and its elimination, predictive models became more complex and precise, eventually culminating in the widely used Coburn-Forster-Kane (CFK) model developed by Coburn et al. (1965).

Through these developments, the modeling approach has been diverse, ranging from empirical data-fitting to rationally derived formulations. These will be dealt with separately in the following sections. First, it is instructive to review some of the earlier findings that provided the foundation for subsequent model development. Unless otherwise stated, all experimental results cited herein refer to human subjects.

Haldane (1895) reported that during the uptake of CO, about half of the quantity breathed was absorbed and that the rate of this absorption was proportional to the respiratory exchange rate per unit of body mass. Haldane estimated that the time taken to attain equilibrium with the ambient CO was 2.5 h and that desorption proceeded more slowly than absorption. In addition, the relative affinity of CO for hemoglobin, resulting in carboxyhemoglobin (COHb; see Chapter 8, this volume, for details on the binding mechanism), was estimated as 140 times that of O_2. This work was followed by the determination of the diffusing lung capacity for CO (D_LCO) by Krogh and Krogh (1910) in subjects at rest and at work. Adopting the single-breath method, these investigators reported values of about 20 and 30 ml STPD·min^{-1}·Torr^{-1}, respectively.

After three decades of little further attention to CO, an intense and productive period of experimentation began. Using radioactive CO, Tobias et al. (1945) found that less than 0.1% of CO oxidized to CO_2, thus establishing that most of the CO absorbed in the body was preserved. Roughton and Root (1945) deduced that 30–40% of the absorbed CO combined with myoglobin-like pigments, but as COHb reduced, these extravascular stores released CO into the blood for washout. In a related study, Roughton (1945) demonstrated that increased levels of both exercise and inhaled O_2 enhanced the elimination of CO. Forbes et al. (1945) confirmed the benefit of O_2 breathing by observing a decrease in the rate of CO uptake under hyperoxic conditions. These findings were consistent with those of Lilienthal and Pine (1946), who reported a rate of CO uptake that was inversely proportional to the level of O_2 breathed. Their data also suggested that D_LCO was lowered with increased O_2. In another study, Lilienthal et al. (1946) determined that the Haldane ratio (M) of affinities of CO and O_2 for hemoglobin was 204, higher than originally estimated.

The level of detail in these studies prompted investigators to construct tables and/or mathematical formulae for the prediction of CO uptake. One such chart that is often cited was provided by Forbes et al. (1945) who based their predictions on the resultant COHb of subjects exposed to different concentrations of CO at sea level and at various levels of exercise. This chart, shown in Figure 1, displays many features on the uptake of CO: (a) that the uptake is exponential in nature, (b) the initial rate of formation of COHb is proportional to the ambient level of CO, and (c) the final (equilibrium) value of COHb increases less proportionally with increasing ambient level of CO. This figure marks the emergence of empirical models as covered in the following section.

2 EMPIRICAL MODELS

The curves shown in Figure 1 were generated with the following formula derived by Forbes et al. (1945):

$$\frac{dCO}{dt} = k \cdot P_ACO \cdot \left(\frac{COHb\%(\infty) - COHb\%(t)}{COHb\%(\infty) - COHb\%(0)} \right) \tag{3.1}$$

where k is a constant which increases with the activity level of the individual, P_ACO is the pressure of CO in the inspired air, and t = ∞ and 0 refer to the percent blood saturation of COHb at equilibrium and at the start of exposure, respectively. The equilibrium value was obtained from the Haldane relationship:

$$\frac{COHb\%(\infty)}{100 - COHb\%(\infty)} = \frac{M \cdot P_ACO}{P_CO_2} \tag{3.2}$$

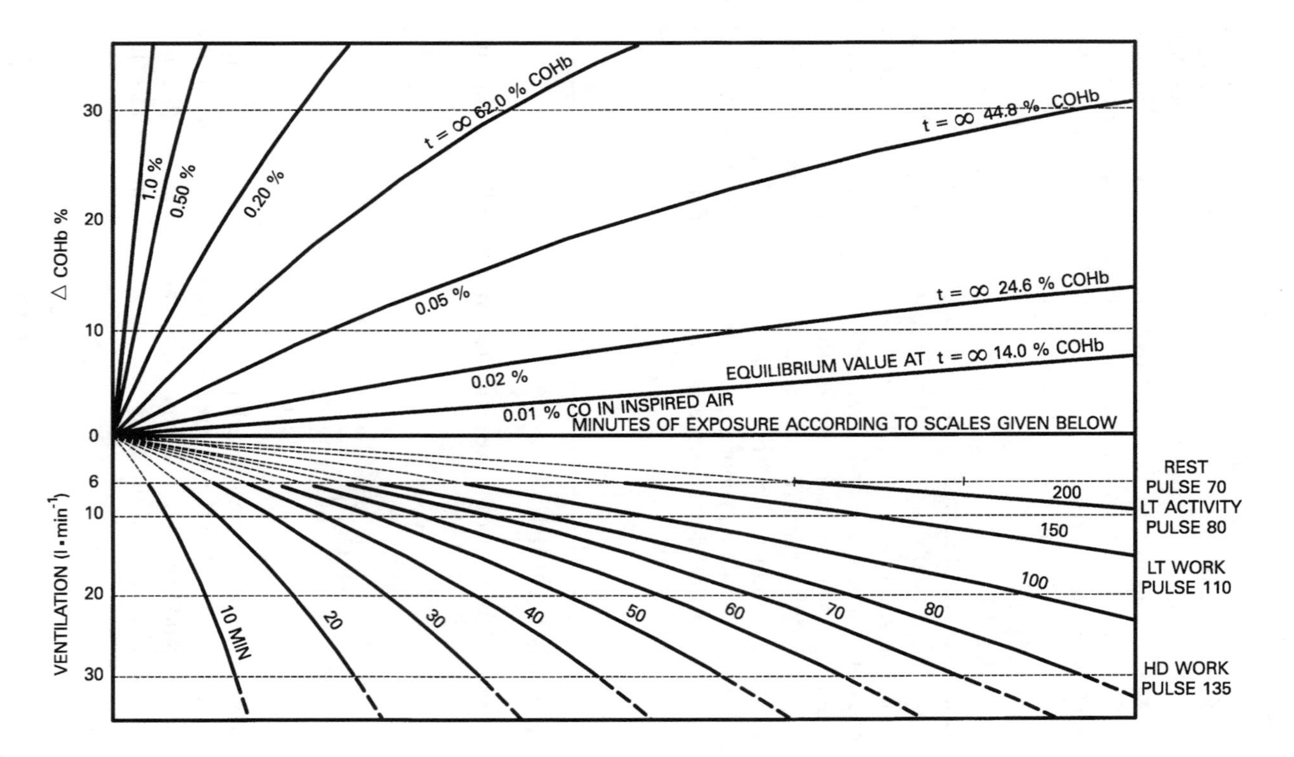

FIGURE 1
Uptake of CO in percentage of the inspired air for various concentrations and rates of ventilation. Upon choosing the desired ventilation rate on the lower ordinate scale and extending to the desired time of exposure, a vertical line drawn upward and intersecting the CO level of interest yields the predicted increase in COHb along the upper ordinate scale. (From Forbes et al., *Am. J. Physiol.,* 143, 594, 1945. With permission.)

where M and P_CO_2 were assumed to have values of 210 and 98 Torr, respectively (see Glossary for explanation of common model variables in these and all subsequent equations). For certain applications, the above two equations suffice to give reasonable estimates of the resultant COHb following an exposure to CO. Goldsmith et al. (1963) essentially applied similar expressions in their prediction of COHb during simulated exposures to fluctuating CO levels in an attempt to replicate ambient urban conditions.

Peterson and Stewart (1970) regressed the following equation of the resultant COHb from data of resting individuals exposed to CO ranging from 25 to 512 ppm for durations of 0.5 to 24 h:

$$\text{COHb\%} = \left(\frac{(\text{ppmCO})^{0.858} \cdot t^{0.630}}{197} \right) \cdot 10^{0.00094 t'} \tag{3.3}$$

where t′ is the postexposure time (experimental values of COHb ranged from about 2 to 25%). The above equation applied to individuals having a low initial COHb value (< 1%). To predict the change in COHb from a higher starting value and to also address the response to acute levels of CO exposure, Stewart et al. (1973) measured COHb values in resting subjects following exposures of up to 35,600 ppm CO for 45 s and from these regressed the following equation:

$$\Delta\text{COHb\%} = (\text{ppmCO})^{1.036} \cdot V_{air} \cdot 10^{-4.48} \tag{3.4}$$

where V_{air} is the volume (liters) of CO-contaminated air breathed.

Ott and Mage (1978) extended the work of Goldsmith et al. (1963) of predicting COHb values under fluctuating ambient urban conditions. Their model is based on the following simple, first order, linear differential equation (similar in form to that used by Forbes (1945)):

$$\tau \cdot \frac{d\text{COHb\%}}{dt} + \text{COHb\%} = 0.5 + \alpha \cdot \text{ppmCO} \tag{3.5}$$

where τ is the time constant of CO uptake, 0.5 represents the endogenous level (%) of COHb, and α is an empirically derived constant. Values of 149.4 min and 0.15 were assumed for τ and α, respectively. Equation 3.5 was developed for ambient levels of CO not exceeding 100 ppm. Venkatram and Louch (1979) applied a similar expression for the purpose of testing/setting CO standards of exposure under similar environmental conditions.

A property of the above differential equation is that for a given dose of CO (i.e., ppmCO·min), the resultant COHb after exposure is higher if the dose is received in less time than if received over a longer period. Furthermore, the former predicted COHb remains higher even during washout. For example, a dose of 600 ppmCO·min received as 10 ppmCO over 60 min leads to 0.996% COHb, while 50 ppmCO over 12 min leads to 1.117 and 1.080% COHb at the end of exposure and at 60 min. This predicted asymmetry was the subject of a later experimental investigation using the CFK model as described in the next section.

The most recent empirical formulae to emerge were developed by Chung (1988) for men at rest and at a light activity level, respectively given by:

$$\mathrm{COHb\%} = \frac{(\mathrm{ppmCO} - 0.1)\cdot t^{0.957} - 62.3}{3.18\cdot t^{0.957} + 2540} \tag{3.6}$$

and

$$\mathrm{COHb\%} = \frac{(\mathrm{ppmCO} - 0.1)\cdot t^{0.713} - 12.0}{0.0785\cdot t^{0.713} + 600} \tag{3.7}$$

These equations were developed for ambient levels of CO not exceeding 1000 ppm. Furthermore, they cannot be applied for very large t, otherwise COHb predictions extrapolate to unreasonably high levels.

Figure 2 shows the predicted COHb for several of the empirical models outlined above. The high predictions based on the model of Ott and Mage (1978) can partly be explained by the relatively low time constant of CO uptake (149.4 min), which corresponds to a level of activity higher than that assumed by Forbes et al. (1945). Yet, it is peculiar that while the predictions of Forbes et al. (1945) and Peterson and Stewart (1970) agree very closely, the former assumes a light activity level while the latter is based strictly on resting conditions. The comparisons are confounded further by the relatively low predictions based on Chung (1988), which, in this example, also assumed a resting condition. Some of these disparities may be attributed to the fact that most empirical equations are accurate only in the domain in which they were derived. Nevertheless, the wide disparity among these predictive equations clearly demonstrates the need for a more rational approach — that is, one based on physical and physiological principles. The next section will introduce such an approach through the development of the CFK model.

3 CFK MODEL

3.1 DEVELOPMENT

As will be seen, two primary physiological variables that govern the rate of CO uptake and its elimination are the ventilation rate and the diffusing

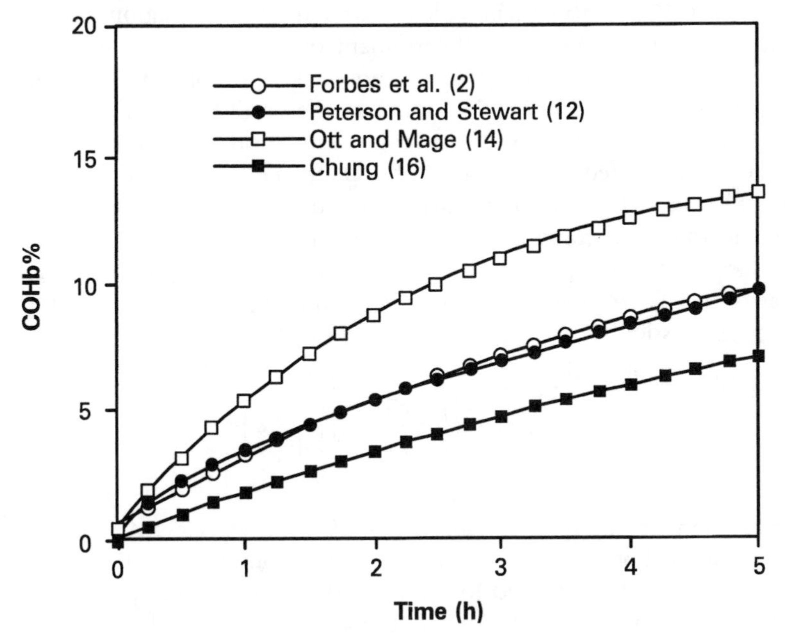

FIGURE 2
Empirically predicted increases in COHb for exposure to 100 ppm CO based on models by Forbes et al. (Equations 3.1 and 3.2 in the present text, and $k = 0.68\%\cdot Torr^{-1}\cdot min^{-1}$ based on Figure 1 of Forbes et al., 1945), Peterson and Stewart (1970) (Equation 3.3), Ott and Mage (1978) (Equation 3.5), and Chung (1988) (Equation 3.6).

lung capacity of CO. Several early investigations that focused on these variables were instrumental in the development of the CFK model and will be reviewed briefly.

Lilienthal and Pine (1946) expressed the uptake of CO as proportional to the dose of CO received times the minute ventilation rate:

$$\Delta COHb\% = 0.05 \cdot V_E \cdot P_A CO \cdot \Delta t \tag{3.8}$$

At about the same time, Pace et al. (1946) also expressed CO uptake in similar terms except that blood volume was included:

$$\Delta COHb\% = \frac{\dot{V}_E \cdot ppmCO \cdot \Delta t}{4.65 \cdot V_b} \tag{3.9}$$

Both expressions yield similar predicted uptakes of CO; for example, choosing $\dot{V}_E = 10$ l STPD·min^{-1}, ppmCO = 1000, $\Delta t = 30$ min, and $V_b = 5500$ ml (average value reported by Pace et al.), leads to ΔCOHb values of 11.4 and 11.7%, respectively.

Bates (1952) subsequently demonstrated that ventilation rate alone, however, can be a misleading determinant of CO uptake. In subjects that hyperventilated, the amount of CO removed was proportionally less than in subjects that breathed normally. Actual uptake of CO was shown by Hatch (1952) to be dependent on the alveolar ventilation rate. Further work by Hatch demonstrated that the rate of CO uptake was proportional to the sum of reciprocals of alveolar ventilation and the diffusing lung capacity of CO, and that each component contributed about the same.

Burrows and Harper (1958) determined the value of the diffusing lung capacity of CO from end-tidal (alveolar) sampling of CO through the following expression:

$$D_LCO = \left(\frac{P_ICO}{P_ACO} - 1\right) \cdot \left(\frac{\dot{V}_A}{P_B}\right) \tag{3.10}$$

The resultant steady-state measure of D_LCO is more relevant to predicting the rate of CO uptake under common exposure conditions than is the single-breath method of Krogh and Krogh (1910). Earlier, Roughton and Forster (1957) determined that D_LCO was a composite of the pure diffusing capacity of the alveolar membrane (D_M) and the specific conductance for CO in whole blood (Θ) expressed by:

$$\frac{1}{D_LCO} = \frac{1}{D_M} + \frac{1}{(\Theta \cdot V_C)} \tag{3.11}$$

where V_C is the volume of blood in the lung capillaries exposed to the alveolar gas. This expression has a significant influence on the rate of CO elimination under hyperoxic conditions, as will been seen later.

The precursor to the CFK model was the work conducted by Forster et al. (1954), who derived separate expressions for CO uptake during breath-holding and during steady-state. Several key assumptions were made that were subsequently applied in the later development. These, for the steady-state approximation, were (a) a negligible blood circulation time compared to the exposure time, (b) total CO stores in the body confined to blood and myoglobin, (c) homogeneity in lung ventilation-perfusion and ventilation-diffusion, and (d) continuous lung ventilation. Forster et al. (1954) then applied a mass balance to describe the rate of CO uptake, which became the basis of the CFK model.

Another consideration leading to the development of the CFK model was the endogenous production of CO in the body. Coburn et al. (1963) reported that this production in healthy nonsmoking males was 0.007 $ml \cdot min^{-1}$ and resulted in an average COHb value of 1.22%. A follow-up study (1964) concluded that this source of CO was an *in vivo* endproduct of heme catabolism.

Finally, Coburn et al. (1965) presented a model of CO uptake, which was subsequently labeled as the CFK, based on many of the findings reviewed above. In addition to the assumptions made by Forster et al. (1954), it was further assumed that the inspiratory and expiratory minute volumes were equal and that CO exchange occurred only through the lungs. By equating the rate of storage of CO in the body to the difference in CO concentrations between the inspired and expired gases, and taking the endogenous production of CO into account, Coburn et al. (1965) derived the following rate of formation of COHb:

$$\frac{d[COHb]}{dt} = \frac{\dot{V}_{CO}}{V_b} + \frac{1}{V_b \cdot \beta} \cdot \left(P_I CO - \frac{[COHb] \cdot P_C O_2}{[O_2Hb] \cdot M} \right) \tag{3.12}$$

where

$$\beta = \frac{1}{D_L CO} + \frac{(P_B - 47)}{\dot{V}_A} \tag{3.13}$$

Percentage carboxyhemoglobin saturation is related to [COHb] by (Peterson and Stewart, 1975):

$$COHb\% = \frac{100 \cdot [COHb]}{1.38 \cdot [THb]} \tag{3.14}$$

where 1.38 (ml STPD·g^{-1} Hb) represents the hemoglobin capacity for O_2 and CO.

If O_2Hb is assumed to be constant (i.e., unaffected by the presence of COHb), then Equation 3.12 becomes a linear differential equation with the following solution:

$$[COHb] = [COHb]_o \cdot e^{-A \cdot t} + \frac{C}{A} \cdot \left(1 - e^{-A \cdot t}\right) \tag{3.15}$$

where the subscript o refers to the initial value,

$$A = \frac{P_C O_2}{V_b \cdot \beta \cdot M \cdot [O_2Hb]} \tag{3.16}$$

and

$$C = \frac{\dot{V}_{CO}}{V_b} + \frac{P_I CO}{V_b \cdot \beta} \tag{3.17}$$

If, on the other hand, O_2Hb is corrected for the presence of COHb, for example, through the following approximate relationship (Peterson and Stewart, 1975):

$$\left[O_2Hb\right] = 1.38 \cdot \left[THb\right] - \left[COHb\right] \tag{3.18}$$

then Equation 3.12 becomes nonlinear. Strategies to solve the nonlinear version have been proposed by several investigators (Marcus, 1980; Muller and Barton, 1987; Peterson and Stewart, 1975; Tyuma et al., 1981). Certain of these and other solutions to test the CFK model are reviewed in the next section.

3.2 VALIDATION

Following the development of the CFK model, Coburn (1970) conducted a careful examination of body stores for CO. It appears that 10 to 15% of total CO stores are taken up by myoglobin located primarily in muscle. Coburn found no evidence for a significant change in the ratio of COMb to COHb for COHb levels up to 12%. It also appears that lumping COMb with COHb is an acceptable approximation when predicting CO uptake, as assumed in all studies discussed below where V_b is interpreted as the effective blood volume.

Peterson and Stewart (1970) found the CFK model provided a better fit to data of CO uptake than did certain empirical formulas derived from very diverse exposures. This study was followed by another (Peterson and Stewart, 1975) where the CFK model was validated for individuals at rest and at exercise exposed to CO from 50 to 200 ppm for times ranging from 20 to 315 min. Peterson and Stewart (1975) then applied the CFK model for exposures of up to 1000 ppm CO for 480 min. Their solution of the CFK model involved a trial-and-error method assuming the linear form (Equation 3.15), but changing the O_2Hb value to reflect the increasing value of COHb through Equation 3.18. This method is considerably more accurate than assuming a fixed value of O_2Hb, especially as COHb values increase appreciably. Although a slight underprediction in the rate of change of COHb is incurred with this method, the error is progressively diminished with increasingly small time steps.

Further verifications of the CFK model followed using similar approximations. Joumard et al. (1981) found the CFK model predicted reasonably well the resultant COHb in individuals exposed to low-level ambient urban levels of CO during normal pedestrian activity. Hauck and Neuberger (1984) also verified the CFK model for exposure to variable levels of CO and activity, but under controlled conditions.

Despite the good agreement claimed by the above studies (Hauck and Neuberger, 1984; Joumard et al., 1981; Peterson and Stewart, 1975), it is not certain that the CFK model was correctly applied. This stems from a study conducted by Tikuisis et al. (1987a) who found that the NIOSH (1972) solution of the CFK model overpredicted observed values of COHb. It was shown that the overprediction, by a factor of about 1.2, was due to (a) neglect of the saturation of the inspired CO by water vapor in the lungs and (b) use of BTPS rather than STPD conditions for $\dot{V}_A$, as originally specified by Coburn et al. (1965). There is no evidence that the studies cited above conformed to these requirements.

Given the importance of $\dot{V}_A$ on the prediction of COHb, an experiment was carried out by Tikuisis et al. (1987b) where this variable was measured in resting individuals exposed to CO. $\dot{V}_A$ was measured on the basis of CO_2 elimination and D_LCO was determined using Equation 3.10. Furthermore, to examine whether the asymmetry suggested by Ott and Mage (1978) is significant, the CO was administered five times in succession at either 1500 or 7500 ppm for 5 or 1 min, respectively. Each separate exposure yielded a dose of 7500 ppmCO·min resulting in a total dose of 37,500 ppmCO·min. Figure 3a indicates no significant difference in the resultant COHb between protocols following each exposure and Figure 3b demonstrates very good agreement between the measured and CFK-predicted values of COHb. Interestingly, however, the measured COHb was consistently lower during the actual exposure than that predicted for each exposure. It was hypothesized that this slight overprediction was the consequence of inherently assuming an instantaneous equilibration by not taking blood circulation into account.

This overprediction in COHb appearance during exposure was the subject of another study (Tikuisis et al., 1992) in which 1 ml arterialized blood samples were taken from the hand every 30 s during exposure to varying levels of CO. In this experiment, the first exposure was 3000 ppmCO for 3 min at rest, followed by three sequential exposures at steady-state exercise. Two levels of exercise were performed (low and moderate resulting in $\dot{V}_A$ values of 15.4 and 22.8 l STPD·min^{-1}, respectively, compared to 5.1 l STPD·min^{-1} at rest). CO levels were selected to produce similar COHb values as predetermined by the CFK model. These exposures, in the order administered, were 1000, 3000, and 1500 ppm CO for 3, 1, and 2 min, respectively, during low exercise, and 667, 2000, and 1000 ppm CO, respectively, during moderate exercise. Figure 4 shows that reasonably good agreement was obtained between the measured and predicted COHb 2 min after the end of each exposure. However, during exposure, it was evident that the appearance of COHb was sigmoidal in shape with a prominent overshoot

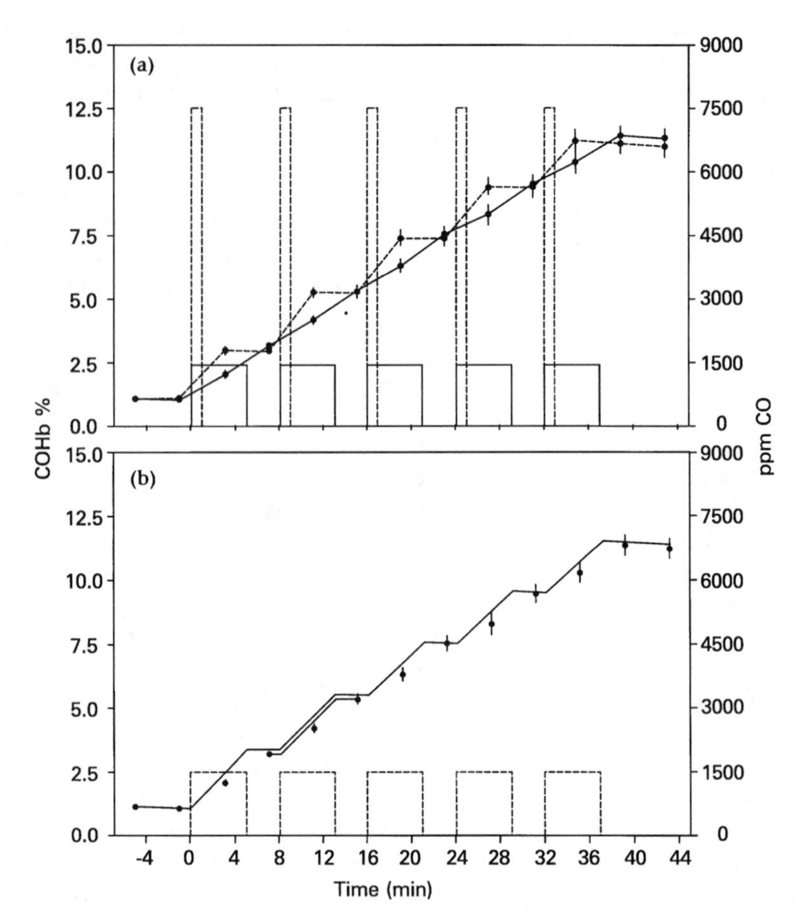

FIGURE 3
(a) Comparison of mean (±SE) of measured COHb% for exposures to 1500 (solid line) and 7500 (dashed line) ppm CO. (b) Comparison of mean (±SE) of measured COHb% for five sequential exposures to 1500 (dashed line) and CFK-predicted values (solid line). (From Tikuisis et al. (1987a). With permission.)

at the end of exposure, especially during the rest condition. To explain these features, a simple blood circulation model was developed, with various channels of blood flow having different circulation times. The pooling of blood at the end of circulation in the shorter channels led to the characteristic overshoot observed in the experiment.

Overshoot was predicted earlier by Godin and Shephard (1972) with a model developed for predicting COHb in the pulmonary capillary blood. In addition to this blood pool, their model included fast and slow blood pools. Overshoot was greatest in the pulmonary capillary blood, less in the fast pool, and nonexistent in the slow pool. Most recently, Benignus et al. (1994) observed overshoot during CO uptake in the arterial blood, while COHb in the venous blood indicated a somewhat delayed sigmoidal shape

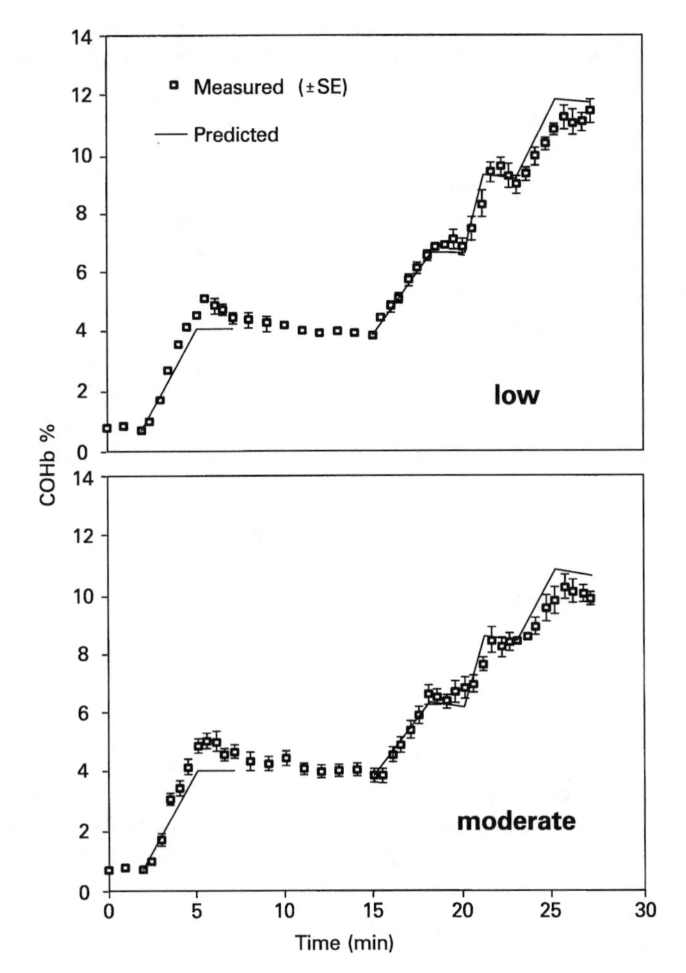

FIGURE 4
Measured and CFK-predicted COHb% for various exposures to CO (see text) against time during rest for the first 7 min, and for low and moderate exercise thereafter. (From Tikuisis et al., *J. Appl. Physiol.*, 72, 1311, 1992. With permission.)

with no overshoot. These observations were modeled quite accurately by Smith et al. (1994), who applied a much more sophisticated blood circulation model than did Tikuisis et al. (1992). All studies, however, agreed that the CFK model-predicted values of COHb matched closely the measured values at some time after exposure, when all blood pools were thoroughly mixed (usually within 2 min).

The CFK model, as presented by Equation 3.12, describes the rate of formation of COHb, although it was originally developed to describe CO uptake. The distinction is more than a simple transformation, because COHb measured at one site in the body does not necessarily represent whole body CO uptake as demonstrated above. However, considering that the CFK model appears to predict the average COHb in the body reasonably well, it can be regarded as a valid predictor of CO uptake as originally intended by Coburn et al. (1965). Caution must be exercised when predictions of COHb for specific body sites are required during exposure to CO, especially those sites with rapid circulation times.

3.3 CO Elimination

Although the examination of the above studies was confined to CO uptake, the CFK model is equally suited to predict CO elimination. Indeed, Coburn et al. (1965) applied the model to demonstrate that CO elimination is enhanced with increased pressure of O_2 breathed (through the increased value of P_CO_2 — see Equation 3.12). The increase in the mean O_2 tension in the pulmonary capillary blood is predicted by (Peterson and Stewart, 1975):

$$P_CO_2 = P_AO_2 - 49 \tag{3.19}$$

where P_AO_2 is the saturated O_2 alveolar pressure given by:

$$P_AO_2 = fO_2 \cdot \left(P_B - 47\right) \tag{3.20}$$

and where fO_2 is the fraction of O_2 in the ambient gas.

While increased O_2 pressure promotes a more rapid rate of CO elimination, the increased rate is not necessarily proportional to O_2 pressure. This arose in a study by Tyuma et al. (1981) whose application of the CFK model against the data of Peterson and Stewart (1970) indicated excellent agreement when air was breathed, but a marked overprediction in CO elimination occurrred when 100% O_2 was breathed at 1 and 3 atm. Tyuma et al. (1981) surmised that the diffusing lung capacity for CO is reduced during hyperoxia. If so, then β would increase (see Equation 3.13) and hence the rate of CO elimination would decrease.

That D_LCO may depend on the inspired pressure of O_2 through the specific conductance of hemoglobin for CO was originally expressed by Roughton and Forster (1957) (see Equation 3.11). Jones et al. (1982) found a decrease in the mean value of D_LCO from about 25.8 to 14.0 ml STPD·min^{-1}·Torr^{-1} in subjects briefly exposed to CO (from <1 to 3000 ppm) in an experiment designed to fix their mean alveolar O_2 fraction at 0.15 and 0.78, respectively; the latter was achieved by subjects breathing 100% O_2 before exposure. This decrease is consistent with the predicted decrease in Θ with increasing pressure of O_2 according to Forster (1987):

$$\frac{1}{\Theta} = 1.30 + 0.0041 \cdot P_AO_2 \tag{3.21}$$

This expression can be used to test the CFK model using previously unpublished CO elimination data from the experiment described by Tikuisis et al. (1992). Recall that this experiment measured CO uptake in subjects at two levels of exercise. After 2 min following the end of the last exposure, the subjects were instructed to continue exercise for an additional 8 min at the same level and were given 100% O_2 to breathe. The decreases in COHb% averaged (±SE) 2.80 (0.15) and 2.98 (0.18) for low and moderate exercise, respectively.

Before the CFK model can be applied, D_M and V_C must be known to determine D_LCO. V_C can be roughly estimated from body mass (wt in kg) using the following linear regression derived from the data of Jones et al. (1982):

$$V_C = -93.25 + 2.743 \cdot \text{wt} \qquad (3.22)$$

The pure diffusing capacity of the alveolar membrane was then determined from D_LCO, measured during the CO exposure and through the use of Equation 3.11. Applying the values of D_M and V_C thus determined for each subject, lead to mean D_LCO values (±SE) of 22.5 (0.9) and 24.5 (1.1) ml STPD·min^{-1}·Torr^{-1} for the low and moderate exercise levels, respectively. These values are almost half of that measured for the CO exposures during exercise. Substitution of the O_2-reduced values of D_LCO during hyperoxia lead to predicted decreases in COHb that agreed with the measured decreases (*p* values of 0.78 and 0.60 at the 0.05 significance level, respectively).

Although not as great as would be predicted with a larger value of D_LCO, the increases in CO elimination are nevertheless substantially greater than if air was breathed and/or the subjects were at rest, as originally reported by Roughton (1945). Even greater rates of CO elimination can be expected under a hyperbaric environment with 100% O_2. The present results strongly support the use of the CFK model for predicting CO elimination provided that the appropriate value of D_LCO is used when subjects are in a hyperoxic environment.

3.4 Sensitivity Analysis

McCartney (1990) conducted an elegant sensitivity analysis of the CFK model that provides both further insight on the uptake of CO and a useful guide on how each variable affects the model prediction. As will be seen, each variable influences the uptake of CO uniquely and at different times during the uptake.

The sensitivity analysis essentially determines the fractional change in the predicted COHb for a fractional change in each independent variable. The resultant normalized sensitivity value is designated as F. If the change in COHb is zero, then the CFK model is not sensitive to that variable. If, on the other extreme F = ±1, then the CFK model is completely sensitive to that variable. Values between 0 and ±1 indicate a partial sensitivity. If the model is sensitive to a variable, then the error in the prediction of COHb is given by F times the fractional error in the value of that variable.

Figure 5 illustrates the temporal sensitivity of the linearized version of the CFK model to several of its key variables. Only two variables affect error at the outset of an exposure — the initial value of COHb and the total blood concentration of hemoglobin [THb]. While the former has no influence shortly after exposure begins, the latter regains its influence because [THb] determines the relationship between COHb and the blood gas concentration of CO. The alveolar ventilation and blood volume also exhibit interesting influences. In these cases, the rate of CO uptake is sensitive to these variables, albeit in opposing directions, but this sensitivity becomes zero once equilibrium is attained. Although not shown, the sensitivity of the CFK model to D_LCO is qualitatively similar to that of $\dot{V}_A$, but at a lower magnitude. Finally, the remaining two variables, the ambient concentration of CO and the Haldane ratio, have an increasing influence with time and affect the equilibrium value of COHb. Greater details on the sensitivity of the CFK model (especially the nonlinear version) to its variables should be sought in McCartney (1990).

3.5 Applications

A number of investigators have applied the CFK model for predicting CO uptake in individuals under a variety of conditions. These include Peterson and Stewart (1975), who tabled the resultant COHb values for exposures from 8.7 to 1000 ppmCO for 1- and 8-h periods. The nominal condition pertained to a resting individual and several subsequent predictions were made where the value of one of the model variables would be altered. Bernard and Duker (1981) confined their application of the CFK model to work environments where activity was well beyond resting and CO levels were potentially very high (up to 200,000 ppm). Their simulation allowed for changes in D_LCO and $\dot{V}_A$ as a function of the O_2 consumption rate. Weir and Viano (1977) applied the CFK model for the prediction of COHb in individuals accidentally exposed to CO and concluded that the model is useful in reconstructing exposure history (that is, to ascertain the ambient CO given the COHb level of the casualty).

Collier and Goldsmith (1983) conducted a detailed study of the equilibrium value of COHb at altitude for low-level exposure to CO. Certain physiological considerations impact the application of the CFK model in this circumstance, such as the reduction of P_CO_2 due to decreasing ambient O_2 pressure. Interestingly, Collier and Goldsmith found that increasing altitude leads to only slightly increasing COHb levels. While altitude decreases O_2Hb, further decreases owing to the presence of CO are marginal compared to

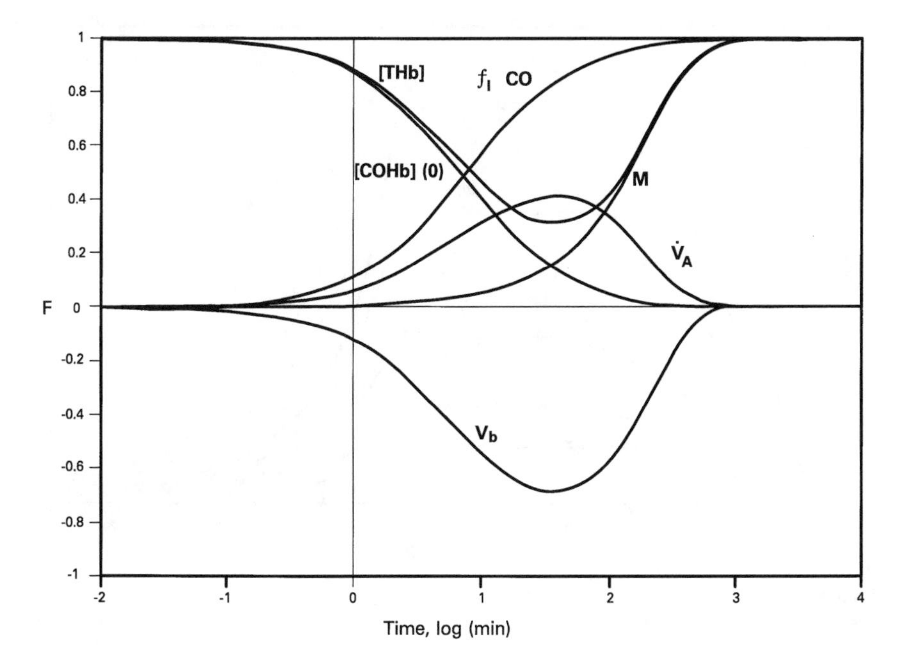

FIGURE 5
Fractional sensitivity of the linearized CFK model prediction of CO uptake to several key variables (f_ICO is the fraction of CO in the ambient air) against time of exposure. F = ±1 indicates 100% sensitivity (see text for details). (From McCartney, M. L., *Am. Ind. Hyg. Assoc. J.*, 51, 169, 1990. With permission.)

similar exposures at sea level. A significant recommendation from their study was that CO exposure standards against health risk at sea level are applicable at altitude if expressed in volumetric terms (i.e., in ppm).

Another interesting and important application of the CFK model was its extension to predicting COHb levels in the human fetus by Hill et al. (1977). Essentially these investigators coupled the CFK model for a mother to one for the fetus where CO transfer between the two was channeled through the placenta. In addition to the many physiological differences between the two bodies, it is noteworthy that the haldane ratio for the fetus was assumed to be 181 vs. 223 for the mother, that the endogenous production of CO per unit volume of blood was assumed to be approximately doubled for the mother compared to the fetus, and that the fetal hemoglobin concentration was assumed to be 15.5 vs. 12.5 $g \cdot ml^{-1}$ blood for the mother. These differences lead to striking differences in the predicted rate of COHb formation between the two. Figure 6 indicates that while fetal COHb lags behind the mother's during initial exposure, fetal levels become higher towards equilibrium for exposures of less than 200 ppm CO. These predictions were subsquently confirmed in sheep (Longo and Hill 1977).

Finally, Benignus (1994) has made available a computer model (EPAPUF) for predicting not only COHb but also the pulmonary variables

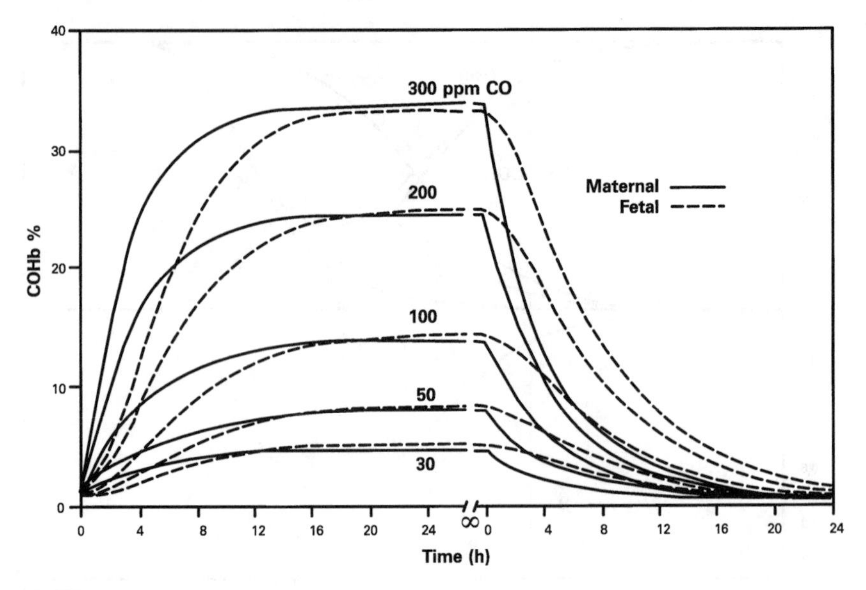

FIGURE 6
Predicted maternal and fetal COHb% during exposure to CO levels from 30 to 300 ppm and washout from equilibrium values. (From Hill et al., *Am. J. Physiol.,* 232, H311, 1977. With permission.)

involved during exposure to CO, O_2, and CO_2. This model, consisting of 35 input and 60 output variables, is designed to allow the simulation of continuous dynamic exposures or the analysis of actual exposures.

3.6 Approximations

There are instances when approximate solutions to the CFK model are warranted for ease of computation. Smith (1990) demonstrated that the linear form (Equation 3.15) provides a good approximation for CO uptake over a broad range of COHb values, but less so for CO elimination. While computational efforts have been minimized with the advent of powerful computers, it is useful to have simple expressions of CO uptake for instructive purposes, quick checks, and possible field application. The following will outline certain of these expressions and their limitations.

When ambient CO levels are sufficiently high, the endogenous production of CO can be neglected. This can be demonstated by considering the equilibrium value of COHb with no ambient CO present (i.e., due to $\dot{V}_{CO}$ only) which, according to Equations 3.12, 3.14, and 3.18, is given by:

$$\mathrm{COHb\%}(\infty) = \frac{100}{1 + \dfrac{P_C O_2}{\dot{V}_{CO} \cdot \beta \cdot M}} \qquad (3.23)$$

Assuming a resting condition in air at 1 atm [from Tikuisis et al., 1987b; $\dot{V}_A$ = 4200 ml STPD·min^{-1} and D_LCO = 20 ml STPD·min^{-1}·Torr^{-1} so that β = 0.22 ml^{-1} STPD·min·Torr (see Equation 3.13), $\dot{V}_{CO}$ = 0.007 ml STPD·min^{-1}, M = 218, and P_CO_2 = 100 Torr], then COHb%(∞) = 0.33. Under exercise conditions, β becomes smaller and COHB%(∞) is reduced. Therefore, an ambient level of CO that increases COHb beyond a few percent can ignore the endogenous contribution of CO.

Further, if the period of exposure to CO is relatively short (such that the increase in COHb does not exceed say 2%), then the influence of a changing value of COHb on its rate of formation can be ignored. Thus,

$$\Delta\left[\mathrm{COHb}\right] \cong \frac{\Delta t}{V_b \cdot \beta} \cdot \left(P_I\mathrm{CO} - \frac{\left[\mathrm{COHb}\right]_o \cdot P_C\mathrm{O}_2}{\left(1.38 \cdot \left[\mathrm{THb}\right] - \left[\mathrm{COHb}\right]_o\right) \cdot M} \right) \tag{3.24}$$

Assuming values of 5500 ml blood and 0.154 g Hb·ml^{-1} blood for V_b and [THb], respectively (McCartney, 1990), then the changes in COHb during a resting exposure to 500 ppm CO over 10 min are 1.35, 1.29, 1.19, 0.94, 0.62, and 0.20% for starting values of 2, 5, 10, 20, 30, and 40%, respectively. Conversions were made using Equation 3.14 and the following relationship:

$$P_I\mathrm{CO} = \mathrm{ppmCO} \cdot \left(P_b - 47\right) \cdot 10^{-6} \tag{3.25}$$

Note the exponentially decreasing increment in COHb as its value approaches equilibrium. The equilibrium value is simply attained by equating Equation 3.12 to zero and ignoring $\dot{V}_{CO}$:

$$\mathrm{COHb\%}\left(\infty\right) = \frac{100}{1 + \dfrac{P_C\mathrm{O}_2 \cdot 10^6}{\mathrm{ppmCO} \cdot \left(P_B - 47\right) \cdot M}} = \frac{100}{1 + \dfrac{643.4}{\mathrm{ppmCO}}} \tag{3.26}$$

where the right-hand side refers to a CO–air environment at 1 atm. Applying the above conditions leads to an equilibrium value of 43.7%.

If the initial value of COHb is very low, then the effect of its presence on the rate of change of COHb can be ignored entirely, that is:

$$\Delta\mathrm{COHb\%} \approx \frac{\Delta t \cdot \mathrm{ppmCO} \cdot \left(P_B - 47\right) \cdot 10^{-4}}{1.38 \cdot \left[\mathrm{THb}\right] \cdot V_b \cdot \beta} \tag{3.27}$$

This expression will overestimate the increase with increasing error as the initial value of COHb is increased. For example, the predicted increase under the above conditions is 1.39%, which is reasonably accurate for low starting COHbs but incorrect for higher values.

Two experiments (Tikuisis et al., 1987b; Tikuisis et al., 1992) described earlier indicated no significant differences in the resultant COHb following exposures to CO of varying ambient levels but over durations that yielded the same dose. This suggests that the integration over time of the ambient CO levels can be used to predict COHb. An acceptable integration time is one that is considerably smaller than the time constant of CO uptake, τ, given by the reciprocal of A (see Equation 3.16). For example, at a COHb value of 2% and assuming the earlier values of the other variables, then τ = 549 min. During heavy exercise (e.g., $\dot{V}_A$ = 30,000 ml STPD·min^{-1} and D_LCO = 60 ml STPD·min^{-1}·Torr^{-1} so that β = 0.04 ml^{-1} STPD·min·Torr), the time constant is proportionally reduced (τ = 102 min).

As a demonstration, consider the following hypothetical exposure profile: 100 ppm CO for 80 min + 800 ppm CO for 10 min + 200 ppm CO for 40 min + 400 ppm CO for 20 min. Each exposure was specifically designed to yield the same dose (i.e., 8000 ppm CO·min). Again, assuming resting conditions (as outlined earlier) and a starting value of 2% COHb, the CFK-predicted value at the end of the entire exposure is 9.58%. If the total dose (32,000 ppm CO·min) was applied uniformly across the 150 min of exposure (i.e., at 213.3 ppmCO), the predicted end value is 9.29%. On the other hand, if the dose was divided in two separate stages such as 100 ppmCO for 80 min + 342.9 ppm CO for 70 min, then the predicted end value of 9.61% agrees more closely to the correct value. Clearly, the shorter the period of averaging (integration), the more accurate the prediction. A reasonable "rule of thumb" is to apply a dose-averaging period of about 10% of the time constant for CO uptake.

4 SUMMARY

This report on modeling the uptake and elimination of CO has traced the development from empirical methods to the rationally derived CFK model. The evidence to date suggests that the latter provides an excellent prediction and should be used whenever possible. Care must be taken to ensure that correct values of the physical and physiological variables of the CFK model are applied. Where predictions for specific sites are required, modifications to the model can and have been made to discriminate between regional differences in COHb appearance (e.g., venous vs. arterial blood and fetal vs. maternal blood). Where the CO exposure profile is erratic, little error is incurred by approximating the exposure by average values over short durations relative to the time constant of CO uptake.

ACKNOWLEDGMENTS

The author gratefully acknowledges the advice of Drs. Brian H. Sabiston and Vernon A. Benignus in the preparation of this manuscript. The author also wishes to thank the American Physiological Society and the American Industrial Hygiene Association for permission to reproduce figures published in their journals. This work was conducted at the Defence and Civil Institute of Environmental Medicine.

GLOSSARY

COHb%	carboxyhemoglobin saturation (%)
O_2Hb%	oxyhemoglobin saturation (%)
[COHb]	concentration of CO bound to hemoglobin (ml STPD·ml^{-1} blood)
[COMb]	concentration of CO bound to myoglobin (ml STPD·ml^{-1} tissue)
[O_2Hb]	concentration of O_2 bound to hemoglobin (ml STPD·ml^{-1} blood)
[THb]	total concentration of hemoglobin (g Hb·ml^{-1} blood)
D_M	pure diffusing lung capacity (ml STPD·min^{-1}·Torr^{-1})
D_LCO	diffusing lung capacity for CO (ml STPD·min^{-1}·Torr^{-1})
M	Haldane ratio
ppmCO	ambient level of CO in parts per million
P_ICO	saturated inspired CO pressure (Torr)
P_ACO	saturated alveolar CO pressure (Torr)
P_AO_2	saturated alveolar O_2 pressure (Torr)
P_CO_2	mean O_2 tension (Torr) in the pulmonary capillary blood
P_B	barometric pressure (Torr)
V_b	effective blood volume (ml)
V_C	blood volume in lung capillaries (ml)
$\dot{V}_E$	minute ventilation (l STPD·min^{-1})
$\dot{V}_A$	alveolar ventilation (ml STPD·min^{-1})
$\dot{V}_{CO}$	endogenous rate of CO production (ml STPD·min^{-1})
t	time (min)
τ	time constant of CO uptake (min)
Θ	specific conductance for CO (ml STPD·min^{-1}·Torr·ml^{-1} blood^{-1})

REFERENCES

Bates, D.V., The uptake of carbon monoxide in health and in emphysema, *Clin. Sci.*, 11, 1, 1952.

Benignus, V.A., A computer model for predicting carboxyhemoglobin and pulmonary parameters associated with exposures to carbon monoxide, oxygen and carbon dioxide, *Aviat. Space Environ. Med.*, 66, 369, 1995.

Benignus, V.A., Hazucha, M.J., Smith, M.V., and Bromberg, P.A., Prediction of carboxyhemoglobin formation due to transient exposure to carbon monoxide, *J. Appl. Physiol.,* 76, 1739, 1994.
Bernard, T.E. and Duker, J., Modeling carbon monoxide uptake during work, *Am. Ind. Hyg. Assoc. J.,* 42, 361, 1981.
Burrows, B. and Harper, P.V., Jr., Determination of pulmonary diffusing capacity from carbon monoxide equilibration curves, *J. Appl. Physiol.,* 12, 283, 1958.
Chung, S.J., Formulas predicting carboxyhemoglobin resulting from carbon monoxide exposure, *Vet. Hum. Toxicol.,* 30, 528, 1988.
Coburn, R.F., The carbon monoxide body stores, *Ann. N.Y. Acad. Sci.,* 174, 11, 1970.
Coburn, R.F., Blakemore, W.S., and Forster, R.E., Endogenous carbon monoxide production in man, *J. Clin. Invest.,* 42, 1172, 1963.
Coburn, R.F., Forster, R.E., and Kane, P.B., Considerations of the physiological variables that determine the blood carboxyhemoglobin concentration in man, *J. Clin. Invest.,* 44, 1899, 1965.
Coburn, R.F., Williams, W.J., and Forster, R.E., Effect of erthrocyte destruction on carbon monoxide production in man, *J. Clin. Invest.,* 43, 1098, 1964.
Collier, C.R. and Goldsmith, J.R., Interactions of carbon monoxide and hemoglobin at high altitude, *Atmos. Environ.,* 17, 723, 1983.
Forbes, W.B., Sargent, F., and Roughton, F.J.W., The rate of carbon monoxide uptake by normal men, *Am. J. Physiol.,* 143, 594, 1945.
Forster, R.E., Diffusion of gases across the alveolar membrane, in *Handbook of Physiology, Vol. 4: Gas Exchange,* Farhi, L. E. and Tenney, S. M., Eds., American Physiological Society, Washington, D.C., 1987, p. 78.
Forster, R.E., Fowler, W.S., and Bates, D.V. Considerations on the uptake of carbon monoxide by the lungs, *J. Clin. Invest.,* 33, 1128, 1954.
Godin, G. and Shepard, R.J., On the course of carbon monoxide uptake and release, *Respiration,* 29, 317, 1972.
Goldsmith, J.R., Terzaghi, J., and Hackney, J.D., Evaluation of fluctuating carbon monoxide exposures, *Arch. Environ. Health,* 7, 33, 1963.
Haldane, J., The action of carbonic oxide on man, *J. Physiol.,* 18, 30, 1895.
Hatch, T.F., Carbon monoxide uptake in relation to pulmonary performance, *Arch. Ind. Hyg. Occup. Med.,* 6, 1, 1952.
Hauck, H. and Neuberger, M., Carbon monoxide uptake and the resulting carboxyhemoglobin in man, *Eur. J. Appl. Physiol.,* 53, 186, 1984.
Hill, E.P., Hill, J.R., Power, G.G., and Longo, L.D., Carbon monoxide exchanges between the human fetus and mother: A mathematical model, *Am. J. Physiol.,* 232, H311, 1977.
Jones, H.A., Clark, J.C., Davies, E.E., Forster, R.E., and Hughes, J.M.B., Rate of uptake of carbon monoxide at different inspired concentrations in humans, *J. Appl. Physiol.,* 52, 109, 1982.
Joumard, R., Chiron, M., Vidon, R., Maurin, M., and Rouzioux, J.-M., Mathematical models of the uptake of carbon monoxide on hemoglobin at low level carbon monoxide levels, *Environ. Health Persp.,* 41, 277, 1981.
Krogh, A. and Krogh, M., On the rate of diffusion of carbonic oxide into the lungs of man, *Skand. Arch. Physiol.,* 23, 237, 1910.
Lilienthal, J.L., Jr. and Pine, M.B., The effect of oxygen pressure on the uptake of carbon monoxide by man at sea level and at altitude, *Am. J. Physiol.,* 145, 346, 1946.
Lilienthal, J.L., Jr., Riley, R.L., Proemmel, D.D., and Franke, R.E., The relationship between carbon monoxide, oxygen and hemoglobin in the blood of man at altitude, *Am. J. Physiol.,* 145, 351, 1946.
Longo, L.D. and Hill, E.P., Carbon monoxide uptake and elimination in fetal and maternal sheep, *Am. J. Physiol.,* 232, H324, 1977.
Marcus, A.H., Mathematical models for carboxyhemoglobin, *Atmos. Environ.,* 14, 841, 1980.
McCartney, M.L., Sensitivity analysis applied to Coburn-Forster-Kane models of carboxyhemoglobin formation, *Am. Ind. Hyg. Assoc. J.,* 51, 169, 1990.

Muller, K.E. and Barton, C.N., A nonlinear version of the Coburn, Forster and Kane model of blood carboxyhemoglobin, *Atmos. Environ.*, 21, 1963, 1987.

National Institute for Occupational Safety and Health, Occupational exposure to carbon monoxide, Report TR-007-72, DHEW/NIOSH, Rockville, MD, 1972.

Ott, W.R. and Mage, D.T., Interpreting urban carbon monoxide concentrations by means of a computerized blood COHb model, *J. Air Pollut. Control Assoc.*, 28, 911, 1978.

Pace, N., Consolazio, W.V., White, W.A., and Behnke, A.R., Formulation of the principal factors affecting the rate of uptake on carbon monoxide by man, *Am. J. Physiol.*, 147, 352, 1946.

Peterson, J.E. and Stewart, R.D., Absorption and elimination of carbon monoxide by inactive young men, *Arch. Environ. Health,* 21, 165, 1970.

Peterson, J.E. and Stewart, R.D., Predicting the carboxyhemoglobin levels resulting from carbon monoxide exposures, *J. Appl. Physiol.*, 39, 633, 1975.

Roughton, F.J.W., The average time spent by the blood in the human lung capillary and its relation to the rates of CO uptake and elimination in man, *Am. J. Physiol.*, 143, 621, 1945.

Roughton, F.J.W. and Forster, R.E., Relative importance of diffusion and chemical reaction rates in determining rate of exchange of gases in the human lung, with special reference to true diffusing capacity of pulmonary membrane and volume of blood in the lung capillaries, *J. Appl. Physiol.*, 11, 290, 1957.

Roughton, F.J.W. and Root, W.S., The fate of CO in the body during recovery from mild carbon monoxide poisoning in man, *Am. J. Physiol.*, 145, 239, 1945.

Smith, M.V., Comparing solutions to the linear and nonlinear CFK equations for predicting COHb formation, *Math. Biosci.*, 99, 251, 1990.

Smith, M.V., Hazucha, M.J., Benignus, V.A., and Bromberg, P.A., Effect of regional circulation patterns on observed COHb levels, *J. Appl. Physiol.*, 76, 1659, 1994.

Stewart, R.D., Peterson, J.E., Fisher, T.N., Hosko, M.J., Baretta, E.D., Dodd, H.C., and Herrmann, A.A., Experimental human exposure to high concentrations of carbon monoxide, *Arch. Environ. Health,* 26, 1, 1973.

Tikuisis, P., Buick, F., and Kane, D.M., Percent carboxyhemoglobin in resting humans exposed repeatedly to 1,500 and 7,500 ppm CO, *J. Appl. Physiol.*, 63, 820, 1987b.

Tikuisis, P., Kane, D.M., McLellan, T.M., Buick, F., and Fairburn, S.M., Rate of formation of carboxyhemoglobin in exercising humans exposed to carbon monoxide, *J. Appl. Physiol.*, 72, 1311, 1992.

Tikuisis, P., Madill, H.D., Gill, B.J., Lewis, W.F., Cox, K.M., and Kane, D.M., A critical analysis of the use of the CFK equation in predicting COHb formation, *Am. Ind. Hyg. Assoc. J.*, 48, 208, 1987a.

Tobias, C.A., Lawrence, J.H., Roughton, F.J.W., Root, W.S., and Gregersen, M.I., The elimination of carbon monoxide from the human body with reference to the possible conversion of CO to CO_2, *Am. J. Physiol.*, 145, 253, 1945.

Tyuma, I., Ueda, Y., Imaizumi, K., and Kosaka, H., Prediction of the carbon monoxy-hemoglobin levels during and after carbon monoxide exposures in various animal species, *Jpn. J. Physiol.*, 31, 131, 1981.

Venkatram, A. and Louch, R., Evaluation of CO air quality criteria using a COHb model, *Atmos. Environ.*, 13, 869, 1979.

Weir, F.W. and Viano, D.C., Prediction of carboxyhemoglobin from transient carbon monoxide exposure, *Aviat. Space Environ. Med.*, 48, 1076, 1977.

CHAPTER 4

Cerebrovascular Effects of Carbon Monoxide

Mark A. Helfaer and Richard J. Traystman

CONTENTS

1 INTRODUCTION

Carbon monoxide (CO) has been the subject of a broad range of investigations over the years. It is present during combustion such as in house fires, as well as in cigarette smoke. In addition, it has also appeared in unexpected areas such as during routine administration of anesthesia

0-8493-4796-3/96/$0.00+$.50

(Fang and Eger, 1994). CO exposure causes significant perturbations in virtually all organs in the body. In this chapter, we will review the experimental data to identify the effects of this molecule on the brain. We will begin with biochemical aspects, then physiological and pathological aspects, and finally clinical (human) effects with attention to behavioral changes associated with exposure to CO.

2 BIOCHEMISTRY

CO binds to and inactivates heme proteins (Penney, 1990). When bound to hemoglobin, the consequence is reduced oxygen carrying capacity with rising carboxyhemoglobin (COHb) levels. This translates into a leftward shift of the hemoglobin dissociation curve, which causes the oxygen to be tenaciously bound to the hemoglobin molecule, which in turn causes tissue hypoxia. There is considerable debate whether this tissue hypoxia is the sole basis for the toxicity ascribed to CO, or whether there are other biochemical effects that mediate the toxicity. The cerebral response to the decreased availability of oxygen is to increase cerebral blood flow, decrease cerebral metabolic rate for oxygen, or increase extraction of substrates (Traystman, 1991). As a general rule, the brain responds by increasing flow in a dose dependent manner during CO administration (Traystman, 1991).

Cytochrome oxidase exists in brain in synaptic as well as nonsynaptic mitochondria. In the anaerobic state, CO is fully and almost entirely irreversibly bound to ferrous cytochrome a3. This binding changes as a function of age in synaptic, but not nonsynaptic, mitochondria (Harmon, 1990). The *in vitro* observation that CO binds to cytochrome a3 has been substantiated *in vivo* in rat cortex by reflectance spectrophotometry (Brown and Piantadosi, 1990). This binding is partially reversed by hyperbaric oxygen administration. By disrupting the final acceptor in the electron transport chain, CO binding could easily cause multiple physiologic perturbations independent of mechanisms dependent solely on hypoxia. Associated with these changes in cytochrome a3, during mixed hypoxia and CO, there is a fall in phosphocreatine and intracellular pH, with preservation of ATP. In these experiments, in normobaric conditions, phosphocreatine and intracellular pH continued to fall 45 min after CO exposure, but improved with hyperbaric (2.5 atm) oxygen with complete recovery at 90 min (Brown and Piantadosi, 1992).

The cytochrome P450 system of enzymes is a heme protein and therefore also binds CO avidly. The liver contains the largest amount of this enzyme and is involved in steroid and fatty acid metabolism. The activity of this enzyme in brain is 4–25% of the activity found in liver (Bhamre et al., 1992), with the concentration highest in brain stem. The implications of CO binding to the P450 system upon cerebral pathophysiology has yet to be demonstrated. Theoretically, inhibition of this enzyme system could result in a more prolonged half-life or higher levels of otherwise detoxified chemicals, which could cause injury to tissue (Anandatheerthavarada et al., 1993).

Nitric oxide synthase is an enzyme that catalyzes the conversion of arginine into citrulline and nitric oxide (NO). NO has been implicated in many systems as one of the endothelial-derived relaxing factors, but has been found in many other tissues, especially neural tissue (Faraci and Brian, 1994). It has been demonstrated that this enzyme has a heme-containing subunit (McMillan et al., 1992) that demonstrates spectra typical of ferroprotoporphyrin IX (Klatt et al., 1992). *In vitro* activity of this enzyme can be inhibited by CO, which could cause local vasoconstriction due to the lack of the tonic release of the vasorelaxation of NO. In addition, because NO is produced in metabolically active neural tissue, it has been postulated that it may be a second messenger, perhaps linking blood flow and metabolism. By disrupting this mechanism of cell-to-cell communication, further injury could be caused (Zhuo et al., 1993). Also, CO itself has similarly been invoked as a putative neurotransmitter and/or a second messenger (Zhuo et al., 1993). The latter hypothesis is based on the fact that CO is produced by the action of the enzyme heme oxygenase, the constitutive form of which has been demonstrated throughout the brain, co-localizing with messenger RNA for soluble guanylyl cyclase (Verma et al., 1993). This pathway is illustrated in Figure 1.

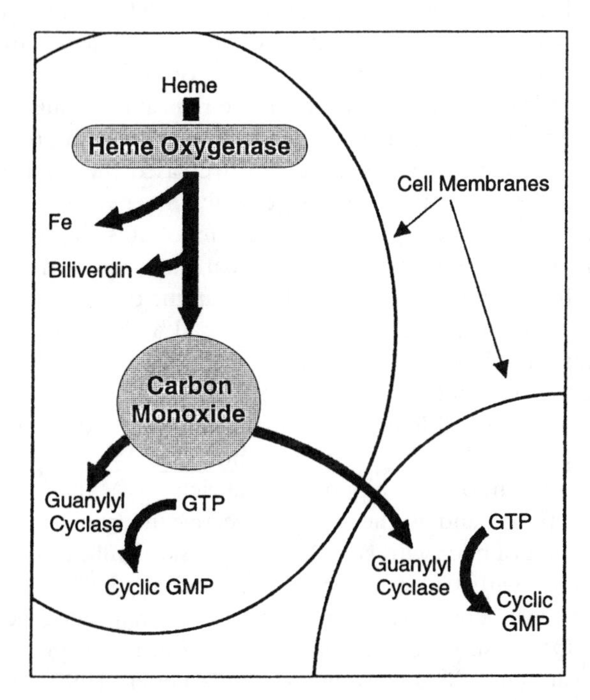

FIGURE 1

Diffuse message. By diffusing through membranes and interacting with the enzyme guanylyl cyclase, carbon monoxide may act as a biological signal. (From Barinaga, M., *Science,* 259, 309, 1993. With permission.)

In addition, zinc protoporphyrin-9 selectively inhibits heme oxygenase and also depletes endogenous guanosine 3′,5′-monophosphate (cGMP) (Verma et al., 1993). These lines of evidence suggest that CO may be a second messenger or a neurotransmitter, which would imply another mechanism by which this very soluble gas could mediate injury in the central nervous system (Barinaga, 1993; Dinerman et al., 1994).

CO has been shown to convert xanthine dehydrogenase to xanthine oxidase in rats (Thom, 1992). The presence of xanthine oxidase in brain has been associated with free radical production and damage, and inhibition of this enzyme has been demonstrated to improve responses to cerebral injury (Kirsch et al., 1992). Evidence of free radical production in brain following CO exposure is supported by decreased ratios of reduced to oxidized glutathione as well as detection of the products of salicylate hydroxylation (Zhang and Piantadosi, 1992). These data are consistent with production of hydroxyl radical production during CO exposure. The site of production of these reactive species of oxygen is postulated to be the mitochondria (Zhang and Piantadosi, 1992). In a rat model, conjugated diene levels, which are a measure of lipid peroxidation in brain, doubled after exposure to CO (Thom, 1990b). By inhibiting the activity of xanthine oxidase, conjugated diene levels did not rise. Treatment of these exposed rats with superoxide dismutase (which scavenges superoxide anion) or deferoxamine (which chelates iron, which contributes to free radical production) decreases conjugated diene levels (Thom, 1992). The conclusion of these studies is that CO exposure could lead to tissue injury by a free radical mechanism initiated by the increase in xanthine oxidase. In addition to the improvement with pharmacologic probes, hyperbaric oxygen decreased the lipid peroxidation associated with CO exposure. This decrease was greater in 3 atm compared with 1 atm (Thom, 1990a). The mechanism of this diminution of lipid peroxidation with administration of hyperbaric oxygen is not known. On the contrary, however, treatment with deferoxamine did not confer histologic improvement in a rat model of hyperglycemia, 1% CO exposure, and right carotid occlusion (MacMillan et al., 1993).

The excitatory amino acids have been implicated as mediators of brain injury in a number of different experimental models (Kirsch et al., 1992). Successive exposures to CO have been demonstrated to cause degeneration in the CA1 region of the hippocampus, rich in N-methyl-D-aspartate (NMDA) receptors, and is the site of selective neuronal vulnerability to injury. Treatment of mice with NMDA antagonists significantly reduces CA1 injury associated with CO exposure (Funatsu et al., 1985). It is therefore likely that this channel is partly responsible for the neurodegeneration associated with CO exposure, and that NMDA antagonists may play a therapeutic role in the treatment of CO poisoning (Ishimaru et al., 1992). Furthermore, these investigators postulate that the delayed amnesia seen after CO poisoning may be due to delayed neuronal death in the CA1 region of the hippocampus as well as a dysfunction in the acetylcholinergic neurons (Nabeshima et al., 1991).

The effects of CO on a wide range of biochemically mediated processes determine its complex role in normal physiologic functioning and pathologic states. Because of the multiple sites of action of CO, the physiologic effects of increased exposure are complex and difficult to predict based upon these well-characterized pathways elucidated by *in vitro* techniques. The physiologic effects of CO are understood more by *in vivo* experiments than by theoretical and logical extension of the biochemistry into the whole animal.

3 PHYSIOLOGY

The brain regulates the amount of oxygen delivered to itself. The effect of CO on cerebral blood flow (CBF) and cerebral oxygen consumption ($CMRO_2$) is determined in part by the relationship between CBF and oxygen availability. In normal adult man, $CMRO_2$ in cerebral hemispheres is 3.5 ml per 100 g per minute (Kety and Schmidt, 1948). Depending upon conditions, about half of this is due to an active component of brain activity such as synaptic transmission, whereas the other half is accounted for by basal requirements such as maintenance of ionic pumps and cell integrity. When activity of either of these components increases, the brain generally will increase CBF as well as increase oxygen extraction, as occurs in other tissues. Because of the effects on cytochrome oxidase, CO may decrease oxygen utilization, and through its effects on hemoglobin, will decrease oxygen delivery. This is in contradistinction from hypoxia alone, where it is solely oxygen delivery that is adversely effected. The level of oxygen tension is determined by many variables illustrated in Figure 2.

Under conditions of hypoxia, CBF increases because of cerebral vasodilation (Iadecola et al., 1994). In addition, the hemodynamic effects of hypoxia result in an elevated mean arterial blood pressure (MABP), which results in a greater driving pressure into the brain, defined as cerebral perfusion pressure (CPP = MABP–ICP), where ICP is intracranial pressure. The mechanism by which CBF increases with either hypoxic hypoxia or CO hypoxia is controversial. One theory is that there is an oxygen sensor (Traystman et al., 1978) in brain. If this sensor responds to PaO_2, then CO hypoxia should not change cerebral vascular resistance (CVR = MABP/CBF), for under these conditions (unlike hypoxic hypoxia), PaO_2 is preserved. Traystman et al. (1978) reported that the rise in CBF during a reduction in arterial oxygen content produced by hypoxic hypoxia (low inspired percentage oxygen) was similar to the rise in CBF associated with CO hypoxia for a similar reduction in oxygen content. During this experiment, MABP increased with hypoxic hypoxia, and decreased with CO hypoxia. It was shown that by denervating the carotid bodies, cerebral vascular resistance fell in the hypoxic hypoxia animals to the same degree as in the CO hypoxia animals. Therefore, it was concluded that it is unlikely that the chemoreceptors play a role in either CO or hypoxic hypoxia and

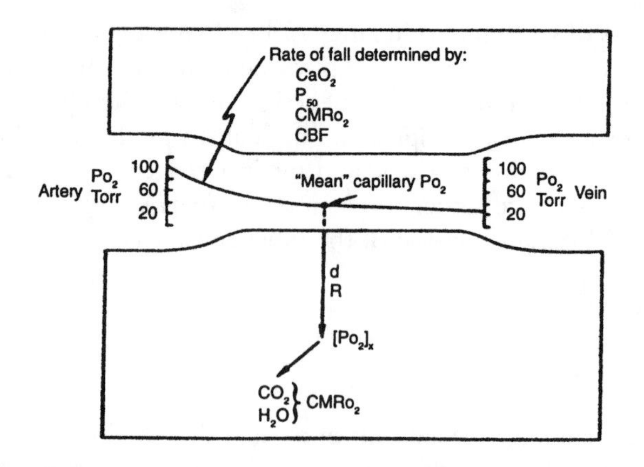

FIGURE 2

Oxygen diffusion in tissue. A simplified, two-dimensional view of a single vessel is shown with the determinants of PO_2 at point x within the tissue ($[PO_2]_x$). Rate of fall of PO_2 within the vessel is determined by arterial oxygen content of (CaO_2), oxyhemoglobin dissociation curve (denoted by P_{50}), cerebral oxygen consumption ($CMRO_2$) and cerebral blood flow (CBF). The ($[PO_2]_x$) is determined by the mean capillary PO_2, the diffusion distance (d), the resistance to diffusion (R) and the rate of oxygen metabolism to CO_2 and H_2O. (From Jones, M. D., Jr. and Traystman, R. J., *Semin. Perinatol.*, 8, 205, 1984. With permission.)

that the brain increases its blood flow in response to its oxygen needs with both hypoxic and CO hypoxia in order to maintain $CMRO_2$. This is independent of whether or not the baroreceptors or chemoreceptors are denervated. The implications for these findings, illustrated in Figure 3, are that if an oxygen sensor exists, it does not reside in either the baroreceptors or the chemoreceptors.

The exception to this is the neurohypophysis, in which CO hypoxia does not cause an increased blood flow (Figure 4), and the increase in blood flow due to hypoxic hypoxia can be blocked by chemoreceptor denervation (Figure 5) (Wilson et al., 1987). Thus, this is a unique brain region that does not respond to CO hypoxia (changes in arterial oxygen tissue delivery) but does respond to hypoxic hypoxia (changes in arterial oxygen tension).

Most of the physiologic experiments that have been discussed have been with very high levels of COHb. Most of the behavioral data, however, have been generated at lower levels of COHb. A dose response curve at low levels (<20%) of CO reveals that the rise in CBF is approximately linear with COHb less than 20%, and no change in $CMRO_2$ (Traystman, 1991). At COHb levels greater than 30%, however, CBF increased (Figure 6) out of proportion to the decrease in oxygen-carrying capacity with an associated fall in $CMRO_2$. At a COHb level of 50%, CBF increased to 200% of baseline, while $CMRO_2$ decreased.

The finding that CBF responses to hypoxic and CO hypoxia are similar is not without controversy. Studies on fetal (Jones et al., 1978) and newborn (Jones et al., 1981) lambs have demonstrated that increases in CBF during hypoxic hypoxia may correlate better with decreased arterial oxygen content

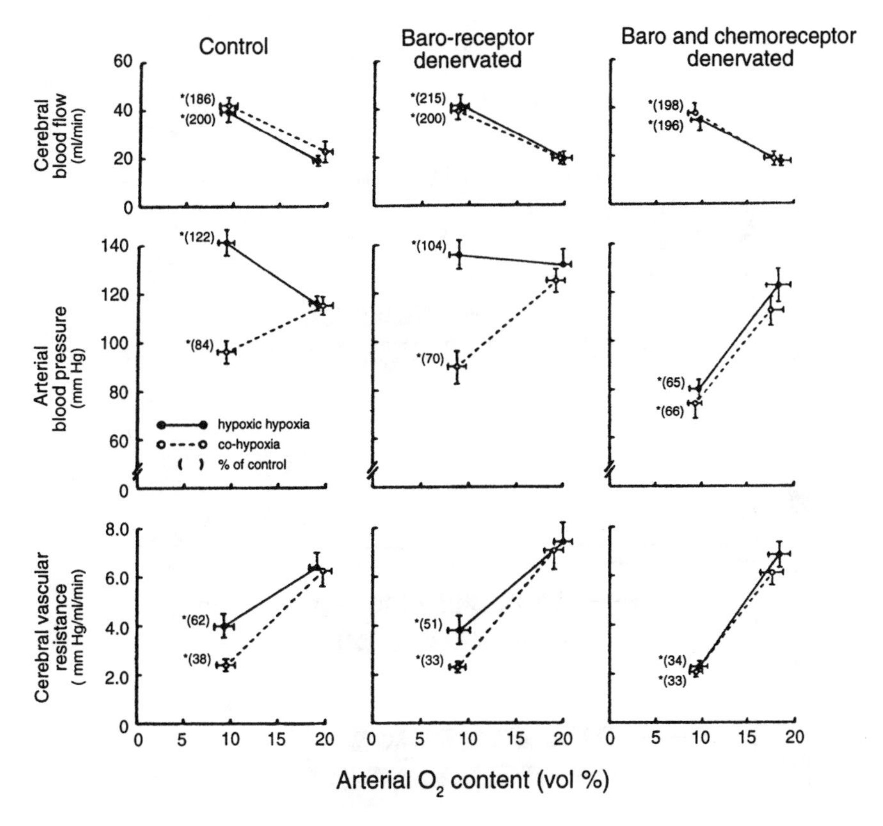

FIGURE 3
Effects of hypoxic and carbon monoxide hypoxemia on cerebral blood flow, mean arterial blood pressure, and cerebral vascular resistance in control, carotid baroreceptor-denervated, and carotid baroreceptor- and chemoreceptor-denervated animals (group 1). Data points and bars represent mean ±SE of 9 animals. Numbers in parentheses are % of control. *$P < 0.05$. (From Traystman, R. J. and Fitzgerald, R. S., *Am. J. Physiol.*, 241, H724, 1981. With permission.)

(CaO_2) than with decreased arterial oxygen tension (PaO_2). During hypoxic hypoxia there is a fall in CaO_2 that translates into a fall in the delivery of oxygen to the microvascular environment. Decreases in oxygen tension, on the other hand, compromise diffusion of oxygen from this environment into the parenchyma. The product of CBF and CaO_2 is cerebral oxygen delivery and is constant as PaO_2 falls because of compensatory rises in CBF. There are age related differences in the cerebral vascular response to hypoxic and CO hypoxia (Koehler et al., 1984). During hypoxic hypoxia, CBF increased in both newborn and adult sheep, in the face of an unchanged $CMRO_2$. During CO hypoxia, CBF increased to a greater extent than the CBF rise during hypoxic hypoxia for a given level of arterial oxygen content. However, in adults, $CMRO_2$ fell by 16% while in newborns, $CMRO_2$ was preserved. It was postulated that the fetal conditions of low arterial oxygen content and left shifted (fetal) hemoglobin–oxygen dissociation curve provided a

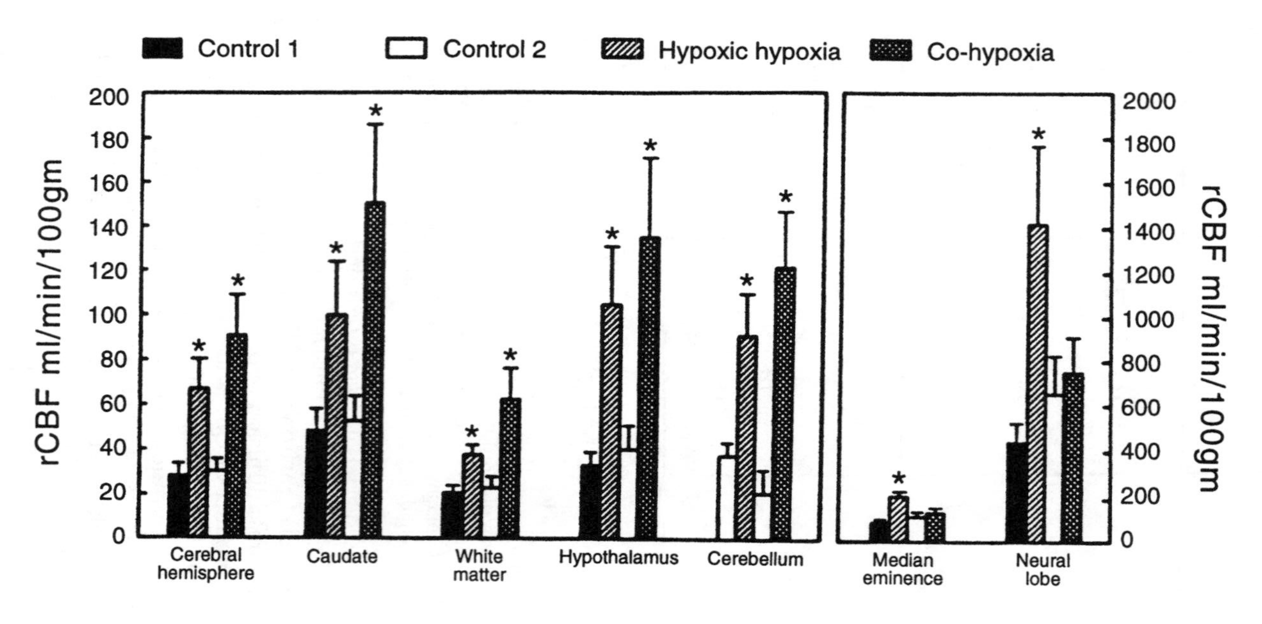

FIGURE 4

Effect of hypoxic hypoxia and carbon monoxide (CO) hypoxia on neurohypophyseal and regional cerebral blood flow (rCBF). Each bar represents mean ±SE of 5 dogs. Both types of hypoxia (diagonal and cross-hatched bars) produced significant increases from control (open and dark bars) in blood flow to all brain regions except neurohypophysis. Both parts of neurohypophysis, median eminence and neural lobe, showed no change from control with CO hypoxia but did have significant flow responses to hypoxia. Note change in vertical axis at right for median eminence and neural lobe blood flow. (From Hanley et al., *Am. J. Physiol.*, 250, H7, 1986. With permission.)

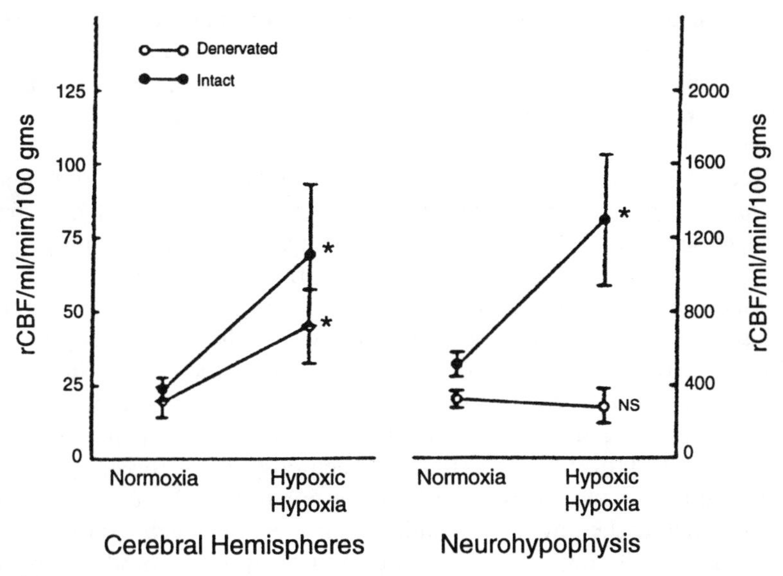

FIGURE 5
Effect of complete chemoreceptor denervation on total cerebral and neurohypophyseal blood flow. Each line represents ±SEM of 6 dogs. (Note change in y axis for neurohypophyseal blood flow.) (From Wilson et al., *Circ, Res.*, 61, II-94, 1987. With permission.)

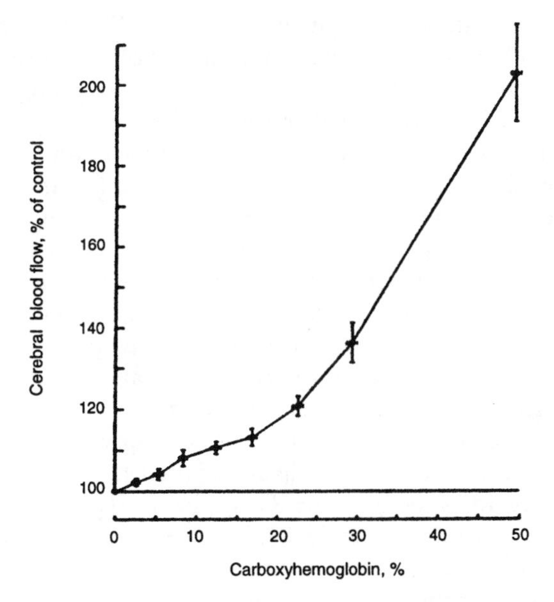

FIGURE 6
Effect of increasing carboxyhemoglobin levels on cerebral blood flow, with special reference to low-level administration (below 20% COHb). Each point represents the mean ±SE of 10 animal preparations. (From Traystman, R. J., in *Air Quality Criteria for Carbon Monoxide,* U.S. EPA, Washington, D.C., 1991)

microcirculatory environment better suited for maintaining $CMRO_2$ (Koehler et al., 1984). In this paradigm, whereas increases in CBF maintain oxygen availability at the microvascular exchange site, overall oxygen transport to the cells becomes relatively more diffusion dependent with CO hypoxia.

Reconciling the differences between these sets of experiments requires reexamining the data from the first set of experiments that demonstrate similar increases in CBF to hypoxic and CO hypoxia. In these experiments, during CO hypoxia, MABP fell in these anesthetized dogs, whereas it was well maintained in unanesthetized sheep (in the second set of experiments). Because cerebral autoregulation is impaired during hypoxia, a drop in perfusion pressure during CO hypoxia may have limited the increase in CBF in the dog. In the dog experiments, cerebral oxygen delivery increased and fractional oxygen extraction decreased during CO hypoxia. These changes may have masked the differences that were demonstrated in the sheep studies. Examining the data in terms of CVR, it is more likely that the sheep studies accurately define the physiologic finding that there is indeed a difference in the cerebrovascular response to hypoxic and CO hypoxia. Another explanation, however, may be the role of the anesthetic in the dog experiments, or the difference in the higher P_{50} (the PaO_2 at which 50% of hemoglobin is saturated) in sheep. This latter question was addressed (Koehler et al., 1983) by exchange-transfusing newborn lambs. Replacement of the high-affinity fetal hemoglobin of newborns with low-affinity adult sheep blood increased the P_{50} by 10 Torr with a concomitant decrease in CBF and cerebral oxygen delivery. This finding alone was important support of the postulated "oxygen sensor" in the brain. With CO hypoxia, however, the hemoglobin dissociation curve shifted back to the left, restoring cerebral oxygen delivery and oxygen extraction to pretransfusion levels. The conclusion of this study was that it was the fall in P_{50} rather than a direct tissue effect that is responsible for the overperfusion in relation to $CMRO_2$ during CO hypoxia.

In sum, these data all support the notion of a cerebral oxygen sensor that is sensitive to changes in tissue oxygen tension. This tissue oxygen tension is determined by the balance of cerebral oxygen delivery, and cerebral oxygen consumption, as well as the position of the oxygen dissociation curve. The sensor that determines cerebrovascular resistance during CO hypoxia has not been defined. On the contrary, a brain CO sensor that mediates the ventilatory response to CO has been identified. Specifically, the tachypnea observed in the conscious intact animals exposed to CO originates in the suprapontine structures, perhaps the diencephalon. This is because the tachypnea is not preserved after midcollicular decerebration. This response is independent of arterial afferents, specifically the carotid bodies (Gautier et al., 1990). It may be that this or an adjacent CO sensor may be responsible for cerebrovascular tone during CO exposure.

Another possibility is that cerebrovascular tone is mediated by local phenomena. Local production of NO causes vasodilation (Faraci and Brian,

1994). It has been postulated and remains controversial whether NO is instrumental in the cerebral vasodilation associated with hypoxia (Faraci and Brian, 1994). Hypoxic challenges result in varying degrees of cerebral vasodilation in the face of NOS inhibition depending on the model, species, and pharmacologic agent employed. On balance, most of the studies demonstrate that hypoxically mediated cerebral vasodilation is minimally effected by NOS inhibition and, therefore, NO is not the sole mediator of vasodilation in the face of cerebral hypoxia. This has been demonstrated by us (Ichord et al., 1994) and reviewed elsewhere (Iadecola et al., 1994). Although the physiologic role of NO during CO exposure has not yet been studied, it is possible that CO exposure could effect NO pathways and may be one of the mechanisms by which CO increases cerebral blood flow.

There are regional differences in tissue injury following severe CO exposure (Figure 7). One explanation for this is the regional CBF response to CO hypoxia. Brain regions in adults with high normoxic blood flows such as caudate, demonstrate a large response to hypoxia. Brain regions with less blood flow, such as the pons, show a relatively lower response (Koehler et al., 1984). CO hypoxia increased CBF to a greater extent than hypoxic hypoxia in all brain regions. In newborn sheep, regional responses differed with each type of hypoxia. With hypoxic hypoxia, the brain stem and the caudate had significantly greater responses compared with other regions such as the cortex. With CO hypoxia, the difference between brainstem and other regions was less marked. In adult sheep, there was no significant interactive effect between the type of hypoxia and the pattern of regional response (Koehler et al., 1984). Of interest is that the neurohypophysis does not respond to alterations in arterial oxygen content. In neurohypophysis, with equivalent reductions in arterial oxygen content, hypoxic hypoxia resulted in a 320% rise in blood flow, whereas during CO hypoxia, no changes were observed (Hanley et al., 1986).

Understanding the pathophysiology of CO poisoning is most clinically germane as it applies to fires. During combustion, other toxic chemicals are released, so there has been research regarding the cerebral effects of these agents in combination with CO. Of particular interest is the combination of CO and cyanide (CN) (Pitt et al., 1979). Comatose states and confusion have been described in a number of experimental paradigms involving the administration of CN. This toxin seems to affect cerebral white matter. Early experiments demonstrated a rise in CBF, ICP, and $CMRO_2$ in response to small doses of CN. It has been demonstrated that both CO and CN increase CBF and maintain $CMRO_2$ constant at low doses. The combination however, additively increased CBF but decreased $CMRO_2$ in a synergistic manner (Pitt et al., 1979). In addition to fires, CO poisoning has gained importance recently in the area of anesthesia. It has been discovered that there is an interaction between some inhalational anesthetics and dry baralyme (a CO_2 absorber) that results in the liberation of CO. This has caused several cases of CO poisoning in patients. The inhalational anesthetics associated with

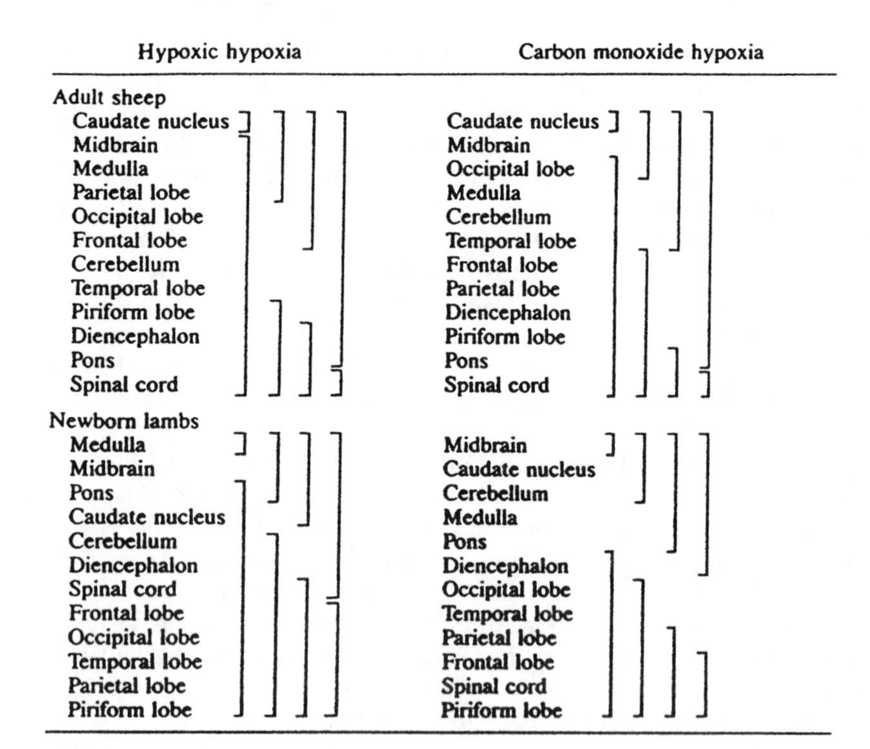

FIGURE 7
Multiple range test was performed within each type of hypoxia and each age group. Each pair of vertical brackets encloses two groups of brain regions that have significantly different responses. (From Koehler et al., *J. Cereb. Blood Flow Metab.*, 4, 115, 1984. With permission.)

this have been desflurane and enflurane, and to a lesser extent, isoflurane. Halothane and sevoflurane seem not to be affected by this interaction. Dryness and, to a lesser extent, elevated temperatures increase the rate of CO production (Fang and Eger, 1994).

In summary, the brain can increase its CBF or oxygen extraction in the face of diminished oxygen availability. The physiologic effects of CO are that, even at low levels, there is a rise in CBF with maintenance of $CMRO_2$ until levels of 30% are reached. The mechanisms responsible for this are uncertain, but may include the effect on the hemoglobin oxygen dissociation curves, effects on tissue oxygen levels, and direct toxic effects on tissue. Areas for future study include examination of a NO-mediated process, along with examination of neural effects that may be mediated by CO acting as a second messenger. Although there are differences in the age- and species-dependent cerebrovascular responses to CO exposure, these differences are small. The combined effects of CN and CO explain some cerebrovascular responses during fires, but further study of cerebrovascular effects of additional compounds released during combustion will be required to further elucidate this pathophysiology.

4 PATHOLOGY

The pathologic consequences of the biochemical and physiologic disruptions associated with CO have been described well. On a microscopic level, the delayed neuronal death in the selectively vulnerable CA1 subfield of the hippocampus has already been discussed (Ishimaru et al., 1991; Newby et al., 1978). In addition, the globus pallidus has been found to be damaged after CO exposure and is dependent upon the fall in local blood flow during the exposure (Ishimaru et al., 1992). Palladial abnormalities including hemorrhagic necrosis have been radiologically demonstrated in humans using magnetic resonance imaging (Silverman et al., 1993; Taylor and Holgate, 1988). Other brain regions that have been described to be damaged during CO exposure have included the thalamus and the centrum semiovale (Tuchman et al., 1990).

In utero exposure of 0.1% CO has been examined in rats at gestational ages of 15 d, which corresponds to the neurodevelopment of 30 weeks in humans. In this model, the caudate nucleus was severely affected (Daughtrey and Norton, 1983) and the pathologic findings increased when the exposure increased from 2 to 3 h. In addition, hemorrhagic infarcts with necrosis in the germinal matrix with debris-filled cavities were noted. The authors suggest that the damage observed in this model is analogous to the periventricular pattern of brain damage observed in premature humans after acute hypoxic events (Daughtrey and Norton, 1982). Other pathologic findings noted after *in utero* exposure of CO in humans are based on autopsy findings in the newborn. These findings include hypoxic ischemic lesions similar to those found in the animal models, decreased brain size and telencephalic dysgenesis (Woody and Brewster, 1990). Juvenile rhesus monkeys exposed to varying levels of CO (0.1–0.3%) for times ranging from 75 to 325 min demonstrate lesions of central white matter. In this study (Ginsberg et al., 1974), the extent of the white matter pathology was correlated best with the degree of metabolic acidosis and arterial hypotension associated with prolonged CO exposure. The authors postulate that this may be the etiology of "CO-leukoencephalopathy," which is often limited to the periventricular area — a susceptible vascular watershed area. The clinical correlates of white matter pathology have been described, and comprise a variety of findings including either scattered or well demarcated spongy and necrotic lesions. Another pathological type of CO leukoencephalopathy is "myelinopathy" marked by demyelination with axonal sparing (Ginsberg et al., 1974).

5 BEHAVIORAL EFFECTS OF CARBON MONOXIDE (CO)

Much of the literature that describes the behavioral effects of CO is confounded by methodological challenges. Among the confounding variables are: the concentration and duration of CO exposure, the tests used to evaluate the effect, the difficulty in "objectifying" some human behavior, and the manner in which the data were handled. Furthermore, there are few studies that simultaneously measure physiologic (i.e., CBF) as well as behavioral variables. As a consequence, there is no clear way in which these biochemical, physiologic, and pathologic effects can be causally linked to behavioral effects of acute CO exposure. What follows is a brief discussion of what tests were performed to evaluate the behavior, and the variable results obtained.

There has been extensive literature concerning the effects of CO on visual pathways. Earlier studies of the effects of COHb on absolute visual thresholds suggested decreased thresholds. Subsequent studies, however, failed to demonstrate a difference in this regard between air and up to 17% COHb (Hudnell and Benignus, 1989). Another test of visual function is the critical flicker frequency, which calculates the frequency at which light flickers appear as a continuous light. Early reports using this methodology did not reveal any effect of CO, but subsequent studies demonstrated that this frequency decreased in a dose-dependent manner from levels of COHb ranging from 4 to 12.7% (Seppanen et al., 1977). When examining visual evoked potentials of patients recovering from CO poisoning, a delay in the P_{100} latency can be demonstrated (He et al., 1993), but this has not been well studied in a dose-dependent acute exposure setting. The overall effects of acute CO exposure on visual function including acuity seems to be difficult to demonstrate (Traystman, 1991).

The effect of COHb on auditory function has been evaluated, and few effects have been demonstrable. Up to 12% COHb did not affect audiograms (Stewart et al., 1970). Furthermore, up to 8.9% COHb did not affect auditory flutter fusion test (Guest et al., 1970). The latter is a test where the frequency of clicks is increased until it is perceived to be a single continuous sound.

Most studies examining fine motor skills fail to demonstrate impairment of function up to 20% COHb (Traystman, 1991). However, a few studies have demonstrated slight impairment of manual dexterity during exposures as low as 7% COHb (Bender et al., 1972). Tracking is a special form of fine motor behavior and hand-eye coordination, and has been demonstrated to be adversely affected by COHB levels between 5 and 20% in some studies (Traystman, 1991). These findings are balanced by other investigators who found no effect on tracking in this range of COHb levels.

Performance of extended low-demand tasks is assessed by the use of tests of vigilance. The literature in this regard is again mixed, with about half of the reports demonstrating an effect, whereas half do not show decrements in vigilance (Traystman, 1991). Impairment of cognition during CO exposures is likewise reported variably depending on the experimental paradigm, and no consistent pattern of response is obvious. The deleterious effect of CO exposure on automobile driving performance has been demonstrated in some studies up to a COHb level of 20%. The effect of CO exposure on EEG has revealed few, if any, effects (Benignus, 1984). Similarly, visual evoked potentials are unaffected by acute exposures to CO up to 22%. Auditory evoked potentials are unaffected by CO when awake, but may increase amplitudes when the test is performed on asleep patients. Rabbits exposed to 1% CO demonstrated a 300% rise in CBF measured by hydrogen clearance with a consequent maintenance of oxygen delivery to brain measured by oxygen delivery index. During this exposure the sensory evoked response voltage decreased by half and recovered only to 80% of baseline with the removal of oxygen. It was therefore concluded that the diminution of somatosensory evoked responses was not due to diminution of oxygen delivery, but rather to some direct effect of CO (Ludbrook et al., 1992). Other investigators found a peak-to-peak amplitude of the N1–P1 increased in a dose-response pattern (He et al., 1993).

In summary, the effects of low dose (<20%) COHb on behavior are variable, so few concrete conclusions can be drawn. It is clear, however, that longer durations of CO exposure are more likely to affect behavior, and levels greater than 20% COHb will likewise have deleterious behavioral effects.

6 CONCLUSIONS

The biochemical, physiologic, and pathologic effects of CO exposure have been reviewed. Although there is a great deal of descriptive information, no clear mechanism has been described that can unify all of these findings. The direct effects of CO on the many heme proteins may mediate some of these findings. Additionally, the effects of CO upon tissue oxygen delivery certainly cause some of the observed effects. It may be that the combination of these two mechanisms may mediate the pathophysiologic effects.

REFERENCES

Anandatheerthavarada, H.K., Shankar, S.K., Bhamre, S., Boyd, M.R., Song, B.J., and Ravindranath, V., Induction of brain cytochrome P-450IIE1 by chronic ethanol treatment, *Brain Res.,* 601, 279, 1993.

Barinaga, M., Carbon monoxide: Killer to brain messenger in one step [news; comment] [see comments], *Science,* 259, 309, 1993.

Bender, W., Goethert, M., and Malorny, G., Effect of low carbon monoxide concentrations on psychological functions, *Staub-Reinhalt. Luft,* 32, 54, 1972.

Benignus, V.A. EEG as a cross species indicator of neurotoxicity, *Neurobehav.Toxicol.Teratol.,* 6, 473, 1984.

Bhamre, S., Anandatheerthavarada, H.K., Shankar, S.K., and Ravindranath, V., Microsomal cytochrome P450 in human brain regions, *Biochem. Pharmacol.,* 44, 1223, 1992.

Brown, S.D. and Piantadosi, C.A., *In vivo* binding of carbon monoxide to cytochrome c oxidase in rat brain, *J. Appl. Physiol.,* 68, 604, 1990.

Brown, S.D. and Piantadosi, C.A., Recovery of energy metabolism in rat brain after carbon monoxide hypoxia, *J. Clin. Invest.,* 89, 666, 1992.
Daughtrey, W.C. and Norton, S., Caudate morphology and behavior of rats exposed to carbon monoxide *in utero, Exp. Neurol.,* 80, 265, 1983.
Daughtrey, W.C. and Norton, S., Morphological damage to the premature fetal rat brain after acute carbon monoxide exposure, *Exp. Neurol.,* 78, 26, 1982.
Dinerman, J., Prabhakar, N., and Snyder, S., Carbon monoxide: A neural messenger molecule, in *Pharmacology of Cerebral Ischemia,* Krieglstein, J. and Overpilcher-Schwenk, H., Eds., Wissenschaftliche Verlagsgesellschaft MbH., Stuttgart, 1994.
Fang, Z.X. and Eger E.I., II, Source of toxic CO explained: -CHF2 anesthetic + dry absorbent, *Anesthesia Patient Safety Foundation Newsletter,* 9, 25, 1994.
Faraci, F.M. and Brian, J.E., Jr., Nitric oxide and the cerebral circulation, *Stroke,* 25, 692, 1994.
Funatsu, K., Yamada, S., Takamuki, K., Nakazawa, Y., and Inanaga, K., Hyperbaric oxygenation treatment of the after-effects of carbon monoxide poisoning, *Dtsch. Med. Wochenschr.,* 110, 140, 1985.
Gautier, H., Bonora, M., and Zaoui, D., Effects of carotid denervation and decerebration on ventilatory response to CO, *J. Appl. Physiol.,* 69, 1423, 1990.
Ginsberg, M.D., Myers, R.E., and McDonagh, B.F., Experimental carbon monoxide encephalopathy in the primate. II. Clinical aspects, neuropathology, and physiologic correlation, *Arch. Neurol.,* 30, 209, 1974.
Guest, A.D.L., Duncan, C., and Lawther, P.J., Carbon monoxide and phenobarbitone: Comparison of effects on auditory flutter fusion threshold and critical flicker fusion threshold, *Ergonomics,* 13, 587, 1970.
Hanley, D.F., Wilson, D.A., and Traystman, R.J., Effect of hypoxia and hypercapnia on neurohypophyseal blood flow, *Am. J. Physiol.,* 250, H7, 1986.
Harmon, H.J., Effect of age on kinetics and carbon monoxide binding to cytochrome oxidase in synaptic and non-synaptic brain mitochondria, *Mech. Aging Dev.,* 53, 35, 1990.
He, F., Liu, X., Yang, S., et al., Evaluation of brain function in acute carbon monoxide poisoning with multimodality evoked potentials, *Environ. Res.,* 60, 213, 1993.
Hudnell, H.K. and Benignus, V.A., Carbon monoxide exposure and human visual detection thresholds, *Neurotoxicol. Teratol.,* 11, 363, 1989.
Iadecola, C., Pelligrino, D.A., Moskowitz, M.A., and Lassen, N.A., Nitric oxide synthase inhibition and cerebrovascular regulation, *J. Cereb. Blood Flow Metab.,* 14, 175, 1994.
Ichord, R.N., Helfaer, M.A., Kirsch, J.R., Wilson, D., and Traystman, R.J., Nitric oxide synthase inhibition attenuates hypoglycemic cerebral hyperemia in piglets, *Am. J. Physiol.,* 266, H1062, 1994.
Ishimaru, H., Katoh, A., Suzuki, H., Fukuta, T., Kameyama, T., and Nabeshima, T., Effects of N-methyl-D-aspartate receptor antagonists on carbon monoxide-induced brain damage in mice, *J. Pharmacol. Exp. Ther.,* 261, 349, 1992.
Ishimaru, H., Nabeshima, T., Katoh, A., Suzuki, H., Fukuta, T., and Kameyama, T., Effects of successive carbon monoxide exposures on delayed neuronal death in mice under the maintenance of normal body temperature, *Biochem. Biophys. Res. Commun.,* 179, 836, 1991.
Jones, M.D., Jr., Sheldon, R.E., Peeters, L.L., Makowski, E.L., and Meschia, G., Regulation of cerebral blood flow in the ovine fetus, *Am. J. Physiol.,* 235, H162, 1978.
Jones, M.D., Jr., Traystman, R.J., Simmons, M.A., and Molteni, R.A., Effects of changes in arterial O_2 content on cerebral blood flow in the lamb, *Am. J. Physiol.,* 240, H209, 1981.
Jones, M.D., Jr., and Traystman, R.J., Cerebral oxygenation of the fetus, newborn, and adult, *Semin. Perinatol.,* 8, 205, 1984.
Kety, J.S. and Schmidt, C.F., The effects of altered arterial tensions of carbon dioxide and oxygen on cerebral blood flow and cerebral oxygen consumption of normal young men, *J. Clin. Invest.,* 484, 492, 1948.
Kirsch, J.R., Helfaer, M.A., Lange, D.G., and Traystman, R.J., Evidence for free radical mechanisms of brain injury resulting from ischemia/reperfusion-induced events, *J. Neurotrauma,* 9, S157, 1992.

Klatt, P., Schmidt, K., and Mayer, B., Brain nitric oxide synthase is a haemoprotein, *Biochem. J.*, 288, 15, 1992.

Koehler, R.C., Traystman, R.J., Rosenberg, A.A., Hudak, M.L., and Jones, M.D., Jr., Role of O_2-hemoglobin affinity on cerebrovascular response to carbon monoxide hypoxia, *Am. J. Physiol.*, 245, H1019, 1983.

Koehler, R.C., Traystman, R.J., Zeger, S., Rogers, M.C., and Jones, M.D., Jr., Comparison of cerebrovascular response to hypoxic and carbon monoxide hypoxia in newborn and adult sheep, *J. Cereb. Blood Flow. Metab.*, 4, 115, 1984.

Ludbrook, G.L., Helps, S.C., Gorman, D.F., Reilly, P.L., North, J.B., and Grant, C. The relative effects of hypoxic hypoxia and carbon monoxide on brain function in rabbits, *Toxicology*, 75, 71, 1992.

MacMillan, V., Fridovich, I., and Davis, J., Failure of iron chelators to protect against cerebral infarction in hypoxia-ischemia, *Can. J. Neurol. Sci.*, 20, 41, 1993.

McMillan, K., Bredt, D.S., Hirsch, D.J., Snyder, S.H., Clark, J.E., and Masters, B.S., Cloned, expressed rat cerebellar nitric oxide synthase contains stoichiometric amounts of heme, which binds carbon monoxide, *Proc. Natl. Acad. Sci. U.S.A.*, 89, 11141, 1992.

Nabeshima, T., Katoh, A., Ishimaru, H., et al., Carbon monoxide-induced delayed amnesia, delayed neuronal death and change in acetylcholine concentration in mice, *J. Pharmacol. Exp. Ther.*, 256, 378, 1991.

Newby, M.B., Roberts, R.J., and Bhatnagar, R.K., Carbon monoxide-and hypoxia-induced effects on catecholamines in the mature and developing rat brain, *J. Pharmacol. Exp. Ther.*, 206, 61, 1978.

Penney, D.G., Acute carbon monoxide poisoning: Animal models: A review, *Toxicology*, 62 123, 1990.

Pitt, B.R., Radford, E.P., Gurtner, G.H., and Traystman, R.J., Interaction of carbon monoxide and cyanide on cerebral circulation and metabolism, *Arch. Environ. Health*, 34, 345, 1979.

Seppanen, A., Hakkinen, V., and Tenkku, M., Effect of gradually increasing carboxyhaemoglobin saturation on visual perception and psychomotor performance of smoking and nonsmoking subjects, *Ann. Clin. Res.*, 9, 314, 1977.

Silverman, C.S., Brenner, J., and Murtagh, F.R., Hemorrhagic necrosis and vascular injury in carbon monoxide poisoning: MR demonstration, *Am. J. Neuroradiol.*, 14, 168, 1993.

Stewart, R.D., Peterson, J.E., Baretta, E.D., Bachand, R.T., Hosko, M.J., and Herrmann, A.A., Experimental human exposure to carbon monoxide, *Arch. Environ. Health*, 21, 154, 1970.

Taylor, R. and Holgate, R.C., Carbon monoxide poisoning: Asymmetric and unilateral changes on CT, *Am. J. Neuroradiol.*, 9, 975, 1988.

Thom, S.R., Antagonism of carbon monoxide-mediated brain lipid peroxidation by hyperbaric oxygen, *Toxicol. Appl. Pharmacol.*, 105, 340, 1990a.

Thom, S.R., Carbon monoxide-mediated brain lipid peroxidation in the rat, *J. Appl. Physiol.*, 68, 997, 1990b.

Thom, S.R., Dehydrogenase conversion to oxidase and lipid peroxidation in brain after carbon monoxide poisoning, *J. Appl. Physiol.*, 73, 1584, 1992.

Traystman, R.J., Cerebrovascular and behavioral effects of carbon monoxide, in *Air Quality Criteria for Carbon Monoxide*, McMullen, T.B. and Raub, J.A. Eds., U.S. Environmental Protection Agency, Washington, D.C., 1991, 74.

Traystman, R.J., Fitzgerald, R.S., and Loscutoff, S.C., Cerebral circulatory responses to arterial hypoxia in normal and chemodenervated dogs, *Circ. Res.*, 42, 649, 1978.

Traystman, R.J. and Fitzgerald, R.S., Cerebrovascular response to hypoxia in baroreceptor- and chemoreceptor-denervated dogs, *Am. J. Physiol.*, 241, H724, 1981.

Tuchman, R.F., Moser, F.G., and Moshe, S.L., Carbon monoxide poisoning: Bilateral lesions in the thalamus on MR imaging of the brain, *Pediatr. Radiol.*, 20, 478, 1990.

Verma, A., Hirsch, D.J., Glatt, C.E., Ronnett, G.V., and Snyder, S.H., Carbon monoxide: A putative neural messenger [see comments] [published erratum appears in *Science* 1994, Jan. 7; 263 (5143):15], *Science*, 259, 381, 1993.

Wilson, D.A., Hanley, D.F., Feldman, M.A., and Traystman, R.J., Influence of chemoreceptors on neurohypophyseal blood flow during hypoxic hypoxia, *Circ. Res.*, 61, II-94, 1987.

Woody, R.C. and Brewster, M.A., Telencephalic dysgenesis associated with presumptive maternal carbon monoxide intoxication in the first trimester of pregnancy, *J. Toxicol. Clin. Toxicol.*, 28, 467, 1990.

Zhang, J. and Piantadosi, C.A., Mitochondrial oxidative stress after carbon monoxide hypoxia in the rat brain, *J. Clin. Invest.*, 90, 1193, 1992.

Zhuo, M., Small, S.A., Kandel, E.R., and Hawkins, R.D., Nitric oxide and carbon monoxide produce activity-dependent long-term synaptic enhancement in hippocampus, *Science*, 260, 1946, 1993.

CHAPTER 5

PULMONARY CHANGES INDUCED BY THE ADMINISTRATION OF CARBON MONOXIDE AND OTHER COMPOUNDS IN SMOKE

Daniel L. Traber and Darien W. Bradford

CONTENTS

0-8493-4796-3/96/$0.00+$.50

1 INTRODUCTION

One of the primary determinants of survival following burn trauma is smoke inhalation injury (Moylan, 1981; Shimazu et al., 1990). Inhalation injury with associated burn injury carries a mortality rate between 45% and 78%, with increasing incidence in the elderly population reaching almost 100% (Heimbach and Waeckerle, 1988). The respiratory tract was involved in 70% of deaths under 12 h after burn and 46% of the patients died beyond 12 h after burn (Zikria et al., 1972b; Zikria et al., 1975). Inhalation injury is characterized by progressive inflammatory changes that result in deterioration of pulmonary function and structure (Shimazu et al., 1987). Carbon monoxide (CO), one of the major products of combustion, when associated with inhalation injury may reach lethal levels (Cohen and Guzzardi, 1983; Lowry et al., 1985). Other products of combustion, including cyanide, aldehydes, acrolein, and hydrogen fluoride, will be discussed later in this chapter (Table 1) (Sheppard, 1989).

2 CLASSIFICATION OF INHALATION INJURY

2.1 Thermal Injury

Direct thermal injury has been shown to cause edema leading to obstruction of the airway above the trachea (Luce et al., 1976; Mellins and Park, 1975; Peters, 1981). Injury from thermal burns below the larynx is extremely uncommon. Mucosal injury generally occurs at air temperatures above 150°C (Crapo, 1984; Herndon et al., 1987; Wald and Balmes, 1990). Heat is

TABLE 1

Chemical Injuries Associated with Products of Combustion

Chemical	Sources	Injury Produced
Aldehydes	Plastics from furniture	Upper airway injury and CNS depression
Acrolein	Acrylics from windows, wood finishes, or wall coverings	Diffuse airway and parenchymal injury
Ammonia	Phenolics and nylon	Upper airway burn
Anhydrides	Chemical, paint, plastics	Airway injury, asthma; pulmonary hemorrhage with high-dose exposure
Carbon dioxide	Closed space fires (as high as 10% CO_2)	Respiratory acidosis, CNS depression
Chlorine	Chemical industry; transportation accidents; gas evolved from mixing chlorine bleach with acid cleaners; swimming pools	Diffuse airway and lung injury
Cyanide	Carpets, upholstery, nylon	Acidosis, shock
	Polyurethane products from isocyanates	Asthma, airway injury, hypersensitivity pneumonitis
Hydrogen chloride	Fabrics, polyvinyl chloride	Mucosal burns and edema, dysrhythmias, shock
Hydrogen fluoride	Teflon from pipes or kitchen utensils	Upper airway injury
Nitrogen dioxide	Nitrocellulose	Lung injury
Sulfur dioxide	Smelters; pulp mills, wineries; oil refineries; power plants	Bronchoconstriction, mucous secretion; airway injury at high concentration

Modified from Sheppard, D., in *Textbook of Pulmonary Diseases, 4th ed.,* Baum, G. and Wolinsky, E., Eds., Little, Brown, Boston, 1989, 840–841. With permission.

energy that is transferred as a result of a temperature difference. Heat energy always flows from a warmer body (higher temperature) to a colder body (lower temperature), until they reach the same temperature. Heat capacity is defined as the quantity of heat necessary to raise a gram of substance 1°C (Weast, 1984). Hot dry air has the heat capacity of about 0.24 cal/g/°C compared to water 1.0 cal/g/°C. Consequently, because of the low heat capacity, even very hot smoke is usually cooled to near 40°C by the time it passes the airway (Moritz et al., 1945). Steam has a heat capacity of 0.48 cal/g/°C, twice that of dry air, but the potential for injury is much higher with steam because as it condenses to water it gives off a considerable amount of calories. Inhalation of steam is a serious burn that is almost universally lethal.

Smoke contains enormous quantities of particulate matter (Silverman, 1978; Trunkey, 1978). Particulate matter generally has a higher specific heat and density in air, which therefore may contribute to heat injury in both the upper and lower airways. Within smoke there are often incomplete products of combustion that are inhaled, and further oxidation is required, releasing large quantities of heat energy that contributes to the injury to the lower airways (Farrow, 1975; Mellins and Park, 1975).

Direct thermal injury usually occurs immediately, and it causes edema, erythema, hemorrhage, and ulceration of the airway (Wald and Balmes, 1990). Fortunately, there are mechanisms that protect the lower airway and lung parenchyma. Air turbulence in the upper airway improves heat exchange from air to mucosa by convection, which reduces the amount of energy absorbed. The normal humidifying mechanisms also serve to protect the respiratory system from heat injury.

There is convincing clinical evidence that hot air is an uncommon source of lower airway and lung parenchymal injury. It is estimated that the frequency of thermal burns below the larynx is generally less than 5% (Fein et al., 1980; Pruitt et al., 1970). It has been reported that occlusion of bronchial artery before inhalation injury will virtually prevent pulmonary edema and markedly reduce the lung lymph flow changes seen after inhalation injury (Abdi et al., 1991; Hales et al., 1989; Traber et al., 1994). The bronchial circulation is a portal system in which the bronchial venous drainage of the extrapulmonic airway enters the azygos vein and the intrapulmonary airway enters the pulmonary circuit at the precapillary level (Charan et al., 1984; Lakshminarayan et al., 1990). Injury to the airway can result in the release of cytotoxins that could then be delivered directly to the pulmonary circuit (Traber et al., 1994). This could possibly explain why damage to the airway creates such profound changes in pulmonary parenchymal function.

2.2 Chemical Injury

There are many toxic water-soluble components associated with smoke. Gases, usually from burning plastics and rubber, include chlorine, sulfur dioxide, ammonia. Fires that occur in an enclosed environment produce incomplete products of combustion, mainly CO. Some of these gases come in contact with the mucosa of the upper airway, which produces epithelial damage. Aldehydes and nitrogen oxides produced by burning wood and cotton are lipid-soluble, and they will destroy lipid cell membranes and denature proteins (Robinson and Miller, 1986; Zikria et al., 1972a). These compounds cause tracheobronchitis with damage to the glottis and supra and infraglottic areas, resulting in massive edema, epithelial debris, and increased capillary permeability, resulting in marked changes of the respiratory physiology.

3 PATHOPHYSIOLOGY

The upper airways are very efficient in cooling, making thermal injury below the vocal cords extremely uncommon. Minimal respiratory epithelial damage may occur in conjunction with areas of tracheal necrosis. Inhalation injury to tracheobronchial areas and release of inflammatory mediators eventually leads to lung damage and edema formation (see Figure 1) (Traber et al., 1988). Complications include loss of compliance, atelectasis, interstitial edema, laryngotracheal bronchitis, pneumonia, and, ultimately, pulmonary failure (Heimbach and Waecherle, 1988). The mechanism responsible for initiating the process that causes the transvacular fluid flux apparently relates to the thermal denaturization of plasma proteins, with a subsequent release of oxygen free radicals and products of arachidonic acid metabolism into the microvascular areas (Jin et al., 1986; Nozaki et al., 1984; Till and Ward, 1987). Parenchymal injury with a resultant increased microvascular fluid flux has been studied in the ovine model of inhalation injury (Herndon et al., 1984; Herndon et al., 1985; Nozaki et al., 1984; Traber et al., 1986). After inhalation injury there was a marked increase in lung lymph flow with only a mild elevation in pulmonary arterial pressure and resistance (Traber and Herndon, 1990). There was also an elevation in lymph-to-plasma protein concentraion ratio, consistent with an elevation in pulmonary microvascular permeability. The changes in lung lymph flow in the sheep model after inhalation injury are accompanied by elevations in extravascular lung water measured by both the double-indicator dilution and gravimetric techniques (Nozaki et al., 1984; Traber et al., 1985). These changes generally have a delayed onset, occurring between 12 and 24 h after injury. These increases in extravascular water correlate with an impairment of gas exchange and increases in lung lymph flow. The oxygen free radicals go on to damage the capillary beds, and thromboxane A_2, a catabolite of arachidonic acid, cause vascular constriction (Traber and Herndon, 1990). The combination of increased microvascular permeability and pressure leads to a rapid and profound edema that can cause complete upper airway obstruction (Pitt et al., 1987).

The tracheobronchial area receives its blood supply mainly from the bronchial artery, which receives 1–2% of cardiac output and supplies approximately 80% of the systemic circulation to the lungs in sheep (Magno and Fishman, 1982). In sheep, after inhalation injury, bronchial blood flow increased 2.5 to 3 times baseline. This may be the reason for the marked hyperemia known to occur after inhalation injury (Kimura et al., 1988). This hyperemic response peaks immediately after injury and returns toward baseline usually within 48 h. The exact mediators of hyperemia have not yet been identified; however, a similar hyperemia is seen after acid aspiration (Basadre et al., 1988). There is strong evidence that the response may be the result of a neuroinflammation. Bronchoconstriction and vasodilation of the bronchial vasculature have been shown to occur after the inhalation of cigarette smoke and are thought to be mediated by neuropeptides (Lundberg et al., 1991). This response, which also occurs after insufflation of cotton smoke, can be markedly reduced by prior administration of capsaicin, which inhibits the release of neuropeptides (Traber et al., 1990). Figure 1 illustrates the pathophysiology of tracheobronchial damage by smoke inhalation in sheep. After a delay of several hours, the inflammatory response is followed by a cellular infiltration with exudate formation (Linares et al., 1989). This exudative material is rich in B-glucuronidase and thromboxane, and it is comsumed by numerous polymorphonuclear neutrophils (PMNs) and lymphocytes (Herndon et al., 1986; Traber et al., 1985). Separation of the basement membrane occurs, leaving behind a bare basement membrane (Abdi et al., 1990; Linares et al., 1989). This tissue damage is perhaps a result of lysosomal enzymes produced by the PMNs and macrophages in the area. Between 12 and 36 h after injury, castlike material may be expectorated.

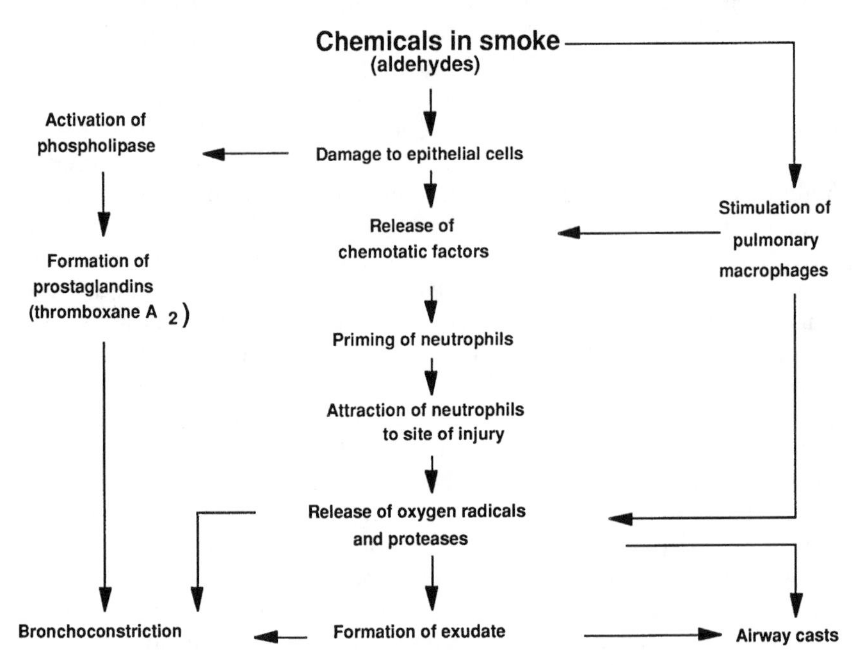

FIGURE 1
The pathophysiology of tracheobronchial damage in sheep.

Smoke inhalation into the tracheobronchial area also produces bronchoconstriction with an increase in respiratory resistance, which could be related to the potent smooth muscle constrictor thromboxane A2 (Petroff et al., 1976; Prien et al., 1988). These changes in resistance may result in hypoxic pulmonary vasoconstriction. Cyclooxygenase inhibitors have shown some reduction in lung injury by blocking thromboxane synthesis (Kimura et al., 1988; Shinozawa et al., 1986). Sheep insufflated with smoke into one lung and air into the contralateral lung showed a marked shunting of blood flow from the injured lung. This has been confirmed by the use of radioactive microspheres (Prien et al., 1987). The changes in pulmonary transvascular fluid flux seen with inhalation injury are markedly reduced by the administration of cyclooxygenase and thromboxane synthesis inhibitors (Kimura et al., 1988; Noshima et al., 1991). Wu et al. (1988) have reported normalization of acute changes in the ventilation/perfusion ratio after inhalation injury by treatment with cyclooxygenase inhibitors.

Injury to tracheobronchial areas depends upon the gaseous and particulate content of the inhaled material (Traber and Herndon, 1990). Acrolein has been found to be present in higher concentrations relative to its toxicity than other compounds (Treitman et al., 1991). In a canine model, acrolein alone produced significant tracheobronchial injury (Venegas et al., 1988). Free radicals present in smoke have also been reported to be edematogenic (Crapo et al., 1984). When nebulized dimethyl sulfoxide, an oxygen free radical scavenger, was given to sheep at the time of smoke inhalation, it greatly reduced the injury and associated mortality (Brown et al., 1988). A similar finding has been found with the intravenous administration of superoxide dismutase (Nguyen et al., 1993).

Clinically, Stone (1979) described the phases of pulmonary injury in children with acute respiratory distress secondary to bronchospasms: laryngeal edema generally occurs within 12 h of insult; from 6 to 72 h, pulmonary edema leading to adult respiratory distress syndrome occurs. Histologically, Linares et al. (1989) described inhalation injury in four distinct although overlapping phases: exudative phase (first 48 h), degenerative phase (from 12 to 72 h), proliferative phase (from 48 h to 7 d), and reparative phase (more than 4 d).

The exudative phase occurs during the first 48 h after injury. The most prominent features include the exudation of protein-containing fluids and the abundance of inflammatory cells (PMNs) (Herndon et al., 1984; Herndon et al., 1986; Traber et al., 1986). Further, prominent interstitial edema followed by infiltration of PMNs leads to alveolar septal enlargement in association with venular and lymphatic dilation. The endothelial lining shows loose junctions or interendothelial gaps through which fluids leak. This leads to the edematous widening of the alveolar interstitium followed by intraalveolar edema and hemorrhage. The type I pneumocytes develop severe intracellular edema with enlarged endoplasmic reticulum cisterns and mitochondrial swelling with large vacuoles. The upper airways show edema and hemorrhages, with PMNs marginated along the small blood vessels. The tracheobronchial bronchial epithelium begins to show cytoplasmic vacuolization with focal necrosis and submucosal edema (Linares et al., 1989).

The degenerative phase occurs within 12 to 72 h after injury. Most striking is the epithelial damage that occurs. Epithelial necrosis occurs along the bronchial tree with shedding and the formation of pseudomembranous casts that are made up of cellular debris, mucus, fibrin, proteinaceous amorphous material, and PMNs (Abdi et al., 1990; Hubbard et al., 1991; Linares et al., 1989). These casts can cause partial or complete obstruction of the bronchial lumen. The tracheobronchial submucosa develops marked congestion and edema. These events continue on to the alveolar wall, where type I pneumocytes develop a distinct necrosis, leading to the exposure of the alveolar epithelial basement membrane. The areas form thick pseudomembranes, loosely adherent to the basal lamina. Hyaline membranes then form from sloughed type I pneumocytes and fibrin, 48 to 72 h after injury. Type II pneumocytes usually remain intact, unlike type I pneumocytes, which suffer irreversible cell damage. Macrophages then appear in the areas of hyaline membrane formation with a proliferation of type II cells (Hubbard et al., 1991; Linares et al., 1989). The proliferative phase occurs from 48 h to 7 d. It is characterized by the proliferation of epithelial cells and macrophage mobilization. Viable basal cells of the upper airways and cells from the ductal glands proliferate to cover damaged epithelium. Type II pneumocytes proliferate to cover the denuded basement membrane leading to the reparative phase (Hubbard et al., 1991).

During the reparative phase (>4 d), pulmonary tissues return to normal or may develop abnormalities such as squamous metaplasia of airway epithelium, septal scarring, or interstitial fibrosis (Barrow et al., 1992; Linares et al., 1989). Small bronchi and bronchioles show a polyploid organizing exudate or show a progressive occlusion of the lumen, developing into fibrous scarring (bronchiolitis obliterans). During this phase, phagocytosis of hyaline membrane remnants and other debris from alveolar macrophages commonly occurs. In many cases, recovery of normal alveolar structure takes place, but this process may lead to an extensive fibroblast proliferation with interstitial collagen formation, leading to subsequent interstitial and intraalveolar fibrosis.

4 INTOXICANTS

4.1 CARBON MONOXIDE (CO)

CO is a combustible, nonirritating, colorless, tasteless, odorless gas that rapidly displaces oxygen from hemoglobin (Gosselin et al., 1976; The *Merck Index,* 1983). It has approximately 250 times the affinity of oxygen. CO intoxication involves compromise of oxygen transport from the lungs to the tissues in the body and possibly an inhibition of the cellular metabolism. The acute effects of inhalation injury include CO toxicity. Decreased O_2 delivery to organs occurs in conjunction with low oxygen concentrations, leading to end organ injury and cell death. Carboxyhemoglobin (COHb) is known to shift the oxyhemoglobin dissociation curve to the left. Patients may receive very high concentrations of oxygen and the dissolved oxygen content becomes elevated; however, there is poor release of oxygen to the tissues. Furchgott and Jothianandan (1991) compared the relaxation induced by nitric oxide (NO), CO, and light. They concluded that CO stimulated cytoplasmic guanylate cyclase (cGMP) of vascular smooth muscle. With this increase in cyclic GMP, relaxation occurs. The extent of relaxation and the increase in cyclic GMP levels were much less with NO. However, NO in the presence of superoxide dismutase (to remove superoxide, which rapidly inactivates NO) is several times more potent than CO in stimulating cytoplasmic guanylcyclase. They further suggest that CO also activates guanylate cyclase by combining with the heme moiety, but its affinity and efficacy are much less than those of NO. We have evaluated the effects of CO in our ovine model (Sugi et al., 1990). The gas was given to achieve the same carboxyhemoglobin levels as we saw in animals with smoke inhalation injury. These levels of CO did not induce a cardiopulmonary response or a change in transvascular fluid flux in the absence of smoke injury. Shimazu et al. (1990) also compared the effects of smoke inhalation injury and CO in a sheep model. Their data were similar to ours with the exception that they saw an increased cardiac output in their model.

4.2 Carbon Dioxide (CO_2)

Levin et al. (1987) found that rats exposed to CO_2 concentrations of up to 15% for 30 min never lost consciousness, and Pryor et al. (1974) found the minimal lethal CO_2 concentration in mice was 40% for 4 h. CO_2 in concentrations up to 10% acts as a respiratory stimulant. Above 10%, the minute volume and respiratory rate decreased (Herpol et al., 1976; Levin et al., 1987; Wong and Alarie, 1982). When sublethal CO_2 levels were added to sublethal concentrations of CO, rats died either during the 30-min exposures, or in the ensuing 24 h. The rate of COHb level depends, in part, on the atmospheric concentration, but also on the rate and depth of breathing, the exposure duration, the blood volume, the barometric pressure, and the diffusivity of CO through the lung (Stewart, 1974). These results indicate that the rate of formation of COHb from exposure to approximately 2500 ppm of CO was 1.5 times greater in the presence of 5.25% CO_2 than in its absence. The increased rate of COHb formation and the maintenance of the equilibrium level at 78% for a greater time period was not sufficient to explain the deaths that occurred with the combined CO and CO_2 for a period of 30 min, because in 60-min experiments, animals with COHb levels of 78% (induced with CO alone) for longer than 30 min did not die (Levin et al., 1987). Levin et al. concluded that the combination of respiratory and metabolic acidosis occurs in the animals when exposed to both CO and CO_2. Recovery was also delayed for at least 30 min after the end of the combined gas exposures, and full recovery did not occur until 60 min after exposure; in contrast, in the CO experiments, recovery occurred within the first 5 min of the postexposure period. The two gases act together by: (1) increasing the rate of formation of COHb, (2) causing a severe degree of acidosis (which is greater than the metabolic acidosis from exposure to CO alone or the respiratory acidosis from CO_2 alone), and (3) prolonging the recovery period from this acidosis after exposure.

4.3 Cyanide (CN)

CN is a colorless combustible gas, less dense than air, with a characteristic odor of burnt almonds (*The Merck Index,* 1983). It has been identified as a product of combustion in polyurethane, silk, wool, polyacrylonitriles, nylon, synthetic rubber, paper nitrocellulose, asphalt, nitrogen-containing polymers, and fire retardants (Becker, 1985; Kirk et al., 1993; Silverman et al., 1988). CN toxicity is often clinically nonspecific; therefore, seriously ill smoke inhalation patients should be suspected of CN poisoning until proven otherwise. CN exerts its toxic effect by inhibiting the reoxidation of reduced cytochrome A_3 in the mitochondrial respiratory chain. Cyanide reacts with Fe (III) of the cytochrome oxidase in the body, preventing the utilization of oxygen by the tissue (blocking of intracellular oxygen transport) (See Figure 1) (Barillo et al., 1986; Einhorn, 1975). This blocks a step in oxidative phosphorylation and prevents the mitochondria from utilizing oxygen (Crapo, 1990). During this process the tricarboxylic acid cycle is arrested, thereby shifting cellular metabolism to anaerobic pathways, leading to lactic acidosis. The half-life of CN is approximately 1 h (Baud et al., 1991). Reports of CO and CN concentration correlations possibly indicate that blood CO concentrations may be considered an index of cyanide poisoning in fire victims. It has been noted that concentrations above 20 ppm in air are considered dangerous, but exposures of 45 to 54 ppm can be tolerated for an hour without serious difficulty (Crapo, 1990). Exposures of 110 to 135 ppm will cause death in less than 1 h and exposures to 280 ppm cause immediate death (Davies, 1986). Inhalation of CN vapors initially causes a reflex stimulation of breathing, which leads to a greater concentration of the gas entering the body. Later, lactic acid production and associated metabolic acidosis stimulates respiratory drive (Morikawa, 1976; Wald and Balmes, 1990). The cause of death is usually secondary to paralysis of the respiratory centers in the brain. CN poisoning has also been associated with increased lactate levels (Einhorn, 1975).

5 IRRITANT GASES

Irritant gases are considered cytotoxic, which is determined by their solubility in water and the duration of exposure (see Table 2) (Crapo, 1990). Table 2 illustrates the relative water solubilities of some common chemical components of smoke. Toxic gases with high solubility dissolve rapidly into water-coated surfaces with which they come in contact. When released into enclosed spaces, upper airway mucosal injury occurs secondarily to mucosal water content of the airway (Schwartz, 1987; Wald and Balmes, 1990). Laryngeal spasm occurs rapidly after injury, which helps to reduce distal exposure. Lower airway injury occurs secondary to higher ventilatory rates, which are driven by increased CO_2 levels, metabolic acidosis, and also exertion.

5.1 Acrolein

Acrolein is a low molecular weight a,b-unsaturated aldehyde that exists in high concentrations in smoke from wood, cotton, and polyethylene (Janssens

TABLE 2

Relative Water Solubilities of Common Components of Smoke

Chemical	Solubility in water
Aldehyde (acrolein)	High
Ammonia	High
Chlorine	High
Hydrogen chloride	High
Sulfur dioxide	High
Oxides of nitrogen (NO_2)	Low
Phosgene	Low

et al., 1994; Kimura et al., 1988; Morikawa, 1976). Acrolein has been detected at life-threatening concentrations in fires. The toxic potency is secondary to the fact that it is a potent thiol-reducing agent, depending primarily on depletion of glutathione levels. Acrolein is known to have mutagenic effects on human bronchial epithelial cells in culture; these effects occur without the known cytopathic effects that occur. Acrolein, when compared with HCl, formaldehyde, or furfuraldehyde, which are all components of smoke, was shown to be edematogenic at an average of 65 min after smoke, whereas HCl produced airway damage but no pulmonary edema (Hales et al., 1992; Terrill et al., 1978). The researchers concluded that acrolein, but not HCl, in synthetic smoke produced a delayed onset, noncardiogenic, and peribronchiolar edema in a roughly dose-dependent fashion. It has been shown that $BW755^c$, a combined lipoxygenase and cyclooxygenase inhibitor, prevented the pulmonary edema after acrolein smoke exposure (Janssens et al., 1994). The latter concluded that thromboxane was probably responsible for the pulmonary hypertension, but not the pulmonary edema after acrolein exposure. This suggests that LTB_4 may be etiologic.

5.2 Ammonia (NH_3)

NH_3 is a colorless gas with a very pungent odor. It is a highly soluble, alkaline, and irritative gas with half the density of air (*The Merck Index*, 1983). NH_3 is used in the manufacturing of chemicals, fertilizers, pharmaceuticals, synthetic fibers, and explosives, and is widely used in refrigeration. Generally considered to be inflammable, mixtures of NH_3 have been known to explode in air. Produced by the combustion of nylon, silk, wool, plastics, and melamine resins, it comes in contact with mucosal membranes, forming ammonium hydroxide (NH_4OH) or hydroxyl ions, which leads to liquefaction necrosis, edema, and sloughing of the airway mucosa (Schwartz, 1987; Wald and Balmes, 1990). In severe cases of exposure, pulmonary edema (delayed onset > 6 h), airway obstruction, and death occurs within 10 min at concentrations above 1000 ppm (Gosselin et al., 1976; Schwartz, 1987; Terrill et al., 1978).

5.3 Chlorine (Cl_2)

Cl_2 is a yellow-green gas with intermediate solubility in water and a density higher than that of air (Sheppard, 1989). Inhalation of chlorine produces diffuse pulmonary injury extending from the oropharynx to the alveoli. In the presence of water, chlorine, a potent oxidant, forms hydrochloric acid (HCl) and hypochlorous acid (HOCl), which is 20 times more toxic than HCl alone (Kaye, 1980; Schwartz, 1987). Chlorine has an odor theshold of 0.2 to 0.4 ppm and is tolerated in concentrations up to 1–2 ppm (*The Merck Index,* 1983). Exposures of 3–6 ppm become irritating, and concentrations of 5–8 ppm cause mild illness (Arena, 1979; Kaye, 1980; Parkes, 1982; Wald and Balmes, 1990). Brief exposures in excess of 40 ppm causes bronchospasm, bronchiolitis, and/or pulmonary edema. Death usually occurs in concentrations in excess of 1000 ppm.

5.4 Hydrogen Chloride (HCl)

Damage from airway acid inhalation appears to be a combination of airway and parenchymal injury that is associated with the release of thromboxane A_2 and prostacyclin (Stothert et al., 1988). Goldman et al. (1991) postulated that peribronchial leukocytes are stimulated by acid to generate circulating mediators (i.e., thromboxane), which in turn activate circulating PMNs and later, the lung endothelium, leading to increased adhesion. This, in turn, leads to delayed lung injury from these sequestered PMNs. Localized acid aspiration induces synthesis of LTB_4 and thromboxane A_2. Inhibition of either leukotriene or thromboxane will limit PMN adhesion and increased lung permeability (Goldman et al., 1992).

Inhalation of HCl is common because it is a component of aspiration of vomitus. We have found that this results in lesions similar to those seen with the inhalation of smoke (Stothert et al., 1990a; 1990b). Hales et al. (1988, 1992), on the other hand, have shown acrolein — but not HCl, formaldehyde, or furfuraldehyde, all of which are components of smoke — to be edematogenic.

5.5 Sulfur Dioxide (SO_2)

SO_2 occurs naturally and also in synthetic rubbers. In the presence of fire these materials generate SO_2, is a very pungent, heavy gas that is extremely toxic (Autian, 1971). The threshold limit value of SO_2 is given as 5 ppm. In contact with water (moisture), SO_2 will form sulfuric acid, which exerts its irritant effects on mucous membranes. Exposure to high concentrations of the gas leads to death, which would probably be secondary to asphyxiation (Einhorn, 1975). Chronic exposure to SO_2 has a greater effect upon those with cardiopulmonary diseases as opposed to the normal population.

5.6 Oxides of Nitrogen

Oxides of nitrogen include nitrous oxide (N_2O), nitric oxide (NO), nitrogen dioxide (NO_2), nitrogen trioxide (N_2O_3), nitrogen tetroxide (N_2O_4), and nitrogen pentoxide (N_2O_5).

NO_2 is a reddish-brown gas with a sweet odor and limited solubility. It is produced by fixation of atmospheric nitrogen during high-temperature combustion (Sheppard, 1989). Outside, it is produced by large fuel-burning point sources and motor vehicles. Indoors, it is produced by gas stoves and space heaters. It is used in the manufacture of explosives and in the chemical, welding, and cleaning industries. The acute injury commonly becomes evident 5 to 72 h later (Gosselin et al., 1976). Symptoms begin with cough, dyspnea, fever, and peripheral blood leukocytosis. NO_2 reacts slowly with water to form HNO_2 and HNO_3. These acids reach their maximal concentrations in the distal airways and are thought to be the principal mediators of injury (Parkes, 1982). NO_2 can cause direct injury to the airway, and alveolar cells with higher concentrations (50 ppm) than normally encountered in the atmosphere (Davies, 1986; Schwartz, 1987; Terrill et al., 1978). Pathological examination revealed extensive damage to the respiratory epithelium and hemorrhagic pulmonary edema (Parkes, 1982). NO_2 has been known to react with blood to form methemoglobin, which is thought to have a minor role in fatal NO_2 exposures (Schwartz, 1987).

5.7 Aliphatic Hydrocarbons

Aliphatic compounds are produced by the thermodegradation of all organic polymers (Autian, 1971). Lower molecular weight compounds produce necrosis, but the biological activity is inversely proportional to molecular weight. When polymers are degraded, unsaturated hydrocarbons may be present that generally have a greater toxic effect than the corresponding saturated compounds.

5.8 Aromatic Hydrocarbons

Benzene and other aromatic compounds have irritating properties as well as systemic toxicity (Autian, 1971). Altering the structure of aromatic compounds may increase or decrease their toxicity. Benzene has the ability to be absorbed not only through inhalation, but through skin. Styrene, a product of polystyrene depolymerization, can produce irritation to mucous membranes and impairment of neurological function at levels above 100 ppm. Both compounds are considered relatively safe at levels below 100 ppm.

5.9 Hydrogen Fluoride (HF)

HF or hydrofluoric acid in its aqueous solution, is one of the strongest and most corrosive inorganic acids known. It is capable of producing severe damage to the lungs, skin, and other soft tissues after only a brief exposure. HF is a corrosive agent in which the hydrogen ion plays an insignificant role (Mistry and Wainwright, 1992). The fluoride ion is the primary agent responsible for the continued destruction seen with acid burns. When hydrofluoric acids react with tissue, many salts are formed that are generally soluble and dissociable. Only calcium and magnesium form neutralizing insoluble fluoride salts that precipitate within the tissues. Cellular calcium then becomes bound and is immobilized; liquefaction necrosis and, ultimately, cell death occurs. It is thought that the local depletion of calcium in the affected tissues leads to release of intracellular potassium from local nerve endings, resulting in the severe pain often associated with acid burns (Flood, 1988).

Inhalation of hydrofluoric acid vapors produces burns with massive swelling of the oropharynx and tracheobronchial tree that is usually related to the delayed pulmonary edema.

5.10 Phosgene

Phosgene (carbonyl chloride, $COCl_2$), is a colorless, water-insoluble gas with three times the density of air (Chu et al., 1982; Dodge, 1977; *The Merck Index,* 1983). It has a sweet, pungent smell similar to newly mown hay and has been used as a strong pulmonary irritant during chemical warfare (Kaye, 1980; Parkes, 1982; Schwartz, 1987). Phosgene is also used in the synthesis of organic compounds and in the separation of metals. In fires, it is produced by the combustion of chlorinated hydrocarbons (e.g., polyvinyl chloride) and is hydrolyzed slowly in the lung to HCl and CO_2 (Schwartz, 1987). Phosgene is considerably more toxic than chlorine, but is thought to contribute little to the toxicity of smoke (Parkes, 1982). Symptoms of cough and chest tightness may be seen with high concentrations of phosgene. The small airway and the alveoli are the primary sites of injury, causing necrosis of the epithelium, pneumonia, and pulmonary edema (Parkes, 1982; *The Merck Index,* 1983; Wald and Balmes, 1990).

6 DIAGNOSIS

The early diagnosis of smoke inhalation injury remains a challenge to the clinician because the characteristic clinical and radiologic signs may not yet be apparent 24 to 48 h after inhalation. Clinical features such as facial burn, singed nasal vibrissae, hoarseness, carbonaceous sputum, and wheezing alert the clinician to the possibly of smoke inhalation injury. Other features include audible wheezes, uvular edema, retrosternal burning, stridor, dyspnea, salivation, and viscid sputum. Determining the type and source (i.e., chemical or smoke) and whether the fire occurred in an enclosed space is most important in the initial evaluation of the patient. In the case of other chemicals, it is important to find out the agents with which the individuals were working. CO poisoning depends on several factors (Table 3) (Klaassen, 1985). Tissues with high oxygen metabolism, such as brain, heart, liver, and kidney, are more sensitive to CO than tissues such as resting skeletal muscle with slower oxygen metabolism. Exposure to significant amounts of CO results in hypoxemia, the signs of which vary from irritability to depression (Peters, 1981). Other symptoms include nausea, vomiting, and headaches. Any suspicion of altered mental status should alert the clinician to the possibility of CO exposure and necessitate administration of high concentrations of oxygen immediately. CO is not metabolized, but excreted by the lungs. It is important to realize that the arterial oxygen tension will remain near normal in CO poisoning, whereas the oxygen content of the blood (O_2 saturation) will fall sharply as oxygen is replaced by CO on the hemoglobin molecule (Prien and Traber, 1988).

TABLE 3

Factors Governing Carbon Monoxide Toxicity

Exposure Level	Uptake	Prevalent Disorders
Concentration of inspired gas	Alveolar minute volume of respiration	Oxyhemoglobin concentration
Duration of exposure	Cardiac output	Atherosclerosis
	Metabolic rate	

Chest X-ray and arterial blood gases are often normal immediately after inhalation of most of these materials. The radiologic findings that may persist for several weeks include the appearance of patchy or diffuse peribronchial infiltrates (Pruitt et al., 1975). Peribronchial cuffing is another early manifestation of interstitial edema and is believed to be caused by direct injury to bronchial mucosa with subsequent swelling. These findings, when positive, are seen in the first 24 h in 90% of patients (Robinson and Miller, 1986). Patients with a significant decrease in the partial pressure of oxygen in arterial blood below 250 mmHg on 100% inspired oxygen have been associated with early respiratory insufficiency (Luce et al., 1976).

Other tests of diagnostic value include fiberoptic bronchoscopy, indirect laryngoscopy, xenon 133 lung scanning, and pulmonary function testing (Agee et al., 1976; Hunt et al., 1975; Moylan et al., 1975; Moylan and Chan, 1978). The endoscopic criteria for an inhalation injury include mucosal erythema, edema, ulceration, hemorrhage, and the presence of carbonaceous sputum (Hunt, 1975). Patients who are in hypovolemic shock may not demonstrate mucosal changes until adequate amounts of fluid have been given. Xenon 133 scanning permits the assessment of parenchymal injuries by measuring the delay in clearance from the lungs of injected or inhaled xenon gas (Peters, 1981). Xenon scanning is a more useful tool for evaluating problems of the terminal airways, which is usually not useful until 3 to 4 h after injury (Levine et al., 1978). Pulmonary function testing can be of value, indicating decreases in lung volume and in flow rates seen with serious inhalation injury, but does not define the extent of the injury.

CN poisoning in the absence of CO is considered a rarity. CN poisoning may or may not be present with carboxyhemoglobin levels exceeding 15%. Smoke containing both CN and CO may cause a cellular-level anoxia by several sequential mechanisms, with possible additive or synergistic effects (Figure 2). When carboxyhemoglobin and CN levels were compared, a very weak correlation was found (Barillo et al., 1986). Rapid screening tests for CN poisoning are commercially available.

CN ⇓

NAD—>FP1(4Fe S)—>CoQ—>(2Fe S)—>cyt b (Fe S) cyt c1—>cyt c—>cyt aa3—>O2

⇑ CO

FIGURE 2
Sequential block of electron transport system. (Abbreviations: NAD = nicotinamide adenine dinucleotide, FMN = flavin mononucleotide, CoQ = coenzyme Q (ubiquinone), cyt b = cytochrome b, cyt c1 = cytochrome c1, cyt c = cytochrome c, cyt a = cytochrome a, cyt a3 = cytochrome a3.)

7 TREATMENT

The treatment of burn patients follows that of any acutely injured patient. Foremost are obtaining and maintaining an airway with adequate delivery of oxygen, and intravenous fluid administration. The decision whether to intubate a burn patient who demonstrates airway obstruction should be relatively easy. The indications for early intubation in a patient with a possible inhalation injury are not as clear. Relative indications for early intubation include central nervous system depression, visible edema, or mucosal injury of the upper airway and circumferential burns of the neck.

Routine early tracheotomy, however, is associated with the development of pulmonary infection and increased morbidity and mortality. Most authors advocate tracheotomy only if endotracheal intubation is impossible or if a patient fails to wean from the ventilator after 1 to 3 weeks. Lund et al. (1985) found that the duration of intubation or tracheotomy is the predominant factor influencing development of late airway sequelae such as tracheal and laryngeal stenosis and granuloma formation. The complications increased in incidence after 3 weeks with either method of intubation. They recommended nasotracheal intubation for up to 3 weeks before performing a tracheotomy.

Treatment for CO exposure should begin with early administration of 100% oxygen by facemask or via endotracheal tube if indicated. Myers et al.

(1981) recommended the use of hyperbaric oxygen for severe cases of suspected CO poisoning, regardless of the time between exposure and presentation, especially when the delay is sufficient to preclude a diagnosis by standard laboratory methods. Raphael et al. (1989) concluded that hyperbaric oxygen was not useful in patients who did not lose consciousness during CO intoxication, irrespective of their carboxyhemoglobin level. The prognosis is poorest for those patients who present with coma.

CN poisoning requires prompt and definitive treatment. There are three treatment modalities for the treatment of CN poisoning:

1. Administer agents that oxidize hemoglobin to methemoglobin. Methemoglobin competes with cytochrome oxidase for CN. Cytochrome oxidase is restored by formation of cyanmethemoglobin. Methemoglobin formation is achieved by inhalation of amyl nitrite or intravenous administration of sodium nitrites (Chen and Rose, 1952).
2. Administer chelating agents that rapidly (1–2 min) form a complex with CN and are then excreted through the kidneys. Cobalt edetate (300–600 mg i.v.) or hydroxycobalamin (4 g i.v.) have been used (Garnier et al., 1981).
3. Enhance detoxification by the mitochondrial enzyme rhodanase to thiocyanate by intravenous administration of thiosulfate (125–250 mg/kg) (Saunders and Himwich, 1950).

Victims with respiratory exposure to hydrofluoric acid must be removed from the area immediately. Administration of 100% oxygen should be started as quickly as possible (Flood, 1988). Administration of 2.5 to 3.0% calcium gluconate solution through a nebulizer has also been recommended in hydrofluoric acid inhalation (Trevino et al., 1983). Pulmonary edema secondary to hydrofluoric acid inhalation should be treated with aggressive pulmonary toilet, diuretics, and aminophylline if indicated.

REFERENCES

Abdi, S., Evans, M.J., Cox, R.A., Lubbesmeyer, H., Herndon, D.N., and Traber, D.L., Inhalation injury to tracheal epithelium in an ovine model of cotton smoke exposure. Early phase (30 minutes), *Am. Rev. Respir. Dis.*, 142, 1436, 1990.

Abdi, S., Herndon, D.N., Traber, L.D., et al., Lung edema formation following inhalation injury: Role of the bronchial blood flow, *J. Appl. Physiol.*, 71, 727, 1991.

Agee, R.N., Long, J., III, Hunt, J.L., et al., Use of 133-xenon in early diagnosis of Injury, *J. Trauma*, 16, 218, 1976.

Arena, J.M. *Poisoning*, 4th ed., Charles C. Thomas, Springfield, IL, 1979.

Autian, J., Toxicological aspects of flammability and combustion, in *Treatise on the Flammability Characteristics of Polymeric Materials*, Einhorn, I.N. and Seader, J.D., Eds., Program Design, Cleveland, 1971.

Barillo, D.J., Goode, R., Rush, B.F.J., Lin, R.L., and Freda, A., Jr., Lack of correlation between carboxyhemoglobin and cyanide in smoke inhalation injury, *Curr. Surg.*, 43, 421, 1986.

Barrow, R.E., Wang, C.Z., Cox, R.A., and Evans, M.J., Cellular sequence of tracheal repair in sheep after smoke inhalation injury, *Lung*, 170, 331, 1992.

Basadre, J.O., Sugi, K., Traber, D.L., Traber, L.D., Niehaus, G.D., and Herndon, D.N., The effect of leukocyte depletion on smoke inhalation injury in sheep, *Surgery*, 104, 208, 1988.

Baud, F.J., Barriot, P., Toffis, V., et al., Elevated blood cyanide concentrations in victims of smoke inhalation, *N. Engl. J. Med.*, 325, 1761, 1991.

Becker, C.E., The role of cyanide in fires, *Vet. Hum. Toxicol.*, 27, 487, 1985.

Brown, M., Desai, M., Traber, L.D., Herndon, D.N., and Traber, D.L., Dimethylsulfoxide with heparin in the treatment of smoke inhalation injury, *J. Burn Care Rehabil.*, 9, 22, 1988.

Charan, N.B., Turk, G.M., and Dhand, R., Gross and subgross anatomy of bronchial circulation in sheep, *J. Appl. Physiol.*, 57, 658, 1984.

Chen, K.K. and Rose, C.L., Nitrite and thiosulfate therapy in cyanide poisoning, *JAMA*, 149, 113, 1952.

Chu, C.S., Chan, M., Ouygan, B.Y., and Zhang, Q.Y., Chemical pulmonary burn injury and chemical intoxication, *Burns Incl. Therm. Inj.*, 9, 111, 1982.

Cohen, M.A. and Guzzardi, L.J., Inhalation of products of combustion, *Ann. Emerg. Med.*, 12, 628, 1983.

Crapo, R.O., Causes of respiratory injury, in *Respiratory Injury: Smoke Inhalation and Burns*, Haponik, E.F. and Munster, A.M., Eds., McGraw-Hill, New York, 1990, 47.

Crapo, J.D., Barry, B.E., Chang, L.Y., and Mercer, R.R., Alterations in lung structure caused by inhalation of oxidants, *J. Toxicol. Environ. Health*, 13, 301, 1984.

Davies, J.W., Toxic chemicals versus lung tissue — An aspect of inhalation injury revisited. The Everett Idris Evans memorial lecture — 1986, *J. Burn Care Rehabil.*, 7, 213, 1986.

Dodge, R.R., Smoke inhalation, in Seminars in Chest Medicine, Clark, R., Taussig, L.M., Weese, W.C., Eds., *Ariz. Med.*, 34, 749, 1977.

Einhorn, I.N., Physiological and toxicological aspects of smoke produced during the combustion of polymeric materials, *Environ. Health Perspect.*, 11, 163, 1975.

Farrow, C.S., Smoke inhalation in the dog: Current concepts of pathophysiology and management, *VM/SAC*, 70, 404, 1975.

Fein, A., Leff, A., and Hopewell, P. C., Pathophysiology and management of the complications resulting from fire and the inhaled products of combustion: Review of the literature, *Crit. Care Med.*, 8, 94, 1980.

Flood, S., Hydrofluoric acid burns, *Am. Fam. Physician*, 37, 175, 1988.

Furchgott, R.F. and Jothianandan, D., Endothelium-dependent and -independent vasodilation involving cyclic GMP: Relaxation induced by nitric oxide, carbon monoxide and light, *Blood Vessels*, 28, 52, 1991.

Garnier, R., Bismuth, C., and Riboulet-Delmas, G., Poisoning from fumes from polystyrene fire, *Br. Med. J.*, 283, 1610, 1981.

Goldman, G., Welbourn, R., Klausner, J. M., et al., Neutrophil accumulations due to pulmonary thromboxane synthesis mediate acid aspiration injury, *J. Appl. Physiol.*, 70, 1511, 1991.

Goldman, G., Welbourn, R., Kobzik, L., Valeri, C.R., Shepro, D., and Hechtman, H.B., Synergism between leukotriene B4 and thromboxane A2 in mediating acid-aspiration injury, *Surgery*, 111, 55, 1992.

Gosselin, R.E., Hodge, H.C., and Smith, R.P., *Clinical Toxicology of Commercial Products: Acute Poisoning*, 4th ed., Williams & Wilkins, Baltimore, 1976.

Hales, C.A., Barkin, P.W., Jung, W., et al., Synthetic smoke with acrolein but not HCl produces pulmonary edema, *J. Appl. Physiol.*, 64, 1121, 1988.

Hales, C.A., Barkin, P., Jung, W., Quinn, D., Lamborghini, D., and Burke, J., Bronchial artery ligation modifies pulmonary edema after exposure to smoke with acrolein, *J. Appl. Physiol.*, 67, 1001, 1989.

Hales, C.A., Musto, S.W., Janssens, S., Jung, W., Quinn, D.A., and Witten, M., Smoke aldehyde component influences pulmonary edema, *J. Appl. Physiol.*, 72, 555, 1992.

Heimbach, D.M. and Waeckerle, J.F., Inhalation injuries, *Ann. Emerg. Med.*, 17, 1316, 1988.

Herndon, D.N., Langner, F., Thompson, P., Linares, H.A., Stein, M., and Traber, D.L., Pulmonary injury in burned patients, *Surg. Clin. North Am.*, 67, 31, 1987.

Herndon, D.N., Traber, D.L., and Linares, H.A., Effects of smoke inhalation on airway blood flow and edema formation, *Circ. Shock*, 16, 45, 1985.

Herndon, D.N., Traber, D.L., Niehaus, G.D., Linares, H.A., and Traber, L.D., The pathophysiology of smoke inhalation injury in a sheep model, *J. Trauma*, 24, 1044, 1984.

Herndon, D.N., Traber, L.D., Linares, H.A., et al. Etiology of the pulmonary pathophysiology associated with inhalation injury, *Resuscitation*, 14, 43, 1986.

Herpol, C., Minne, R., and Van Outryve, E., Biological evaluation of the toxicity of gases produced under fire conditions by synthetic materials. Part 1: Methods and preliminary experiments concerning the reaction of animals to simple mixtures of air and carbon dioxide or carbon monoxide, *J. Combust. Sci. Technol.*, 12, 217, 1976.

Hubbard, G.B., Langlinais, P.C., Shimazu, T., Okerberg, C.V., Mason, A.D., and Pruitt, B.A., The morphology of smoke inhalation injury in sheep, *J. Trauma*, 31, 1477, 1991.

Hunt, J.L., Agee, R.N., and Pruitt, B.A., Fiberoptic bronchoscopy in acute injury, *J. Trauma*, 15, 641, 1975.

Janssens, S.P., Musto, S.W., Hutchinson, W.G., et al., Cyclooxygenase and lipoxygenase inhibition by BW-755C reduces acrolein smoke-induced acute lung injury, *J. Appl. Physiol.*, 77, 888, 1994.

Jin, L.J., LaLonde, C., and Demling, R.H., Lung dysfunction after thermal injury in relation to prostanoid and oxygen radical release, *J. Appl. Physiol.*, 61, 103, 1986.

Kaye, S., *Handbook of Emergency Toxicology*, 4th ed., Charles C. Thomas, Springfield, IL, 1980.

Kimura, R., Traber, L.D., Herndon, D.N., Linares, H.A., Lübbesmeyer, H.J., and Traber, D.L., Increasing duration of smoke exposure induces more severe lung injury in sheep, *J. Appl. Physiol.*, 64, 1107, 1988.

Kimura, R., Traber, L., Herndon, D., Niehaus, G., Flynn, J., and Traber, D.L., Ibuprofen reduces the lung lymph flow changes associated with inhalation injury, *Circ. Shock*, 24, 183, 1988.

Kirk, M.A., Gerace, R., and Kulig, K.W., Cyanide and methemoglobin kinetics in smoke inhalation victims treated with the cyanide antidote kit, *Ann. Emerg. Med.*, 22, 1413, 1993.

Klaassen, C.D., Nonmetallic enviromental toxicants: Air pollutants, solvents and vapors, and pesticides, *The Pharmacological Basis of Therapeutics*, Goodman, G.A., Goodman, L.S., and Rall, T.W., Eds., Macmillan, New York, 1985, 1628.

Lakshminarayan, S., Kowalski, T.F., Kirk, W., and Butler, J., The effect of bronchial venous pressure on pulmonary edema in the dog, *Respir. Physiol.*, 82, 317, 1990.

Levin, B.C., Paabo, M., Gurman, J.L., Harris, S.E., and Braun, E., Toxicological interactions between carbon monoxide and carbon dioxide, *Toxicology*, 47, 135, 1987.

Levine, B.A., Petroff, P.A., Slade, C.L., and Pruitt, B.A., Prospective trials of dexamethasone and aerosolized gentamicin in the treatment of inhalation injury in the burned patient, *J. Trauma*, 18, 188, 1978.

Linares, H.A., Herndon, D.N., and Traber, D.L., Sequence of morphologic events in experimental smoke inhalation, *J. Burn Care Rehabil.*, 10, 27, 1989.

Lowry, W.T., Juarez, L., and Petty, C.S., Studies of toxic gas production during actual structural fires in the Dallas area, *J. Forensic Sci.*, 30, 59, 1985.

Luce, E.A., Su, C.T., and Hoopes, J.E., Alveolar-arterial oxygen gradient in the burn patient, *J. Trauma*, 16, 212, 1976.

Lund, T., Goodwin, C.W., McManus, W.F., et al., Upper airway sequelae in burn patients requiring endotracheal intubation or tracheostomy, *Ann. Surg.*, 201, 374, 1985.

Lundberg, J.M., Alving, K., Karlsson, J.A., Matran, R., and Nilsson, G., Sensory neuropeptide involvement in animal models of airway irritation and of allergen-evoked asthma, *Am. Rev. Respir. Dis.*, 143, 1429, 1991.

Magno, M.G. and Fishman, A.P., Origin, distribution, and blood flow of bronchial circulation in anesthetized sheep, *J. Appl. Physiol.*, 53, 272, 1982.

Mellins, R.B. and Park, S., Respiratory complications of smoke inhalation in victims of fires, *J. Pediatr.*, 87, 1, 1975.

Mistry, D.G. and Wainwright, D.J., Hydrofluoric acid burns, *Am. Fam. Physician*, 45, 1748, 1992.

Morikawa, T., Acrolein, formaldehyde, and volatile fatty acids from smoldering combustion, *J. Combust. Toxicol.*, 3, 135, 1976.

Moritz, A.R., Henriques, F.C., and McLean, R., The effect of inhaled heat on the air passages and lungs: An experimental investigation, *Am. J. Pathol.*, 21, 311, 1945.

Moylan, J.A. and Chan, C.K., Inhalation injury — An increasing problem, *Ann. Surg.*, 188, 34, 1978.

Moylan, J.A., Adib, K., and Birnbaum, M., Fiberoptic bronchoscopy following thermal injury, *Surg. Gynecol. Obstet.*, 140, 541, 1975.

Moylan, J.A., Inhalation injury: A primary determinant of survival following major burns, *J. Burn Care Rehabil.*, 2, 78, 1981.

Myers, R.A., Snyder, S.K., Linberg, S., and Cowley, R.A., Value of hyperbaric oxygen in suspected carbon monoxide poisoning, *JAMA*, 246, 2478, 1981.

Nguyen, T. T, Herndon, D.N., Cox, C.S., et al., Effect of manganous superoxide dismutase on lung fluid balance after smoke inhalation injury, *Proc. Am. Burn Assoc.*, 25, 31, 1993.

Noshima, S., Fujioka, K., Isago, T., Traber, L.D., Herndon, D.N., and Traber, D.L., The effect of a thromboxane synthetase inhibitor, OKY-046, on cardiopulmonary function after smoke inhalation injury, *FASEB J.*, 5, A371, 1991.

Nozaki, M., Guest, M.M., Hirayama, T., and Larson, D.L., Heat-activated permeability factor derived from erythrocytes and its significance in burns, *J. Burn Care Rehabil.*, 5, 30, 1984.

Parkes, W.R., *Occupational Lung Disorders*, 2nd ed., Butterworth, London, 1982.

Peters, W.J., Inhalation injury caused by the products of combustion, *Can. Med. Assoc. J.*, 125, 249, 1981.

Petroff, P.A., Hander, E.W., Clayton, W.H., and Pruitt, B.A., Pulmonary function studies after smoke inhalation, *Am. J. Surg.*, 132, 346, 1976.

Pitt, R.M., Parker, J.C., Jurkovich, G.J., Taylor, A.E., and Curreri, P.W., Analysis of altered capillary pressure and permeability after thermal injury, *J. Surg. Res.*, 42, 693, 1987.

Prien, T. and Traber, D.L., Toxic smoke compounds and inhalation injury — A review, *Burns Incl. Therm. Inj.*, 14, 451, 1988.

Prien, T., Traber, D.L., Richardson, J.A., and Traber, L.D., Early effects of inhalation injury on lung mechanics and pulmonary perfusion, *Intensive Care Med.*, 14, 25, 1988.

Prien, T., Traber, L.D., Herndon, D.N., Stothert, J.C., Jr., Lübbesmeyer, H.J., and Traber, D. L., Pulmonary edema with smoke inhalation, undetected by indicator-dilution technique, *J. Appl. Physiol.*, 63, 907, 1987.

Pruitt, B.A., Erickson, D.R., and Morris, A., Progressive pulmonary insufficiency and other pulmonary complications of thermal injury, *J. Trauma*, 15, 369, 1975.

Pruitt, B.A., Flemma, R.J., DiVincenti, F.C., Foley, F.D., Mason, A.D., and Young, W.G., Pulmonary complications in burn patients. A comparative study of 697 patients, *J. Thorac. Cardiovasc. Surg.*, 59, 7, 1970.

Pryor, A.J., Fear, F.A., and Wheeler, R.J., Mass life fire hazard: Experimental study of the life hazard of combustion products in structural fires, *J. Fire Flammability/Combustion Toxicol. Suppl.*, 1, 191, 1974.

Raphael, J.C., Elkharrat, D., Jars Guincestre, M.C., et al. Trial of normobaric and hyperbaric oxygen for acute carbon monoxide intoxication, *Lancet*, 2, 414, 1989.

Robinson, L. and Miller, R.H., Smoke inhalation injuries, *Am. J. Otolaryngol.*, 7, 375, 1986.

Saunders, J.P. and Himwich, W.A., Properties of the trans-sulfurase responsible for conversion of cyanide to thiocyanate, *Am. J. Physiol.*, 163, 404, 1950.

Schwartz, D.A., Acute inhalational injury, *Occup. Med.*, 2, 297, 1987.

Sheppard, D., Noxious gases: Pathogenetic mechanisms, in *Textbook of Pulmonary Diseases*, Baum, G. and Wolinsky, E., Eds., Little, Brown, Boston, 1989, 831.

Shimazu, T., Ikeuchi, H., Hubbard, G.B., Langlinais, P.C., Mason, A.D., and Pruitt, B.A., Smoke inhalation injury and the effect of carbon monoxide in the sheep model, *J. Trauma*, 30, 170, 1990.

Shimazu, T., Yukioka, T., Hubbard, G.B., Langlinais, P.C., Mason, A.D., and Pruitt, B.A., A dose-responsive model of smoke inhalation injury. Severity-related alteration in cardiopulmonary function, *Ann. Surg.*, 206, 89, 1987.

Shinozawa, Y., Hales, C., Jung, W., and Burke, J., Ibuprofen prevents synthetic smoke-induced pulmonary edema, *Am. Rev. Respir. Dis.*, 134, 1145, 1986.

Silverman, H.M., Smoke inhalation, *Principles and Practice of Emergency Medicine*, Schwart, G.R., Safar, P., and Stone, J.H., Eds., W.B. Saunders, Philadelphia, 1978.

Silverman, S.H., Purdue, G.F., Hunt, J.L., and Bost, R.O., Cyanide toxicity in burned patients, *J. Trauma*, 28, 171, 1988.

Stewart, R.D., The effect of carbon monoxide on man, *J. Fire Flammability/Combustion Toxicol. Suppl.*, 1, 167, 1974.

Stone, H.H., Pulmonary burns in children, *J. Pediatr. Surg.*, 14, 48, 1979.

Stothert, J.C., Basadre, J.O., Gbaanador, G.B., et al., Airway aspiration of hydrochloric acid in sheep, *Circ. Shock*, 30, 237, 1990a.

Stothert, J.C., Gbaanador, G.B., Basadre, J., Flynn, J., Traber, L., and Traber, D., Bronchial blood flow and eicosanoid blockade following airway acid aspiration, *J. Trauma*, 30, 1483, 1990b.

Stothert, J.C., Jr., Herndon, D.N., Lübbesmeyer, H.J., et al., Airway acid injury following smoke inhalation, *Prog. Clin. Biol. Res.*, 264, 409, 1988.

Sugi, K., Theissen, J.L., Traber, L.D., Herndon, D.N., and Traber, D.L., Impact of carbon monoxide on cardiopulmonary dysfunction after smoke inhalation injury, *Circ. Res.*, 66, 69, 1990.

Terrill, J.B., Montgomery, R.R., and Reinhardt, C.F., Toxic gases from fires, *Science*, 200, 1343, 1978.

The Merck Index: An Encyclopedia of Chemicals, Drugs and Biologicals, Merck and Co., Rahway, NJ, 1983.

Till, G.O. and Ward, P.A., Oxygen radicals and lipid peroxidation in experimental shock, *Prog. Clin. Biol. Res.*, 236A, 235, 1987.

Traber, D.L., Herndon, D.N., Stein, M.D., Traber, L.D., Flynn, J.T., and Niehaus, G.D., The pulmonary lesion of smoke inhalation in an ovine model, *Circ. Shock*, 18, 311, 1986.

Traber, D.L., Linares, H.A., Herndon, D.N., and Prien, T., The pathophysiology of inhalation injury — A review, *Burns Incl. Therm. Inj.*, 14, 357, 1988.

Traber, D.L., Schlag, G., Redl, H., and Traber, L.D., Pulmonary edema and compliance changes following smoke inhalation, *J. Burn Care Rehabil.*, 6, 490, 1985.

Traber, D.L. and Herndon, D.N., Pathophysiology of smoke inhalation, in *Respiratory Sequelae of Burns*, Haponik, E.F. and Munster, A.M., Eds., McGraw Hill, New York, 1990, 61.

Traber, D.L., Bradford, D.W., and Traber, L.D., Acute lung injury: Experimental approaches to treatment, in *Yearbook of Intensive Care and Emergency Medicine 1994*, Vincent, J.-L., Ed., Springer-Verlag, Berlin, 1994, 441.

Traber, L.D., Herndon, D.N., Turner, J., Sant Ambrogio, G., and Traber, D.L., Peptide mediation of the bronchial blood flow elevation following inhalation injury, *Circ. Shock*, 31, 13, 1990.

Treitman, R.D., Burgess, W.A., and Gold, A., Air contaminants encountered by fire fighters, *Am. Ind. Hyg. Assoc. J.*, 41, 796, 1991.

Trevino, M.A., Herrmann, G.H., and Sprout, W.L., Treatment of severe hydrofluoric acid exposures, *J. Occup. Med.*, 25, 861, 1983.

Trunkey, D.D., Inhalation injury, *Surg. Clin. North Am.*, 58, 1133, 1978.

Venegas, J.G., Yamada, Y., Custer, J., and Hales, C.A., Effects of respiratory variables on regional gas transport during high-frequency ventilation, *J. Appl. Physiol.*, 64, 2108, 1988.

Wald, P.H. and Balmes, J.R., Respiratory effects of short-term high intensity toxic inhalations: Smoke, gases, and fumes, *J. Intensive Care Med.*, 2, 260, 1990.

Weast, R.C., *CRC Handbook of Chemistry and Physics*, 65th ed., CRC Press, Boca Raton, FL, 1984.

Wong, K.L. and Alarie, Y., A method for repeated evaluation of pulmonary performance in unanesthetized, unrestrained guinea pigs and its application to detect effects of sulfuric acid mist, *Toxicol. Appl. Pharmacol.*, 63, 72, 1982.

Wu, W., Barie, P., and Halebian, P., Attenuation of ventilation-perfusion mismatching with ibuprofen in inhalation lung injury, *Proc. Am. Burn Assoc.*, 20, 105, 1988.

Zikria, B.A., Budd, D.C., Floch, F., and Ferrer, J.M., What is clinical smoke poisoning? *Ann. Surg.*, 181, 151, 1975.

Zikria, B.A., Ferrer, J.M., and Floch, H.F., The chemical factors contributing to pulmonary damage in "smoke poisoning," *Surgery*, 71, 704, 1972a.

Zikria, B.A., Weston, G.C., Chodoff, M., and Ferrer, J.M., Smoke and carbon monoxide poisoning in fire victims, *J. Trauma*, 12, 641, 1972b.

CHAPTER 6

Effects of Carbon Monoxide Exposure on Developing Animals and Humans

David G. Penney

CONTENTS

0-8493-4796-3/96/$0.00+$.50

1 INTRODUCTION

As is well known, carbon monoxide (CO) is an insidious poison, being as it is odorless, colorless, tasteless, and noncorrosive. It readily binds to hemoglobin, preventing oxygen binding and hence decreasing the oxygen transport capability of the cardiovascular system. It also binds to other iron-containing compounds in the body and under certain circumstances may exert its effects strongly and directly at the cellular level — that is, as a histotoxic mechanism (see Chapter 8, this volume).

Because CO quickly crosses the placental membranes (Longo, 1977), it can act as a potent poison to the developing fetus. Although the earliest studies of possible short- and long-term deleterious developmental effects of CO were more than a half century ago, many gaps still remain in our understanding of CO's effects and the risks posed by maternal exposure to CO. Moreover, the studies available in the literature are often less than satisfactory in answering pressing medico-legal questions for a variety of reasons: (1) extrapolation from animal to the human is always problematic, (2) use of many different animal species, (3) exposure to CO at different concentrations, (4) exposure to CO may be chronic or acute, (5) differences in maturity of the fetuses of different species at various stages relative to the human, and others. Most of what we know about the effects of prenatal CO exposure has come from animal studies; information from human studies is mainly anecdotal.

Developmental toxicity may be defined as involving the death of the developing organism, structural anomalies, altered growth, and functional deficiencies resulting from exposure to a toxic substance that occurs while the organism is undergoing rapid development. Immature organisms may be especially vulnerable because a toxic exposure insufficient to produce maternal/adult toxicity may adversely affect the fetus or neonate. Toxic responses to early exposure may occur early in life, later in life, or may reappear under conditions of stress or ill health.

There are various reasons why fetal and/or developing organisms may be especially vulnerable to toxic substances such as CO: (1) usually low or very much lower partial pressure of oxygen available, (2) differences in uptake and elimination of CO from fetal blood, and (3) the possibility that tissue hypoxia may differ between the fetus and adult even at similar COHb concentrations.

This chapter attempts to synthesize most of the available information on the embryonic, fetal, and general developmental effects of CO exposure. The emphasis is on long-term effects of CO — effects persisting post-birth. In carrying out this charge, several excellent past reviews in this area should be pointed out. There will be no attempt to recapitulate their discussion. With one exception, this review will not discuss a related body of work, postnatal CO exposure.

Discussion will begin with the general effects of CO on birth weight, duration of pregnancy, etc., proceeds to the cardiovascular effects, but since the central nervous system is the major target organ of this poison, most of the discussion is spent on the neural and behavioral effects of CO.

2 ANIMAL STUDIES

Animal studies of prenatal exposure to CO have centered on two organ systems, the circulation and the central nervous system. Summaries of many older studies of the general effects of CO on animals are seen in Table 1. It is known that a variety of cardiovascular and hematopoietic changes result from chronic CO exposure, and when it occurs early they may become lasting ones (Penney, 1990). Acute CO exposure also produces profound cardiovascular alterations (Penney, 1988). Drs. Helfaer and Traystman reviewed the cerebrovascular effects of CO in Chapter 4 of this volume.

As early as 1933, Wells (1933) reported abortions, resorptions, and the abnormal growth of survivors in mice exposed to 5900 or 15,000 ppm CO for 5–8 min every second day during gestation. Williams and Smith (1935) observed decreased litter size and preweaning survival in rats from mothers treated with 3400 ppm CO for 1 h per day. Other investigators have found increased fetal mortality (Astrup et al., 1972; Choi and Oh, 1975; Dominick and Carson, 1983; Rosenkrantz et al., 1986; Singh et al., 1993; Singh and Scott, 1984), decreased fetal body weight (Astrup et al., 1972; Fechter and Annau, 1976; Garvey and Longo, 1978; Schwetz et al., 1979; Singh et al., 1993), increased resorptions (Schwetz et al., 1979), and skeletal anomalies (Astrup et al., 1972; Schwetz et al., 1979; Singh et al., 1993). CO had no effect on implantation in a study by Singh and Scott (1984).

Tachi and Aoyama (1983) found that intermittent exposure of female rats to 1000–1600 ppm CO for various periods of time after conception resulted in a decrease in fetal body weight of 20% on gestation days 14 and 21. CO exerted no effect on the number of fetuses or corpora lutea. These findings were confirmed in a later study (Tachi and Aoyama, 1986).

In rats exposed to 1100–1200 ppm, for 2 h per day throughout gestation, it was observed that fetal body weight and the weight of placentas was decreased compared to pair-fed and ad lib fed controls. There was no effect on litter size. The pregnant females receiving CO showed decreased food consumption compared with ad lib controls and increased hematocrit (Leichter, 1993).

In female rats exposed to 100 ppm CO for days 1–16, 4–12, 10–22 or 18–22 days of gestation, placental weight was increased 11–13% where exposure continued until term. The placental hypertrophy that takes place presumably acts to improve oxygen transport. Fetal body weight was reduced 6–8% when exposure occurred for all or the second half of gestation. There

TABLE 1

General Effects of Prenatal Exposure to Carbon Monoxide in Animals

Species	Maternal Treatment	Maternal Toxicity	COHb(%)	Development Effects	Ref.
Mouse	5900 or 15000 ppm for 5-8 min every second day of gestation.	Acutely - unconsciousness	—	Abortions, resorptions, and abnormal growth of survivors.	Wells, 1933
Mouse	125, 250, 500 ppm on gestation days 7–18.	—	—	500 ppm CO: Increased fetal mortality, no effect on number of implantation sites.	Singh and Scott, 1984
Mouse	65, 125, 250 or 500 ppm exposure, during gestation days 8–18; feeding 27, 16, 8, or 4% protein diets.	None	—	All CO levels and the 8% and 4% protein diets decreased fetal body wt. Incidence of dead or resorbed fetuses was higher in all CO-exposed and protein deficient groups. As dietary protein decreased, CO exposure increased the incidence of fetal deaths. Gross malformations included brachygnathia with protruding tongue, microstomia, microcephaly, open mouth, open eyes, etc.	Singh et al., 1993
Rat	3400 ppm for 60 min/day for 3, 6, or 8.3 mo.	Decreased body wt., appetite, and muscle tone; lack of grooming.	60–70	Deceased litter size, preweaning survival.	Williams and Smith, 1935
Rat	750 ppm for 3 h/day on gestation days 7, 8, or 9.	Unreported	—	Absorptions, stillbirths, skeletal anomalies, and decreased fetal body wt. and crown-rump length.	Choi and Oh, 1975
Rat	0, 30, or 90 ppm or 13% oxygen in nitrogen on gestation days 3–20.	Decrease in successful pregnancies.	4.8–8.8	13% oxygen: 12% decrease in body wt.; 90 ppm CO: 14% increase in brain wt., 24% decrease in lung wt., decrease in brain serotonin.	Garvey and Longo, 1978
Rat	1000–1600 ppm, 81 min, twice daily, from conception for various periods of gestation.	Slight decrement in body wt. gain.	—	Decrease fetal body wt. 20% on gestation days 14 and 21. No effect on number of fetuses or corpora lutea.	Tachi and Aoyama, 1983

Rat	1000–6000 ppm twice/day for 2 h, 40 min total from gestation days 0–6, 7–13, 14–20, 0–20.	—	—	Decreased fetal wt. at gestation day 21.	Tachi and Aoyama, 1986
Rat	100 ppm exposure, at various stages of gestation.	COHb 10–14%		Placental wt. increased 11–13% where exposure continued till term. Fetal body wt. reduced 6–8% with exposure for all or the second half of gestation; no change where CO exposure ceased before term.	Lynch and Bruce, 1989
Rat	1100–1200 ppm exposure, 2 h per day, throughout gestation.	Increased hematocrit. Food consumption decreased compared with ad lib controls.		Body wts. of fetuses and placentas decreased when compared to pair-fed and ad lib controls. No effect on litter size.	Leichter, 1993
Rabbit	90 or 180 ppm from mating until the day before parturition.	—	8–9, 16–18	At 90 ppm: 9.9% mortality of neonates, 13% decrease in body wt.; at 180 ppm: 35% mortality of neonates, 11% decrease in birth wt., and increase in malformations.	Astrup et al., 1972
Mouse Rabbit	7 or 24 h/day of 250 ppm, on gestation days 6–15 for mice and on gestation days 6–18 for rabbits.	Transient increase in body wt. for mice in 7 h/day group.	10–11 (mice); 13–15 (rabbits)	Mice: Increase in resorptions and body wt. with 7 h/day exposure, decrease in body wt. and crown rump length with 24 h/day exposure; both exposures increased skeletal anomalies. Rabbits: Increase in body wt. and crown-rump length with 7 h/day exposure.	Schwetz et al., 1979
Rabbit	12 “puffs” of 2700–5400 ppm daily, from gestation days 6–18.	Decreased maternal respiration rate; signif. maternal death rate.	—	Larger number of fetal deaths.	Rosenkrantz et al., 1986
Pig	150–450 ppm for 48–96 h between gestation days 108–110.	—	—	Linear increase in number of stillbirths, which became signif. when maternal COHb exceeded 23%.	Dominick and Carson, 1983

Modified from McMullen, T.B. and Raub, J.A., Eds., *Air Quality Criteria for Carbon Monoxide,* U.S. EPA, Washington, D.C., 1991. With permission.

was no change where CO exposure ceased before term. No effect on fetal survival was noted (Lynch and Bruce, 1989).

A number of studies therefore make it clear that CO exposure is fetotoxic.

2.1 Effect on Implantation

While many of the studies reviewed earlier in this chapter concern the effects of CO on fetal development, few have examined the initial days of pregnancy — that is, before the blastocyst implants in the uterine wall. Only one study has reported on this (Singh and Scott, 1984).

The effects of CO inhalation were determined on the day of blastocyst implantation (Day 5 of pregnancy). To this end, nulliparous rats (Sprague-Dawley, Charles River Breeding Labs.), 200–250 g body weight, were maintained in individual cages in a controlled environment: room temperature (20–25°C), humidity (40–60%) and photoperiod (14 h light: 10 h dark; lights on at 2:00 a.m.). They were provided free access to Purina Lab Chow and water. Vaginal smears were taken daily to stage the estrous cycle. Pro-estrous animals were placed overnight with a male; spermatozoa in the vaginal lavage the following morning indicating mating (Day 1 of pregnancy). In the rat, copulation occurs prior to follicle rupture, so that eggs encounter spermatozoa and are fertilized shortly after ovulation. Under the photoperiod used, copulation occurred the evening of pro-estrous and ovulation occurred between 12:00 and 2:00 a.m. Because the time of ovulation is closely synchronized by photoperiod, the time of fertilization is relatively constant.

Vaginal smears were taken between 9:00 and 10:00 the morning following breeding. Sperm-bearing animals were randomly assigned to either control (air) as treatment (CO) groups. The treatment groups were exposed to 200 ppm CO at 10:00 a.m. on Day 1 of pregnancy. This CO concentration was previously shown (Penney et al., 1983) to have marked effects on fetal cardiovascular development in the rat. Continuous CO exposure took place in large plastic bags containing standard wire rat cages, as previously described (Penney et al., 1993). Maintenance of food, water, and clean litter required removal of rats from the test atmosphere for 2–3 min daily.

For detection of blastocyst implantation sites, the rats were lightly anesthetized with ether, an external jugular vein exposed and 0.25 ml of Evans Blue dye (1% in saline) injected intravenously. Fifteen minutes later, the rats were euthanized by ether-overdose, the uteri removed, cleaned of adventitia, placed on white blotting paper, extended to normal length and examined under an illuminated magnifier at 1.75×. Extravascular dye accumulation in areas of blastocyst-induced increased capillary permeability reveals implantation sites visible as discrete blue bands. Such bands were counted and the mean number of implantation site/cornu calculated.

Exposure to 200 ppm CO for the initial 5 d after mating significantly reduced maternal body weight gain (Table 2) and maternal water and food consumption (Table 3), as compared with controls maintained in room air. Nevertheless, there was no effect on blastocyst implantation, whether examined in the right horn, left horn, or as a total (Table 4). In rats in which pregnancy was allowed to proceed for a longer period, implantation sites, fetuses, and resorptions were determined. No significant differences were noted in this regard between the CO-exposed and the AIR groups (Table 5). While the fetal and placental weights in the CO group appeared considerably smaller, the differences were not significant.

TABLE 2

Maternal Cumulative Body Weight Gain (g) During 200 ppm Carbon Monoxide Exposure

	Days Post-Mating			
Group (*n*)	2	3	4	5
CO (18)	–2.67 ±1.14	1.72 ±1.27	5.06 ±1.27	8.50 ±1.16
AIR (15)	7.80 ±1.61	10.67 ±1.49	14.00 ±1.51	15.73 ±1.56
	++	++	++	++

Note: Values are means ± standard error of the mean. Comparing treatment and control groups: ++, $P<0.01$.

2.2 Cardiovascular Effects

The cardiovascular system is one of the two organ systems on which CO has its greatest effect (the other being the nervous system), chronic exposure producing prominently polycythemia and cardiac hypertrophy. Summaries of many older studies of the cardiovascular effects of CO in animals are seen in Table 6.

Hoffman and Campbell (1977) reported an increased heart weight/body weight ratio on the 5th day post-birth in rats exposed to 230 ppm CO throughout gestation. Hematocrit and hemoglobin concentration were also increased at that time. Using several CO concentrations (60, 125, 250, 500 ppm), Prigge and Hochrainer (1977) confirmed these findings, noting that wet heart weight was increased also. In this case, hematocrit and hemoglobin were decreased in those rats exposed to 250 and 500 ppm. Penney et al. (1980) also found increased heart wet weight and heart weight/body weight ratio in rats exposed to 200 ppm CO *in utero*. These

TABLE 3

Maternal Daily Water and Food Consumption (g) During Carbon Monoxide Exposure

Water

	Days Post-Mating			
Group (*n*)	2	3	4	5
CO (18)	18.7	28.7	29.3	31.7
	±1.6	±1.2	±1.0	±0.9
AIR (15)	36.3	35.5	35.9	37.4
	±1.3	±2.0	±1.7	±1.4
	+++	+	+	++

Food

	Days Post-Mating			
Group (*n*)	*2*	*3*	*4*	*5*
CO (18)	9.4	13.5	15.7	15.4
	±0.6	±0.5	±0.4	±0.4
AIR (15)	19.6	17.5	19.4	18.6
	±0.7	±0.4	±0.7	±0.4
	+++	++	+	++

Note: Values are means ± standard error of the mean. Comparing treatment and control groups: +, $P<0.05$; ++, $P<0.01$; +++, $P<0.001$.

TABLE 4

Effects of Carbon Monoxide on Implantation

Group (*n*)	Right Horn	Left Horn	Total
CO (11)	4.09	5.09	9.18
	±0.78	±0.97	±1.57
AIR (9)	6.00	5.33	11.33
	±0.67	±1.00	±1.43

Note: Values are means ± standard error of the mean.

TABLE 5

Effects of Carbon Monoxide on Fetal Development

Group (*n*)	Sites	Fetuses	Resorptions	Fetal Wt. (g)	Plac. Wt. (g)
CO (7)	12	13.3	.83	1.63	.51
	±2.1	±0.8	±1.27	±1.16	±0.05
AIR (6)	9.2	9.2	0	2.78	.79
	±2.9	±2.9		±0.57	±0.28

Note: Values are means ± standard error of the mean.

results were confirmed the same year by Fechter et al. (1980), who exposed rats to 150 ppm CO throughout gestation. Interestingly, however, they reported a decreased heart dry weight and no change in cardiac DNA content.

In 1983, Penney et al. (1983) carried out an extensive survey of the effects of fetal CO exposure from days 5–22 of gestation, using 157, 166, and 200 ppm CO. They found in all three cases that ventricular wet weight increased, as did heart weight/body weight ratio, and heart dry weight. Neonatal erythrocyte count, placental weight, and placental weight/body weight ratio were increased. Identical results were obtained in studies of fetuses examined daily during the final 4 d of gestation at 200 ppm CO. Cardiac hypertrophy at birth was not due to elevated myocardial water content. Heart DNA content was increased at 157 and 200 ppm CO in neonates and fetuses. Cardiac hydroxyproline concentration, as an index of collagen, was unaffected. The M subunit composition of lactate dehydrogenase was elevated, from 4 d before birth until birth, at 200 ppm CO. Heart myoglobin concentration and content were elevated. Clearly prolonged maternal CO inhalation exerts significant effects on fetal body, heart, and placental growth, and on various blood and heart constituents.

In follow-up studies, Penney and collaborators (Clubb et al., 1986) explored the possible changes in heart cellular composition that might result from chronic fetal CO exposure. In rats exposed to 200 ppm CO from gestational day 7 until birth, it was found that along with increases in heart wet weight and heart weight/body weight ratio, total cardiomyocyte number was increased in the right ventricle. The results of this study suggest that increased hemodynamic load due to CO exposure during the fetal period results in cardiac hypertrophy because of increased myocyte hyperplasia.

Two other groups of rats in this study (Clubb et al., 1986) either continued exposure to CO (500 ppm) for 28 d as neonates, or were newly introduced to CO (500 ppm) immediately post-birth. Postnatal exposure in either case resulted in polycythemia and massive cardiac hypertrophy at age 28 d, and an increased total cardiomyocyte number in the left ventricle. It

TABLE 6

Cardiovascular Effects of Perinatal Exposure to Carbon Monoxide in Rats

Exposure	COHb (%)	Body Wt.	Heart Wt.	Blood	Other	Ref.
150 ppm, gestation days 1–21	15	—	Wet wt. increased; heart wt/body wt ratio increased; dry wt. decreased.	—	Nucleic acid unchanged; no signif. differences at postnatal days 4–21.	Fechter et al., 1980
230 ppm, gestational day 2–postnatal day 21.	24	Decreased	Heart wt./body wt. increased on postnatal day 5.	Hematocrit and hemoglobin increased on postnatal day 5.	—	Hoffman and Campbell, 1977
60, 125, 250, 500 ppm.	—	Decreased	Wet wt. increased; heart wt./body wt. increased.	Hemtocrit and hemoglobin decreased at 250–500 ppm.	—	Prigge and Hochrainer, 1977
157, 166, 200 ppm, gestational days 5–22.	24.9	Decreased	Wet ventricular wt. increased; heart wt./body wt. increased; dry heart wt. increased	—	Increased LDH M subunit; increased heart DNA content.	Penney et al., 1983
200 ppm	27.8	Decreased	Wet wt. increased; heart wt./body wt. increased.	—	No lasting effects of prenatal exposure.	Penney et al., 1980
30, 90 ppm	4.8-8.8	—	—	—	—	Garvey and Longo, 1978
200 ppm from gestational day 7–postnatal day 1.	—	Decreased at postnatal day 28	Wet wt. and heart wt./body wt. increased at birth.	—	Increased myocyte number in right ventricle.	Clubb et al., 1986
200 ppm from gestational day 7–postnatal day 28.	—	—	Wet wt. and heart wt./body wt. increased at birth.	—	Prenatal CO increased myocyte number in right ventricle; postnatal CO increased myocyte number in left ventricle.	Clubb et al., 1986

200 ppm from postnatal day 1–28.	—	Decreased at postnatal day 21–28	wet wt. and heart wt./body wt. increased at birth.	—	Increased myocyte number in left ventricle.	Clubb, 1986
500 ppm at postnatal days 1–32.	—	—	wet wt. and heart wt./body wt. increased in adulthood.	Hematocrit increased at some ages in adulthood.	Heart rate increased 10–15% in adulthood.	Penney et al., 1984
300 and 700 ppm on postnatal days 1–32.	ca. 30% and 50%	Decreased in adulthood only in 700 ppm group.	Wet wt. and heart wt./body wt. increased in adulthood only in 700 ppm group.	Hematocrit increased in adulthood in females.	Increased ventricular DNA content in adulthood.	Penney et al., 1988
500 ppm on postnatal days 1–32.	—	Decreased during CO exposure.	Heart wt./body wt. increased during exposure.	—	Increased myocyte number in left ventricle.	Clubb et al., 1989
500 ppm on postnatal days 1–32.	—	Decreased during CO exposure.	Wet wt. and heart wt./body wt. increased during CO exposure.	Hematocrit increased during CO exposure.	Exercise in adulthood of neonatally CO-exposed rats excessively increased heart wt., ventricular dilatation, and cardiomyocyte length.	Penney et al., 1989
500 ppm on postnatal days 1–32.	—	—	—	—	Coronary blood flow substantially increased in adulthood as compared to controls.	Boluyt et al., 1991

TABLE 6 (continued)

Cardiovascular Effects of Perinatal Exposure to Carbon Monoxide in Rats

Exposure	COHb (%)	Body Wt.	Heart Wt.	Blood	Other	Ref.
500 ppm on postnatal days 1–30.	—	—	—	—	In adulthood, increase in number of small arteries in all heart regions. Large blood vessels, both arteries and veins, enlarged in most heart regions. Increased small, large and total artery cross-sectional area.	Penney et al., 1993

Modified from McMullen, T.B. and Raub, J.A., Eds., *Air Quality Criteria for Carbon Monoxide,* U.S. EPA, Washington, D.C., 1991. With permission.

appears that early CO exposure, whether pre- or postnatal has the potential for sharply altering cardiac cellular development.

Early studies by Penney and his group showed that inhalation of carbon monoxide (500 ppm) by rats for 4 weeks after birth induces cardiac hypertrophy far more extreme than that in similarly treated adults (Penney et al., 1974). In following these animals into adulthood (3½ months of age), it was noted that heart mass failed to regress to the level of air-reared controls (Penney et al., 1980), confirming an earlier observation by other investigators (Asmussen and Paulsen, 1953). Subsequent experiments showed that this "persistent cardiac hypertrophy" (PCH) remains at a statistically significant level for more than one third of the maximal life span of the rat, occurs in both sexes, and is greater in right ventricle than in left ventricle (Penney et al., 1982; Penney et al., 1984). PCH was found not to be associated with altered myocardial water or collagen contents (Penney et al., 1982), differences in body weight, polycythemia, systemic vascular hypertension (Penney et al., 1984), pulmonary vascular hypertension (Penney et al., 1992; Tucker and Penney, 1993), altered ventricular wall compliance (Penney et al., 1984), or heart failure (Banerjee et al., 1983). Other experiments showed that PCH is increased with increasing CO concentration in the range of 350–700 ppm and is greater in females than in males at any given concentration (Penney et al., 1988). The degree of PCH (in adulthood) is directly related to degree of juvenile (i.e., initial) cardiac hypertrophy (Penney et al., 1988). Manipulation of litter size also alters the PCH produced by CO, it being considerably greater in adults reared as pups that had inhaled CO in litters of 4, as compared to those exposed to CO in litters of 16 (Penney et al., 1987). Thus, although increased nutrition in itself irreversibly augments body and heart growth, nutrition also amplifies the effect of neonatal CO exposure. Neonatal hyperthyroidism abolishes PCH (Dubeck et al., 1989).

Rats with PCH as the result of neonatal CO exposure also display an elevated resting heart rate, both in the conscious and the unconscious states. We have called this phenomenon "persistent tachycardia" (Penney et al., 1984). The explanation for this effect remains elusive. Although cardiac output is normal in such rats, elevated heart rate may be compensating a marginally subnormal stroke volume. As with PCH, resting heart rate is also increased by rearing in small litters (Penney et al., 1987).

As indicated above, perinatal CO exposure also results in "cellular remodeling" of the heart — that is, ventricular cardiomyocyte number is increased. Other studies confirm this notion. For example, it was noted that ventricular DNA content becomes elevated during postnatal CO exposure (Penney and Weeks, 1979). This condition persists into adulthood (Penney et al., 1982), and the extent to which it occurs is greater the greater the CO exposure concentration used (Penney et al., 1988). The increase in ventricular DNA content is directly related to the degree of PCH. Rearing neonates in small litters during exposure to CO further increases DNA content, and the difference from rats reared in large litters with and without concurrent exposure to CO persists into adulthood (Penney et al., 1987). Thus, increased postnatal nutrition enhances DNA content in previously CO-exposed rats, just as it does PCH and persistent tachycardia. Neonatal T4 treatment abolishes the persistent elevation of ventricular DNA content (Penney et al., 1987), apparently by abbreviating the period of myocyte hyperplasia.

As discussed above, direct measurements of cardiomyocyte numbers have been made in rats exposed to 500 ppm CO as neonates. Total ventricular cell number at age 15 d in one study (Clubb et al., 1989) was 37.7 ±0.8 million in the CO group as compared to 34.6 ±0.8 million in the controls. At age 200 d, the total cell number was 36.1 ±2.1 million as compared to 32.4 ±1.5 million, respectively. During the last mitotic division of cardiomyocytes in rats, the cells undergo karyokinesis without cytokinesis, producing instead of more uninucleate daughter cells, single larger binucleate cells. This process occurs 4–12 d after birth. Studies of CO-exposed neonates show that it is delayed several days (Clubb et al., 1986), although the percentage of binucleate myocytes eventually achieved (90–95%) is not altered by CO. It is probably through this delay in binucleation — and the additional complete mitotic divisions that occur in some fraction of the cardiomyocyte population — that additional muscle cells arise.

There are two other aspects of cellular remodeling resulting from neonatal CO exposure: decreased cardiomyocyte volume and increased cardiomyocyte length. In rats exposed to 500 ppm CO, cell volume in the right ventricle was 23,832 ±628 um^3 compared to 25,328 ±1435 um^3 in the controls; while in the left ventricle (plus interventricular septum) volume was 27,644 ±1,017 um^3 vs. 32,225 ±1,118 um^3, respectively (Clubb et al., 1989). The length of binucleate myocytes from the right ventricle and the left ventricle of 5-month-old rats were increased 5–14.5 um (Penney et al., 1989). Other studies have found increased muscle cell length in acutely induced hypertrophy in adult animals (Anversa et al., 1979); however, ours is the first observation of "stretched" muscle cells in adulthood resulting from a neonatal stressor.

Several studies have been carried out to assess the functional capacity of the CO-remodeled heart. In one study for example, swim-training of previously neonatally CO-exposed rats was found to elicit significantly different responses than either adult CO exposure or swim-training alone or in combination: (1) a greater degree of cardiac hypertrophy, (2) thinner relative ventricular wall thickness, and (3) longer binucleate cardiomyocyte length (Penney et al., 1989). We suggested that the larger number of smaller muscle cells permitted greater organ hypertrophy. Increased cell "stretching" is consistent with the wall thinning and increased heart dimensions we observed in the swim-trained rats that had been exposed to CO.

In a second study, previously CO-exposed adult rats were aortic-constricted (Boluyt et al., 1991). In an isolated "working-heart" preparation, coronary blood flow was found to be substantially increased as compared to controls at the same workload. This may reflect a CO-induced persistent alteration of coronary vessel anatomy.

Based on clues from earlier studies, the coronary vasculature was examined in adult rats that had been exposed to 500 ppm CO as neonates (Penney et al., 1993). Morphometric analysis revealed a significant increase in the number of small arteries in all heart regions. The large blood vessels, both arteries and veins, were significantly enlarged in most heart regions. CO exposure significantly increased small, large, and total artery cross-sectional area. These results suggest profound and persistent changes in coronary vessel architecture can result from chronic neonatal CO exposure.

2.3 Cell and Molecular Changes in Heart

Neonatal heart development involves the neonatal myocyte "transition" that changes heart growth from primarily hyperplastic to primarily hypertrophic, as well as the remodeling of the nonmyocyte space through extracellular matrix formation and capillary angiogenesis. Expression of growth factors and their receptors during this period that are capable of acting intra- and intercellularly may be important in initiating and modulating the proliferation and maturation of the cardiomyocyte and those processes outside the cell. The ability of chronic CO exposure to perturb these control systems during the neonatal period was examined in a rat model.

All male litters (randomly reconstituted to 10 each) of 1-day-old Sprague-Dawley rats were purchased from Charles River Breeding Laboratories. Half of the litters were chosen to inhale CO continuously within large plastic bags inflated by an air-CO mixture (Penney, 1984), while control litters remained in room air.

Ventricles were obtained from the animals at 2, 7, 14, 39 and 49 d post CO or AIR treatment. Northern blot hybridization analysis of specific gene transcripts was performed using electrophoretically separated total RNA, transferred by capillary blotting to nylon filters, UV cross-linked, and probed with {a-^{32}P}-labeled cDNA inserts.

The following cDNA probes were used: rat IGF-I and IGF-II genes and the Type-2 IGF/mannose-6-phosphate receptor (ATCC); FGF-receptor (Flg); cardiac and skeletal troponin I (c/s-TNI); cardiac troponin C (c-TNC); mouse transforming growth factor-β1 and-β3; rat creatine kinase-M and-B; and Types 1, 3, and 4 human collagen genes (ATCC).

Ventricular transcript levels for the insulinlike growth factor IGF-II and the Type-2 IGF receptor both decreased in abundance during development and were not markedly altered by CO exposure (Table 7). An exception was the Type-2 IGF receptor, which, while remaining at low levels in the AIR rats, completely disappeared in the CO rats at age 39 d.

Transcript levels for the Type-l FGF receptor (Flg) were expressed equally in young CO and AIR rats, but was reduced in the older CO rats relative to the controls.

TABLE 7

Changes in Various Ventricular Transcripts and Morphological Features in One-Day-Old Rats Exposed to 500 ppm Carbon Monoxide for Up to 49 Days

	Days Post-Birth				
Transcript	2	7	14	39	49
IGF-II	⇔	⇔			
Type-2 IGF	⇔	⇔		⇓	⇓
Type-I FGF receptor (Flg)	⇔	⇔		⇓	⇓
TGF-β1	⇔	⇑	⇑	⇔	⇔
TGF-β3	⇔	⇑		⇓	⇔
Collagen type 1	⇔	⇔	⇔	⇔	⇔
Collagen type 3	⇔	⇔	⇔	⇔	⇔
MHC-β	⇔	⇔	⇓		
MHC-∂		⇔	⇓	⇓	⇓
Troponin C	⇔	⇔	⇔	⇔	⇔
Actin			⇓	⇓	⇓
ANF	⇔	⇑			⇑
CK-M			⇓	⇓	⇓
CK-B	⇑	⇑	⇔	⇔	⇔
Capillary density, right ventricle				⇑, 30–40%	
Capillary density, left ventricle				⇑, 15%	

Transcript levels for TGF-β1 (transforming growth factor) and TGF-β3 were higher in 7-and 14-day-old rats exposed to CO. TGF-β1 remained elevated in the CO rats through day 39 and then declined at 49 d. TGF-β3 levels were generally lower than TGF-β1 in both groups, yet the CO-exposed rats showed a higher level than the controls at age 7 d and the AIR group a higher level than the CO group at age 39 d.

Transcript levels for collagen Types 1 and 3 (extracellular matrix genes) were not markedly altered by CO exposure. A modest increase in Type 1 collagen abundance was seen with CO at age 7 d.

In terms of changes in ventricular myosin, which have been previously noted in conjunction with cardiac hypertrophy and altered myocyte growth *in vitro*, we observed the normal transition in isoform predominance from MHC-β to MHC-∂ in both groups of rats. In contrast, MHC-∂ levels were markedly decreased in the CO-exposed rats, with no concomitant increase in MHC-β. MHC-β was decreased at age 14 d in the CO rats.

Transcript levels for the cardiac and skeletal forms of troponin C (TNC) and troponin I (TNI) were not altered by CO exposure. The normal changeover from the skeletal to the cardiac isoforms was observed. Cardiac actin was decreased in the CO-exposed rats after the 14th day.

While ANF transcripts showed a developmental decline in both groups, expression was greater in the CO rats at age 7 d. The 49-day-old rats showed a pronounced reexpression of the ANF gene as compared with the AIR group.

Although similar in the CO and AIR rats at ages 2 and 7 d, CK-M transcripts (late fetal and neonatal form) were depressed by CO exposure at ages 14 through 49 d. CK-B transcripts (embryonic and fetal form) were elevated in the CO group at 2 and 7 d, remaining at detectable levels in both groups through 49 d.

Capillary density of the CO-exposed rats was increased most in the right ventricle as compared to the AIR rats (>30–40%) and to a lesser extent in the left ventricle (>±15%).

This study indicates that CO exposure of the neonate has a limited and selective influence on the expression of several growth factors and growth factor receptors in the heart, which may be associated with cardiomyocyte proliferation (Engelmann and Penney, 1991). Notably, expression of ANF and other "fetal" myocyte characteristics were extended. In contrast, more marked stimulation of extracellular matrix genes occurred, which is in agreement with morphological and physiological findings. This suggests that the normal autocrine/paracrine growth regulatory mechanisms are modified. These data suggest that the "transitional" period of heart development is a period of limited plasticity as it pertains to myocyte proliferation, whereas ventricular remodeling, as evidenced by extracellular matrix gene expression, is more readily adaptive to CO exposure and the associated hemodynamic and "humoral" influences. A marked increase in right ventricular capillary density was also noted.

2.4 Neural Effects

The central nervous system is a major target organ of CO. For a recent review of the neural and behavioral responses to CO, mainly in adult animals and humans, see the chapter by Penney and White in *The Vulnerable Brain and Environmental Risks: Special Hazards from Air and Water* (Penney, 1994). Summaries of older studies of the neural effects of CO are seen in Table 8.

Exposure of chicken embryos to CO for 3 h on the 7th day of development caused hemorrhages in developing forebrain areas (Daughtrey et al., 1983). Less branching of the dendritic tree and reduced nuclear size of the paleostriatum primitivum neurons was observed in embryos examined 1 d before hatching.

In pregnant rats exposed to 1000 ppm CO for 3 h on gestation day 15, damage to the germinal matrix overlying the developing caudate nucleus was observed 24 h later (Daughtrey and Norton, 1982). The damage consisted of hemorrhagic infarcts with necrosis, and cavities with cellular debris. It is similar to damage observed in premature humans after acute hypoxic episodes. The damage seen *in utero* was also seen in newborn, and in individuals up to 7 months of age (Daughtrey and Norton, 1983). This damage consisted of ectopic swellings of caudate tissue into the lateral ventricles. The incidence of such damage was 20% in rats exposed for 2 h

TABLE 8

Neural Effects of Prenatal Exposure to Carbon Monoxide in Animals

Species	Mater./Neonat. Treatment.	Mater./Embry. Toxicity	COHb (%)	Development Effects	Ref.
Chicken	4000 ppm exposure, 7 d of age, for 3 h	High fetal death rate with greater than 3 h exposure.	>51	Hemorrhage in forebrain area. Reduced branching of dendritic tree and reduced nuclear size of the paleostriatum primitivum seen in embryos 1 d before hatching.	Daughtrey et al., 1983
Rat	1000 ppm exposure, 3 h on gestation day 15.	—	50	24 h after exposure, damage to the germinal matrix overlying the developing caudate nucleus was observed, consisting of hemorrhagic infarcts with necrosis and cavities with cellular debris.	Daughtrey and Norton, 1982
Rat	75, 150, and 300 ppm exposure, throughout gestation.	No effect on maternal body wt. gain.	—	Cerebellar wts. decreased with increasing CO conc. Norepinephrine and serotonin conc. in the pons medulla decreased directly with increasing CO concentation at 21 d post-birth. Norepinephrine also increased with CO conc. in neocortex	Storm and Fechter, 1985a
Rat	150 ppm exposure, throughout gestation.	—	—	Cerebellar norepinephrine concentration and content elevated during the 2nd–6th postnatal weeks.	Storm and Fechter, 1985b
Rat	75, 150 or 300 ppm exposure, conception through 10 d post-birth.	—	—	Cerebellar wt. and content of gamma-aminobutyric acid decreased 10 and 21 d post-birth. Total cerebellar high-affinity tritiated GABA uptake decreased in 21-day-old rats exposed to 300 ppm.	Storm et al., 1986
Rat	75, 150, and 300 ppm exposure, throughout gestation.	Slight decline in maternal body wt. gain at highest CO conc.	11.5,18.5, 26.8	Cerebellar wt. and neurochemical markers of GABAergic neurons reduced. Numbers of fissures and degenerating Purkinje and granule cells in the cerebral cortex decreased. DNA content of the neostriatum increased as the result of 300 ppm CO exposure, and dopamine content in rats receiving 150 and 300 ppm CO. Norepinephrine and 5-hydroxytryptamine slightly decreased in the pons medulla at 21 d of age.	Fechter, 1987

Rat	Acute 150–1000 ppm exposure, from day 17 of gestation.	—	—	Hippocampal ornithine decarboxylase activity doubled after 4 h at 500 ppm and tripled at 600 ppm.	Packianathan et al., 1993
Rat	150 ppm exposure, throughout gestation.	—	—	Pathological signs suggestive of Wallerian degeneration in sciatic nerve at 16 mos. Not seen at younger age, or in rats exposed to CO postnatally.	Carratu et al., 1993a
Rat	75 and 150 ppm exposure, throughout gestation.	—	—	Inactivation kinetics of transient sodium current in sciatic nerve slowed, and there was a negative shift in the sodium equilibrium potential. Sodium inactivation was not modified in older rats.	Carratu et al., 1993b
Rat	75 and 150 ppm exposure, throughout gestation.	—	—	Splenic macrophage phagocytosis of *Candida albicans* and superoxide radical release reduced in 15- and 21-day-old pups.	Giustino et al., 1993
Cat	2000–3000 ppm exposure, 76–150 min.	—	37–64	Most vulnerable areas were the cerebral white matter and brain stem, followed by the basal ganglia and thalamus, and then the cerebral cortex.	Okeda et al., 1986
Rhesus monkey	1000–3000 ppm exposure; near term, 1–3 h	Fetal hypoxia; bradycardia, hypotension, metabolic acidosis.	>60	Of 9 newborn, 4 were normal, 1 showed moderate dysfunction, and 4 severe damage. One newborn showed hypotonia, lethargy, a poor suck response, and apneic spells. Autopsy revealed hemorrhagic necrosis affecting the globus pallidus and putamen bilaterally. Four severely damaged newborn had increased intracranial pressure with splitting of the cranial sutures and prominent retinal hemorrhages. Neurologically, they displayed nystagmus, opisthotonus, and intermittent extensor spasms. These had hemorrhagic necrosis of the cerebral cortex, the basal ganglia, and the thalamus of both hemispheres.	Ginsberg and Myers, 1974

Modified from McMullen, T.B. and Raub, J.A., Eds., *Air Quality Criteria for Carbon Monoxide,* U.S. EPA, Washington, D.C., 1991. With permission.

as fetuses, and 70% following a 3-h exposure, but behavioral tests failed to demonstrate motor function decrement.

Female Long Evans rats were exposed to 75, 150, and 300 ppm CO during all of gestation (Fechter, 1987). Weight of the cerebellum was reduced, as well as certain neurochemical markers of GABAergic neurons. The cerebral cortex showed decreased numbers of fissures and degenerating Purkinje and granule cells. DNA content of the neostriatum was increased as the result of 300 ppm CO exposure, as was dopamine content in rats receiving 150 and 300 ppm CO. In rats examined at age 21 d, norepinephrine and 5-hydroxytryptamine tended to be decreased in the pons medulla.

In rats exposed to 75, 150, or 300 ppm CO throughout gestation, Storm and Fechter (1985a) reported that cerebellar weights decreased with increasing CO concentration in pups at age 21 d. Norepinephrine and serotonin concentration in the pons medulla decreased directly with increasing CO concentation at 21 d post-birth, and norepinephrine increased in the neocortex and tended to increase in hippocampus at age 42 d. CO exposure had no effect on maternal body weight gain or on long-term body weight of the offspring.

The same group in a closely related study (Storm and Fechter, 1985b), found that cerebellar norepinehrine concentration and content were elevated during the 2nd through 6th postnatal weeks in pups from pregnant female rats exposed to 150 ppm CO throughout gestation.

In a third study on the influence of prenatal CO exposure on brain transmitters, Storm et al. (1986) found that cerebellar weight and the content of gama-aminobutyric acid was decreased 10 and 21 d post-birth in rats that had been exposed to 75, 150, or 300 ppm CO from conception through age 10 d. Total cerebellar high-affinity tritiated GABA uptake was decreased in 21-day-old rats exposed to 300 ppm. The cerebella of rats exposed to 300 ppm CO had fewer fissures.

Singh et al. (1993), in a study of combined fetal mouse CO exposure and protein deficiency, found that fetal body weight was decreased at all CO levels and when the 8% and 4% protein diets were used. Dams were exposed to 65, 125, 250, or 500 ppm exposure, during gestation days 8–18, and were fed 27%, 16%, 8%, or 4% protein diets. The incidence of dead or resorbed fetuses was higher in all CO-exposed and protein deficient groups. As dietary protein decreased, CO exposure led to an increasing incidence of fetal deaths. Exposure to CO increased the mean percentage of litters with gross malformations, especially those of the jaw. The gross malformations included brachygnathia with protruding tongue, microstomia, microcephaly, open mouth, open eyes, etc. CO is teratogenic in protein deficient mice and this teratogenicity is enhanced by an increase in protein deficiency in the maternal diet.

The areas most vulnerable to maternal CO poisoning in the cat in late fetal development are the cerebral white matter and brain stem, followed by the basal ganglia and thalamus, and then the cerebral cortex (Okeda et al., 1986). In earlier stage fetuses, the frequency and severity of brain changes are less, and the cerebral white matter and basal ganglia are most often involved. It was found that the grade of fetal brain damage correlated with maternal acidosis but not with hypotension; moreover, the late fetal stage brain was more easily damaged than is the maternal brain.

Of nine newborn Rhesus monkey fetuses exposed to 1000–3000 ppm CO over 1–3 h, four were normal, one showed moderate dysfunction, and four had severe neurologic damage (Ginsberg and Myers, 1974). Bradycardia, hypotension, and metabolic acidosis developed in response to maternal CO exposure. One newborn showed hypotonia, lethargy, a poor suck response, and apneic spells. Autopsy revealed hemorrhagic necrosis affecting the globus pallidus and putamen bilaterally. Four severely damaged newborn had increased intracranial pressure with splitting of the cranial sutures and prominent retinal hemorrhages. Neurologically, they displayed nystagmus, opisthotonus, and intermittent extensor spasms. These monkeys had hemorrhagic necrosis of the cerebral cortex, the basal ganglia, and the thalamus of both hemispheres.

Carratu et al. (1993b) exposed male Wistar rats to 75 ppm and 150 ppm CO *in utero*, from days 0 through 20 of gestation. As 40-day-old rats, the inactivation kinetics of transient sodium current in sciatic nerve fibers were significantly slowed. Also, the voltage–current relationship of these animals showed a negative shift of sodium equilibrium potential. As still older rats, sodium inactivation was not modified, indicating that the CO-induced changes are reversible.

In a second study of prenatal exposure to 150 ppm CO, this group (Carratu et al., 1993a) found pathological signs in sciatic nerve suggestive of Wallerian degeneration in rats at age 16 months. These changes were characterized by: (1) widening of the nodal gaps and more marked visualization of Schmidt-Lanterman incisures, (2) disruption of the myelin sheath in both paranodal and internodal regions, and (3) fragmentation of nerve fibers with formation of chains of ovoids. These changes were not seen in younger prenatally exposed rats or in rats exposed to the same CO concentrations postnatally.

In rats exposed to 75 and 150 ppm CO throughout gestation, it was noted that splenic macrophage phagocytosis of *Candida albicans* and superoxide radical release was reduced in 15- and 21-day-old pups (Giustino et al., 1993). This suggests that exposure to low concentrations of CO during fetal development is capable of inducing immunological changes in offspring.

Hippocampal ornithine decarboxylase activity was seen to double at 500 ppm and triple at 600 ppm in fetal rats briefly exposed to CO (Packianathan et al., 1993). They found a similar response in rats treated with hypoxic hypoxia, suggesting its similarity to CO hypoxia.

Based on the above discussion, exposure of the developing fetus to CO produces pathological changes to the brain (Ginsberg and Myers, 1974; Okeda et al., 1986; Daughtrey and Norton, 1982; Daughtrey et al., 1983) and peripheral nerves (Carratu et al., 1993a; Carratu et al., 1993b), and alterations in brain growth (i.e., weight) (Storm and Fechter, 1985a; Storm et al., 1986), neurotransmitter concentration and/or content (Storm and Fechter, 1985a, 1985b; Storm et al., 1986), GABAergic neurons (Fechter, 1987), enzyme activity (Packianathan et al., 1993), and immunity (Giustino et al., 1993).

2.5 Behavioral Effects

Dr. Vernon Benignus (Chapter 10, this volume) discusses the behavioral effects of CO exposure in adult animals and humans. This section will detail the behavioral modifications resulting from prenatal CO exposure. Summaries of the available studies in this area are found in Table 9.

Exposure to 125 ppm CO during gestation days 7–18 in the mouse was associated with an impaired righting reflex on the day of birth, and impaired negative geotaxis 10 d after birth (Singh, 1986). Impaired serial righting was noted 14 d after birth following 65 ppm and 125 ppm CO exposure. In another mouse study, exposure to CO throughout gestation, producing 6–11% COHb, resulted in increased errors in running a heat-motivated maze (Abbatiello and Mohrman, 1979).

Fechter and Annau (1976, 1977) reported decreased locomotor responses to L-dopa in open field tests 1 and 4 d post-birth in rats that had received 150 ppm CO throughout gestation. They later noted decreased negative geotaxis on postnatal day 3 and decreased homing behavior on postnatal days 3–5 in fetal rats exposed to the same CO concentration (Fechter and Annau, 1980; Fechter et al., 1980).

The response to 1000 ppm CO for 2 or 3 h on gestation day 15 in rats was found to be an increase in maze exploratory activity as measured 30 d post-birth (Daughtrey and Norton, 1983).

With regard to memory impairment, Mactutus and Fechter (1984) reported decreased acquisition and retention at 24 h of two-way active avoidance in rats age 30 d that had been exposed to CO throughout gestation.

Mactutus and Fechter (1985) found no effect on two-way avoidance acquisition with moderate or difficult task requirements at postnatal day 120 in rats that had been exposed to 150 ppm CO prenatally, but minimal and pronounced decreased retention at 24 h and 28 d, respectively, and decreased acquisition and retention of two-way avoidance at postnatal days 300–360. This suggests that the observed performance impairment results from a memory deficit and not an effect on sensory, motor, or motivational factors.

In rats exposed to approximately 1200 ppm CO for 2.6 h per day from conception till day 21 of gestation, Tachi and Aoyama (1990) noted a delay in the appearance of the upper and lower incisors. They also observed that the righting and free-fall righting reflexes and negative geotaxis were delayed by 1 d. CO exerted no effect on duration of pregnancy.

In utero exposure of rats to 150 ppm from 0 to 20 d of gestation resulted in a reduction in the minimum frequency of ultrasonic calls made by pups removed from the nest, and a decrease in the responsiveness to a challenge dose of diazepam, but did not affect locomotor activity or D-amphetamine-induced hyperreactivity (Di Giovanni et al., 1993). Also, adult rats so exposed as fetuses exhibited alterations in the acquisition of an active avoidance task.

Morris et al. (1985) observed that piglets previously exposed to 250 ppm CO as fetuses from gestation day 109 on, took longer to nurse for the first time, and had lower performance in tests of orientation, maneuvering, and investigation of environment. Piglets exposed to 200 ppm CO showed no such effects. They suggest that the threshold for such behavioral effects of CO is between 200 and 250 ppm.

CO exposure (COHb = 16.5–27.8%) in fetal sheep produces tachycardia, and decreases the incidences of breathing activity and rapid eye movement in a dose-dependent manner (Koos et al., 1988). The changes were similar to those produced by hypoxic hypoxia and anemia.

Clearly, fetal exposure to CO produces many persistent alterations in behavior.

3 HUMAN STUDIES

Relatively few studies of the fetal effects of CO exposure are available. Unfortunately, most of those available are anecdotal-case studies that provide little insight into the mechanisms underlying observed dysfunction. Koren et al. (1991) reviewed the older literature, finding more than a dozen reports beginning in 1929 of human fetal/infant effects of CO poisoning during pregnancy (Table 10). Table 11 presents a summary of some recent published CO poisoning cases involving pregnancy.

Norman and Halton (1990) reviewed the literature through 1989, finding 60 cases of CO poisoning during pregnancy. They grouped the cases into those where exposure occurred during the first trimester, and those where exposure occurred during the second and third trimesters. They noted that anatomical malformations showed a marked preponderance in cases where exposure took place in the first 13 weeks of pregnancy. In contrast, functional alterations occurred as the result of CO exposure over a wide developmental age range. Autopsies of stillbirths showed that the central nervous system was a target for CO-induced damage at any time during pregnancy. In general, unconsciousness or coma in the pregnant victim was found to bode ill for fetal outcome. They conclude that CO has teratogenic and embryotoxic potential when exposures are sufficient to cause significant increase in maternal COHb levels and encourage the aggressive use of oxygen treatment in the pregnant patient.

A prospective, multicenter study of acute CO poisoning during pregnancy was recently carried out, involving 40 cases in Ontario, Canada and a number of U.S. states (Koren et al., 1991). Exposure occurred during the

TABLE 9

Behavioral Effects of Prenatal Exposure to Carbon Monoxide in Animals

Species	Mater./Neonat. Treatment	Mater./Embry. Toxicity	COHb (%)	Development Effects	Ref.
Mouse	Exposure throughout gestation.	—	6–11	Increased errors in heat-motivated maze at postnatal day 40.	Abbatiello and Mohrman, 1979
Mouse	0, 65, 125 ppm on gestation days 7–18.	None.	—	125 ppm CO: Impaired righting reflex on postnatal day 1 and impaired negative geotaxis on postnatal day 10; 65 and 125 ppm CO: Impaired serial righting on postnatal day 14.	Singh, 1986
Rat	150 ppm throughout gestation.	No effect on litter size or fetal mortality.	12.2–14.0	3.3% decrease in birth wts. and decrease in preweaning wts.; decreased locomotor responses to L-dopa in open field on postnatal days 4 and 14; increased rate of habituation on postnatal day 14.	Fechter and Annau, 1976
Rat	150 ppm throughout gestation.	No effect on litter size or neonatal mortality.	15	4.9% decrease in birth wt. and decrease in preweaning wts.; decreased response to L-dopa in open field at postnatal days 1 and 4; increased rate of habituation of activity on postnatal day 14.	Fechter and Annau, 1977
Rat	150 ppm throughout gestation (cross fostering for weight measures).	—	—	7.6% decrease in birth wts. and decreased preweaning wts.; decreased negative geotaxis on postnatal day 3; decreased homing behavior on postnatal days 3–5.	Fechter and Annau, 1980; Fechter et al., 1980
Rat	1000 ppm, 2 or 3 h on gestation day 15.	Acutely: Loss of righting reflex, followed by coma; no effect on litter size.	ca. 50	26% increase in exploratory activity in maze at postnatal day 30.	Daughtrey and Norton, 1983

Rat	Exposure throughout gestation.	No effect on maternal wt. gain, gestation length or litter size.	15.6	Decreased acquisition and 24 h retention of two-way active avoidance at postnatal day 30.	Mactutus and Fechter, 1984
Rat	Exposure throughout gestation.	—	15.6	No effect on two-way avoidance acquisition with moderate or difficult task requirements at postnatal day 120, but minimal and pronounced decreased retention at 24 h and 28 d, respectively; decreased acquisition and retention of two-way avoidance at postnatal days 300–360.	Mactutus and Fechter, 1985
Rat	1200 ppm exposure, 81 min, twice per day, days 0–20 of gestation.	No effect on duration of pregnancy.	—	Appearance of upper and lower incisors delayed; righting reflex, free-fall righting and negative geotaxis delayed 1 d.	Tachi and Aoyama, 1990
Rat	150 ppm exposure, throughout gestation.	—	—	Reduction in minimum frequency of ultrasonic calls made by pups removed from the nest; decrease in responsiveness to a challenge dose of diazepam, but no effect on locomotor activity or D-amphetamine-induced hyperreactivity. Adults exhibited alterations in acquisition of an active avoidance task.	Di Giovanni et al., 1993
Pig	200 and 250 ppm exposure, after gestation day 109.	—	19.8 and 22.4, respectively	At 250 ppm: longer to suckle the first time; compromised performance in tests of ability to orient, maneuver, and investigate environment.	Morris et al., 1985

Modified from McMullen, T.B. and Raub, J.A., Eds., *Air Quality Criteria for Carbon Monoxide,* U.S. EPA, Washington, D.C., 1991. With permission.

TABLE 10

Neurologic Disorders Resulting from Maternal Carbon Monoxide Poisoning

Neurologic Disorders	Date Cited
Softening in the basal ganglia	1929
Hydrocephalus internus	1935
Microcephaly and tetraplegia	1940
Retardation of psychomotor development	1947
Extensive damage to the globus pallidus, striatum, red zone of substantia nigra, and lateral nucleus of the thalamus area of the cortex, manifested a diffuse loss of neurons, and polymicrogyria affected the frontal and anterior central regions	1949
Mental retardation	1951
Injury to putamen and globus pallidus, injury to cerebral cortex	1956
Convulsions and behavior indicative of brain toxicity	1956
Slight retardation in development, strabismus	1956
Cyanotic, no reflexes	1956
Bilateral status marmoratus of the putamen, medial globus pallidus and subthalamic nucleus suffered neural loss	1957
Injury to putamen and globus pallidus, injury to cerebral cortex, extensive damage to centrum semiovale	1959
Multicystic cavitary degeneration of the white matter, cortex and basal ganglia showed a total loss of nerve cells	1962
Injury to putamen and globus pallidus and cerebral cortex, extensive damage to centrum semiovale	1963
Symmetrical temporal microgyria, and massive hemispheric destruction	1967
Mental retardation, strabismus	1977
Mental retardation, athetosis, spasticity	1977

Modified from Koren et al., *Repro. Toxicol.*, 5, 397, 1991. With permission.

first trimester in 12, second trimester in 14, and in the third trimester in 14. Poisonings resulted from malfunctioning furnaces ($n = 23$), malfunctioning hot water heaters ($n = 7$), car fumes ($n = 6$), methylene chloride ($n = 3$), and yacht engine fumes ($n = 1$). Outcome was adversely affected in three of five pregnancies that involved severe toxicity — two stillbirths and one cerebral palsy. All five occurred during the second or third trimester. All pregnancies with less severe poisonings resulted in normal infants.

A retrospective case-control study in Colorado (Alderman et al., 1987) found no association between higher carbon monoxide exposure during the last 3 months of gestation and higher odds of low birth weight after adjusting for the confounding effects of maternal race and education.

Hyperbaric oxygen has been used successfully for the treatment of severe CO poisoning in the pregnant patient (Hollander et al., 1987; Van Hoesen et al., 1989).

Measurement of COHb in 134 pregnant women who smoked revealed a linear relationship between the maternal blood COHb and the fetal blood COHb, with an r value of 0.80 (Bureau et al., 1982). Fetal COHb was approximately 5% higher than maternal COHb at each level of maternal

COHb. There was a significant correlation between the number of cigarettes smoked per day and fetal COHb and maternal COHb.

Leonard et al. (1989) using 19 mother–infant pairs, determined the fetal:maternal COHb concentration ratio to be 1.40 when maternal COHb dropped no more than 10% during labor. When maternal COHb dropped by more than 10% due to hyperventilation, the fetal:maternal COHb concentration ratio was 1.83.

3.1 Thyroid Function in the Pregnant Smoker

Cigarette smoking during pregnancy is reported to have a number of deleterious consequences (e.g., low birth weight, prematurity, learning disorders) (Longo, 1977; Naeye, 1979; Rush, 1980). Ericsson et al. (1987) reported that babies born to smoking mothers have elevated serum concentrations of thyroglobulin in the cord blood, without significant changes in serum thyroid-stimulating hormone (TSH), suggesting cigarette smoke has a direct effect on the thyroid gland of the fetus. Others (Meberg and Marstein, 1986) found a moderate decrease in cord blood TSH, suggesting hyperfunction of the thyroid gland. Interestingly, it was observed (Chanoine et al., 1991) that smoking during pregnancy in areas of borderline iodine intake is correlated with thyroid enlargement in newborn. This increase was secondary to a decrease in birth weight. The Chanoine group also found a positive correlation between cord serum thyroglobulin and SCN levels.

Along with CO, the main subject of interest in this volume, hydrogen cyanide (CN) is another toxic component of cigarette smoke. In the body, CN is detoxified into thiocyanate (SCN), a goitrogenic compound. Decreased thyroid function was first reported in smokers resident to areas of marginal iodine intake (Lagasse et al., 1982). Even in areas with adequate iodine intake substantial decreases in serum T3 and T4 concentrations were found in heavy smokers compared with nonsmokers (Sepkovic et al., 1984). The incidence of goiter was found to be slightly increased in smokers vs. ex-smokers or never-smokers (Christensen et al., 1984; Melander et al., 1981). On the other hand, others have found significantly elevated serum T_3 in smokers, although T_4 and TSH levels were unchanged (Karakaya et al., 1987).

Because pregnancy causes increased stress on thyroid function even in the presence of adequate iodine intake, factors that block its activity might be expected to depress thyroid hormone levels and elevate TSH levels. Thus, the pregnant population is an ideal one in which to investigate the effect of cigarette smoking and CN uptake on thyroid function.

A prospective study of cigarette smoking and thyroid function in the pregnant woman and newborn was carried out at a community hospital. Between February 1993 and April 1994, pregnant clinic patients reporting a smoking habit were recruited and enrolled at the time of their first prenatal visit, along with nonsmoking pregnant clinic patients. Patients with a history

TABLE 11

Effects of Human Exposure to Carbon Monoxide During Early Development, from the Recent Literature

Exposure	COHb (%)	Acute Symptoms	Chronic Symptoms	Ref.
Accidental exposure at 8 weeks of age.	24.5	Dizziness, headache, chest tightness.	None.	Copel et al., 1982
Accidental exposure at 33 weeks of age.	23.7	Headache, vomiting, dry mouth, fatigue, stomach cramps fetal heart tones absent.	Fetal demise, stillborn birth.	Cramer, 1982
Accidental exposure at 13 weeks of age.	60	Convulsion, hypotonic, unconscious.	Recovery of minor neurologic deficits by 6 weeks.	Venning et al., 1982
"Light" exposure.	4–27	Hyper-reflexia, auditive memory impairment and spatial orientation problems.	Auditive and visual impairment.	Klees et al., 1985
"Medium" exposure.	6–36	Coma, unconscious, normal.	Anxiety or emotional instability, memory impairment, spatial/temporal disorganization and perceptual problems, minimal or no effect.	Klees et al., 1985
"Heavy" exposure.	37	Coma, language and motor regression, violent anger/nervousness.	Persistent emotional instability.	Klees et al., 1985
Accidental exposure at 21 d of age.	>15	Lethargy, vomiting.	None	O'Sullivan, 1983
28 pediatric exposures.	15	Asymptomatic	Headaches, memory deficit, school performance decline.	Crocker and Walker, 1985
	16.7	Nausea/headache		
	19.8	Vomiting		
	18.6	Lethargy		
	24.5	Visual symptoms/syncope		
	36.9	Seizures		

Accidental expsosure at 37 weeks of age.	58.2	Unconsciousness adult respiratory distress syndrome grand mal seizure.	None.	Margulies, 1986
Accidental exposure at 38 weeks of age.	~45	Sluggishness, lethargy	None.	Hollander et al., 1987
Accidental exposure at 16–30 weeks of age (3 cases).	9.6–39	Nausea, vomiting, headache, dizziness.	None.	Caravati et al., 1988
Accidental exposure at 13–38 weeks of age (3 cases)	5–32	Nausea, vomiting, headache, dizziness flushed skin unconsciousness, coma respiratory rales adult respiratory distress syndrome fetal heart tones absent.	Fetal demise, stillborn birth, multiple morphologic anomalies.	Caravati et al., 1988
Accidental exposure at 37 weeks of age	47.2	Headache, nausea, vomiting chest pain unresponsiveness	None.	Van Hoesen et al., 1989
Accidental exposure at 28 weeks of age	7	Unconsciousness combative, confused peripheral cyanosis bilateral alveolar infiltrates fetal heart tones absent	Fetal demise, stillborn birth.	Farrow et al., 1990
Occupational exposure at 6 weeks of age	5.4	Weakness, nausea, confusion.	Psychomotor development delayed, hypotonia, no speech, telencephalic dysgenesis	Woody and Brewster, 1990
Accidental exposure at 10 weeks of age and afterward	14	Headache, dizziness	Cleft lip and palate, low-set ears, hypoplastic external genitalia, bilateral retinal colobomata, multiple cardiac defects, early postnatal death.	Hennequin et al., 1993

Modified from McMullen, T.B. and Raub, J.A., Eds., *Air Quality Criteria for Carbon Monoxide,* U.S. EPA, Washington, D.C., 1991. With permission.

of thyroid disease, or who used cocaine, marijuana, or other illegal drugs, were excluded. The following data were collected at the time of the first visit: name, age, history, height, weight, blood pressure, current medications, goiter incidence, smoking history, marijuana and other illegal drug use, location of abode, type of heating plant, make and year of automobile, job description and location, and almonds in diet. Blood samples were drawn at the time of the second visit (16–20 weeks) and at delivery, along with routine samples for hematology, hepatitis, drug screen, etc. Umbilical cord (newborn) blood was also sampled at delivery. Samples were processed and stored as required, and analyzed for the thyroid hormones and for COHb, CN, and SCN.

Forty smokers and 22 nonsmokers were initially recruited; 27 smokers and 18 nonsmokers completed the study (Table 12). No significant differences in measures of thyroid function were observed initially or at the time of delivery between women smoking an average of just over one pack per day, and nonsmokers (Table 13). The same was true of the newborns, born to smokers and nonsmokers (Table 14). Smokers showed significantly higher levels of COHb and SCN than nonsmokers at the first visit and at the time of birth (Table 12). Smokers had significantly higher blood hemoglobin than nonsmokers in initial measurements. Newborns of smokers had significantly higher levels of SCN than newborns of nonsmokers (Table 13). There were no significant differences between smokers and nonsmokers in terms of auto age, housing, abode heat source, employment, or ingestion of almonds.

Even though the women smokers and their newborn had higher SCN levels, thyroid function was not significantly altered. This suggests that the SCN levels remained sub-threshold for the metabolic reactions responsible for thyroid hormone production. This study suggests that in the setting of adequate iodine intake, no significant relationship exists between cigarette smoking and thyroid function in pregnant women and their newborn, although pregnant women who smoke do have elevated thiocyanate and COHb, and show evidence of polycythemia.

TABLE 12

Demographic Data

Group (*n*)	Age (yrs.)	Race (B/W)	Body Wt. (lb.)	Height (in.)	Preg. Stage (wks.)	Grav.	Para.	BPs (mm Hg)
Smokers	25.2	3/18	161.8	64.5	15.6	3.04	1.15	122.2
(22)	±1.2		±8.3	±0.5	±1.4	±0.30	±0.26	±2.0
Nonsmokers	20.4	8/10	157.3	63.5	20.1	2.00	±0.44	114.9
(18)	±0.7		±10	±0.7	±2.1	±0.30	±0.20	±2.5
	++							

Note: Values are means ± standard error of the mean. Comparing treatment and control groups: ++, $P<0.01$.

TABLE 13

Maternal Blood Values

Second Visit

Group (*n*)	COHb (%)	Hb (g/L)	CN (ug/L)	SCN (ug/L)	T4 RIA (ug/dL)	T3 RIA (ng/dL)	TSH (mIU/ml)
Smokers (27)	3.99	12.74	5.3	6.2	10.5	174.8	1.59
	±.36	±.20	±3.8	±.8	±.4	±5.7	±.24
Nonsmokers (18)	1.01	11.97	.4	2.5	11.9	184.6	±1.63
	±.17	±.27	±.3	±.5	±.6	±9.4	±.18
	++	+		++			

Delivery

Group (n)	COHb (%)	Hb (g/L)	CN (ug/L)	SCN (ug/L)	T4 RIA (ug/dL)	T3 RIA (ng/dL)	TSH (mIU/ml)
Smokers (27)	4.05	10.59	.23	8.2	11.3	176.2	2.52
	±0.49	±0.35	±0.05	±0.8	±0.5	±7.4	±.41
Nonsmokers (18)	1.83	10.89	.1	2.7	12.2	185.9	±2.10
	±0.29	±0.27	±0	±0.2	±0.6	±6.3	±0.15
	++			++			

Note: Values are means ± standard error of the mean. Comparing treatment and control groups: +, $P<0.05$; ++, $P<0.01$.

TABLE 14

Baby Blood Values

Group (*n*)	COHb (%)	CN (ug/L)	SCN (ug/L)	T4 RIA (ug/dL)	T3 RIA (ng/dL)	TSH (mIU/ml)
Smokers (22)	7.20	.89	9.08	10.3	44.5	9.8
	±0.54	±0.51	±2.59	±0.6	±2.9	±1.6
Nonsmokers (18)	6.25	.1	2.8	10.6	52.5	±7.6
	±0.35	±0	±0.2	±0.5	±3.8	±0.8
			++			

Note: Values are means ± standard error of the mean. Comparing treatment and control groups: ++, $P<0.01$.

ACKNOWLEDGMENTS

Thanks go to Jerald A. Mitchell, Ph.D., for his collaboration in carrying out the rat embryo implantation study, and to Gary L. Engelmann for his valuable collaboration in making possible the heart growth factor study. The assistance of Brigitte Ngoyi, M.D., and Melody Abraham, M.D., in doing the human prospective study of thyroid function is much appreciated.

REFERENCES

Abbatiello, E.R. and Mohrman, K., Effects on the offspring of chronic low exposure carbon monoxide during mice pregnancy, *Clin. Toxicol.,* 14, 401, 1979.

Alderman, B.W., Baron, A.E., and Savitz, D.A., Maternal exposure to neighborhood carbon monoxide and risk of low birth weight, *Public Health Rep.,* 102, 410, 1987.

Anversa, P., Olivetti, G., Melissari, M., and Loud, A.V., Morphometric study of myocardial hypertrophy induced by abdominal aortic stenosis, *Lab. Invest.,* 40, 341, 1979.

Asmussen, E. and Paulsen, N.V., Cardiac hypertrophy in CO-treated young rats and their later ability to withstand stress, *Acta Physiol. Scand.,* 29, 307, 1953.

Astrup, P., Trolle, D., Olsen, H.M., and Kjeldsen, K., Effect of moderate carbon monoxide exposure on fetal development, *Lancet,* 2, 1220, 1972.

Banerjee, S.K., Parker, C.J., Jr., and Penney, D.G. Cardiac myosin isozymes, ATPase and taurine in rats of various ages inhaling carbon monoxide, *Fed. Proc.,* 42, 466, 1983.

Boluyt, M., Penney, D.G., Clubb, F.J., and White, T.P., Exposure of neonatal rats to carbon monoxide alters cardiac adaptation to aortic constriction in adulthood, *J. Appl. Physiol.,* 70, 2697, 1991.

Bureau, M.A., Monette, J., Shapcott, O., Pare, C., Mathieu, J.-L., Lippe, J., Blovin, D., Berthiaume, Y., and Begin, R., Carboxyhemoglobin concentration in fetal cord blood and in blood of mothers who smoked during labor, *Pediatrics,* 69, 371, 1982.

Caravati, E.M., Adams, C.J., Joyce, S.M., and Schafer, N.C., Fetal toxicity associated with maternal carbon monoxide poisoning, *Ann. Emerg. Med.,* 17, 714, 1988.

Carratu, M.R., Cagiano, R., De Salvia, M.A., and Cuomo, V., Wallerian degeneration in rat sciatic nerve comparative evaluation between prenatal and early postnatal exposure to carbon monoxide, *Soc. Neurosci.,* 19, 834, 1993a.

Carratu, M.R., Renna, G., Giustino, A., De Salvia, M.A., and Cuomo, V., Changes in peripheral nervous system activity produced in rats by prenatal exposure to carbon monoxide, *Arch. Toxicol.,* 67, 297, 1993b.

Chanoine, J.P., Toppet, V., Bourdoux, P., Spehl, M., and Delange, F., Smoking during pregnancy: A significant cause of neonatal thyroid enlargement, *Br. J. Obstet. Gynaecol.,* 98, 65, 1991.

Choi, K.D. and Oh, Y.K., A teratological study on the effects of carbon monoxide exposure upon the fetal development of albino rats, *Chungang Uihak,* 29, 209, 1975.

Christensen, S.B., Ericsson, U.-B., Janzon, L., and Tibblin, S., Influence of cigarette smoking on goiter formation, thyroglobulin, and thyroid hormone levels in women, *J. Clin. Endocrinol. Metab.,* 58, 615, 1984.

Clubb, F.J., Penney, D.G., Baylerian, M.S., and Bishop, S.P., Cardiomegaly due to myocyte hyperplasia in perinatal rats exposed to 200 ppm carbon monoxide, *J. Mol. Cell. Cardiol.,* 18, 477, 1986.

Clubb, F.J., Jr., Penney, D.G., and Bishop, S.P., Cardiomegaly in neonatal rats exposed to 500 ppm carbon monoxide, *J. Mol. Cell. Cardiol.,* 21, 945, 1989.

Copel, J.A., Bowen, F., and Bolognese, R.J., Carbon monoxide intoxication in early pregnancy, *Obstet. Gynecol.*, 59, 26S, 1982.

Cramer, C.R., Fetal death due to accidental maternal carbon monoxide poisoning, *J. Toxicol.: Clin. Toxicol.*, 19, 297, 1982.

Crocker, P.J. and Walker, J.S., Pediatric carbon monoxide toxicity, *J. Emerg. Med.*, 3, 443, 1985.

Daughtrey, W.C., Newby-Schmidt, M.B., and Norton, S., Forebrain damage in chick embryos exposed to carbon monoxide, *Teratology*, 28, 83, 1983.

Daughtrey, W.C. and Norton, S., Caudate morphology and behavior of rats exposed to carbon monoxide *in utero*, *Exp. Neurol.*, 80, 265, 1983.

Daughtrey, W.C. and Norton, S., Morphological damage to the premature fetal rat brain after acute carbon monoxide exposure, *Exp. Neurol.*, 78, 26, 1982.

Di Giovanni, V., Cagiano, R., De Salvia, M.A., Giustino, A., and Lacomba, C., Neurobehavioral changes produced in rats by prenatal exposure to carbon monoxide, *Brain Res.*, 616, 126, 1993.

Dominick, M.A. and Carson, T.L., Effects of carbon monoxide exposure on pregnant sows and their fetuses, *Am. J. Vet. Res.*, 44, 35, 1983.

Dubeck, J.K., Penney, D.G., Brown, T.P., and Sharma, P., Thyroxine treatment of neonatal rats suppresses normal and stress-stimulated heart cell hyperplasia and abolishes persistent cardiomegaly, *J. Appl. Cardiol.*, 4, 195, 1989.

Engelmann, G.L. and Penney, D.G., Carbon monoxide-induced neonatal cardiomyocyte growth: Molecular analysis, *J. Mol. Cell. Cardiol.*, 23 (Suppl. III), S-53, 1991.

Ericsson, U.B., Ivarsson, S.A., and Persson, P.H., Thyroglobin in cord blood, *Eur. J. Ped.*, 146, 44, 1987.

Farrow, J.R., Davis, G.J., Roy, T.M., McCloud, L.C., and Nichols, G.R., II, Fetal death due to nonlethal maternal carbon monoxide poisoning, *J. Forensic Sci.*, 35, 1448, 1990.

Fechter, L.D., Neurotoxicity of prenatal carbon monoxide exposure, Report No. 12, Health Effects Institute, Baltimore, 1987.

Fechter, L.D. and Annau, Z., Effects of prenatal carbon monoxide exposure on neonatal rats, *Adverse Effects Environ. Chem. Psychotropic Drugs*, 2, 219, 1976.

Fechter, L.D. and Annau, Z., Prenatal carbon monoxide exposure alters behavioral development, *Neurobehav. Toxicol.*, 2, 7, 1980.

Fechter, L.D. and Annau, Z., Toxicity of mild prenatal carbon monoxide exposure, *Science*, 197, 680, 1977.

Fechter, L.D., M., T., Miller, B., Annau, Z., and Srivastava, U., Effects of prenatal carbon monoxide exposure on cardiac development, *Toxicol. Appl. Pharmacol.*, 56, 370, 1980.

Garvey, D.J. and Longo, L.D., Chronic low level maternal carbon monoxide exposure and fetal growth and development, *Biol. Reprod.*, 19, 8, 1978.

Ginsberg, M.D. and Myers, R.E., Fetal brain damage following maternal carbon monoxide poisoning: An experimental study, *Acta Obstet. Gynecol. Scand.*, 53, 309, 1974.

Giustino, A., Cagiano, R., Carratu, M.R., De Salvia, M.A., Panaro, M.A., Jirillo, E., and Cuomo, V., Immunological changes produced in rats by prenatal exposure to carbon monoxide, *Pharmacol. Toxicol.*, 73, 274, 1993.

Hennequin, Y., Bium, D., Vamos, E., Steppe, M., Goedseels, J., and Cavatorta, E., *In utero* carbon monoxide poisoning and multiple fetal abnormalities, *Lancet*, 341, 240, 1993.

Hoffman, D.J. and Campbell, K.I., Postnatal toxicity of carbon monoxide after pre- and postnatal exposure, *Toxicol. Letts.*, 1, 147, 1977.

Hollander, D.I., Nagey, D.A., Welch, R., and Pupkin, M., Hyperbaric oxygen therapy for the treatment of acute carbon monoxide poisoning in pregnancy: A case report, *J. Reprod. Med.*, 32, 615, 1987.

Karakaya, A., Tuncel, N., Alptuna, G., Kocer, Z., and Erbay, G., Influence of cigarette smoking on thyroid hormone levels, *Human Toxicol.*, 6, 507, 1987.

Klees, M., Heremans, M., and Dougan, S., Psychological sequelae to carbon monoxide intoxications in the child, *Sci. Total Environ.*, 44, 165, 1985.

Koos, B.J., Matsuda, K., and Power, G.G., Fetal breathing and sleep state responses to graded carboxyhemoglobinemia in sheep, *J. Appl. Physiol.*, 65, 2118, 1988.

Koren, G., Sharav, T., Pastuszak, A., Garrettson, L.K., Hill, K., Samson, I., Rorem, M., King, A., and Dolgin, J.E., A multicenter, prospective study of fetal outcome following accidental carbon monoxide poisoning in pregnancy, *Reprod. Toxicol.,* 5, 397, 1991.

Lagasse, R., Bourdoux, P., and Courtois, P., Influence of the dietary balance of iodine/thiocyanate and protein on thyroid function in adults and young infants, in *Nutritional Factors Involved in the Goitrogenic Action of Cassava,* Delange, F., Iteke, F.B., and Ermans, A.M., Eds., International Development Research Centre, Ottawa, Canada, 1982.

Leichter, J., Fetal growth retardation due to exposure of pregnant rats to carbon monoxide, *Biochem. Arch.,* 9, 267, 1993.

Leonard, M.B., Vreman, H.J., Ferguson, J.E., II, Smith, D.W., and Stevenson, D.K., Interpreting the carboxyhaemoglobin concentration in fetal cord blood, *J. Dev. Physiol.,* 11, 73, 1989.

Longo, L.D., The biological effects of carbon monoxide on the pregnant woman, fetus, and newborn infant, *Am. J. Obstet. Gynecol.,* 129, 69, 1977.

Lynch, A.-M. and Bruce, N.W., Placental growth in rats exposed to carbon monoxide at selected stages of pregnancy, *Biol. Neonate,* 56, 151, 1989.

Mactutus, C.F. and Fechter, L.D., Moderate prenatal carbon monoxide exposure produces persistent, and apparently permanent, memory deficits in rats, *Teratology,* 31, 1, 1985.

Mactutus, C.F. and Fechter, L.D., Prenatal exposure to carbon monoxide: Learning and memory deficits, *Science,* 223, 409, 1984.

Margulies, J.L., Acute carbon monoxide poisoning during pregnancy, *Am. J. Emerg. Med.,* 4, 516, 1986.

McMullen, T.B. and Raub, J.A., Eds., *Air Quality Criteria for Carbon Monoxide,* EPA-600/8-90/045F, U.S. Environmental Protection Agency, Washington, D.C., 1991.

Meberg, A. and Marstein, S., Smoking during pregnancy: Effects on the fetal thyroid function, *Acta Paediatr. Scand.,* 75, 762, 1986.

Melander, A., Nordenskjold, E., Lundh, B., and Thorell, J., Influence of smoking on thyroid activity, *Acta Med. Scand.,* 209, 41, 1981.

Morris, G.L., Curtis, S.E., and Simon, J., Perinatal piglets under sublethal concentrations of atmospheric carbon monoxide, *J. Animal Sci.,* 61, 1070, 1985.

Naeye, R.L., The duration of maternal cigarette smoking, fetal and placental disorders, *Early Human Dev.,* 3, 229, 1979.

Norman, C.A. and Halton, D.M., Is carbon monoxide a workplace teratogen? A review and evaluation of the literature, *Ann. Occup. Hyg.,* 34, 335, 1990.

O'Sullivan, B.P., Carbon monoxide poisoning in an infant exposed to a kerosene heater, *J. Pediatr.,* 103, 249, 1983.

Okeda, R., Matsuo, T., Kuroiwa, T., Tajima, T., and Takahashi, H., Experimental study on pathogenesis of the fetal brain damage by acute carbon monoxide intoxication of the pregnant mother, *Acta Neuropathol. (Berlin),* 69, 244, 1986.

Packianathan, S., Cain, C.D., Stagg, R.B., and Longo, L.D., Ornithine decarboxylase activity in fetal and newborn rat brain: Responses to hypoxia and carbon monoxide hypoxia, *Dev. Brain Res.,* 76, 131, 1993.

Penney, D.G., Carbon monoxide-induced cardiac hypertrophy, in *Growth of the Heart in Health and Disease,* Zak, R., Ed., Raven Press, New York, 1984, 337.

Penney, D.G., The neural and behavioral effects of carbon monoxide, in *The Vulnerable Brain and Environmental Risks: Special Hazards from Air and Water,* Isaacson, R.L. and Jensen, K.F., Eds., Plenum Press, New York, 1994, 3.

Penney, D.G., Postnatal modification of cardiac development: A review. *J. Appl. Cardiol.,* 5, 325, 1990.

Penney, D.G., A review: Hemodynamic response to carbon monoxide, *Environ. Health Perspect.,* 77, 121, 1988.

Penney, D.G., Barthel, B.G., and Skoney, J.A., Cardiac compliance and dimensions in carbon monoxide-induced cardiomegaly, *Cardiovasc. Res.,* 18, 270, 1984.

Penney, D.G., Baylerian, M.S., and Fanning, K.E., Temporary and lasting cardiac effects of pre- and postnatal exposure to carbon monoxide, *Toxicol. Appl. Pharmacol.,* 53, 271, 1980.

Penney, D.G., Baylerian, M.S., Thill, J.E., Fanning, C.M., and Yedavally, S., Postnatal carbon monoxide exposure: Immediate and lasting effects in the rat, *Am. J. Physiol.,* 243, H328, 1982.

Penney, D.G., Baylerian, M.S., Thill, J.E., Yedavally, S., and Fanning, C.M., The cardiac response of the fetal rat to carbon monoxide exposure, *Am. J. Physiol.,* 244, H289, 1983.

Penney, D.G., Caldwell-Ayre, T.M., Gargulinski, R.B., and Hawkins, B.J., Effect of litter size (i.e., nutrition) on carbon monoxide-induced persistent heart changes, *Growth,* 51, 321, 1987.

Penney, D.G., Clubb, F.J., Jr., Allen, R.C., Jen, C., Banerjee, S., and Hull, J.A., Persistent early heart changes do alter the cardiac response to exercise training in adulthood, *J. Appl. Cardiol.,* 4, 223, 1989.

Penney, D.G., Cook, K., and Sakai, J., Heart growth: Interacting effects of carbon monoxide and age, *Growth,* 38, 321, 1974.

Penney, D.G., Gargulinski, R.B., Hawkins, B.J., Santini, R., Caldwell-Ayre, T.M., and Davidson, S.B., The effects of carbon monoxide on persistent changes in young rat heart: Cardiomegaly, tachycardia and altered DNA content, *J. Appl. Toxicol.,* 8, 275, 1988.

Penney, D.G., Giraldo, A., and VanEgmond, E.M., Chronic carbon monoxide exposure in young rats alter coronary vessel growth, *J. Toxicol. Environ. Health,* 39, 207, 1993.

Penney, D.G., Stryker, A., and Baylerian, M.S., Persistent cardiomegaly induced by carbon monoxide and associated tachycardia, *Am. J. Physiol.,* 56, 1045, 1984.

Penney, D.G., Tucker, A., and Bambach, G.A., Heart and lung alterations in neonatal rats exposed to CO or high altitude, *J. Appl. Physiol.,* 73, 1713, 1992.

Penney, D.G. and Weeks, T.A., Age dependence of cardiac growth in the normal and carbon monoxide-exposed rat, *Dev. Biol.,* 71, 153, 1979.

Prigge, E. and Hochrainer, D., Effects of carbon monoxide inhalation on erythropoiesis and cardiac hypertrophy in fetal rats, *Toxicol. Appl. Pharmacol.,* 42, 225, 1977.

Rosenkrantz, H., Grant, R.J., Fleischman, R.W., and Baker, J.R., Marihuana-induced embryotoxicity in the rabbit, *Fundam. Appl. Toxicol.,* 7, 236, 1986.

Rush, D., Cigarette smoking, nutrition, social status, and perinatal loss: Their interactive relationships, in *Embryonic Fetal Death,* Academic Press, New York, 1980.

Schwetz, B.A., Smith, F.A., Leong, B.K.J., and Staples, R.E., Teratogenic potential of inhaled carbon monoxide in mice and rabbits, *Teratology,* 19, 385, 1979.

Sepkovic, D.W., Haley, N.J., and Wynder, E.L., Thyroid activity in cigarette smokers, *Arch. Intern. Med.,* 144, 501, 1984.

Singh, J., Early behavioral alterations in mice following prenatal carbon monoxide exposure, *Neurotoxicology,* 7, 475, 1986.

Singh, J., Aggison, L., Jr., and Moore-Cheatum, L., Teratogenicity and developmental toxicity of carbon monoxide on protein-deficient mice, *Teratology,* 48, 149, 1993.

Singh, J. and Scott, L.H., Threshold for carbon monoxide induced fetotoxicity, *Teratology,* 30, 253, 1984.

Storm, J.E. and Fechter, L.D., Alteration in the postnatal ontogeny of cerebellar norepinephrine content following chronic prenatal carbon monoxide, *J. Neurochem.,* 45, 965, 1985a.

Storm, J.E. and Fechter, L.D., Prenatal carbon monoxide exposure differentially affects postnatal weight and monoamine concentration of rat brain regions, *Toxicol. Appl. Pharmacol.,* 81, 139, 1985b.

Storm, J.E., Valdes, J.J., and Fechter, L.D., Postnatal alterations in cerebellar GABA content, GABA uptake and morphology following exposure to carbon monoxide early in development, *Dev. Neurosci.,* 8, 251, 1986.

Tachi, N. and Aoyama, M., Effect of cigarette smoke and carbon monoxide inhalation by gravid rats on the conceptus weight, *Bull. Environ. Contam. Toxicol.,* 31, 85, 1983.

Tachi, N. and Aoyama, M., Effect of restricted food supply to pregnant rats inhaling carbon monoxide on fetal weight, compared with cigarette smoke exposure, *Bull. Environ. Contam. Toxicol.,* 37, 877, 1986.

Tachi, N. and Aoyama, M., Postnatal growth in rats prenatally exposed to cigarette smoke or carbon monoxide, *Bull. Environ. Contam. Toxicol.,* 45, 641, 1990.

Tucker, A. and Penney, D.G., Pulmonary vascular responsiveness in rats following neonatal exposure to high altitude or carbon monoxide, *Exper. Lung Res.,* 19, 699, 1993.

Van Hoesen, K.B., Camporesi, E.M., Moon, R.E., Hage, M.L., and Piantadosi, C.M., Should hyperbaric oxygen be used to treat the pregnant patient for acute carbon monoxide poisoning? A case report and literature review, *JAMA,* 261, 1039, 1989.

Venning, H., Roberton, D., and Milner, A.D., Carbon monoxide poisoning in an infant, *Br. Med. J.,* 284, 651, 1982.

Wells, L.L., The placental effects of carbon monoxide on albino rats and the resulting neuropathology, *Biologist,* 15, 80, 1933.

Williams, I.R. and Smith, E., Blood picture, reproduction, and general condition during exposure to illuminating gas, *Am. J. Physiol.,* 110, 611, 1935.

Woody, R.C. and Brewster, M.A., Telencephalic dysgenesis associated with presumptive maternal carbon monoxide intoxication in the first trimester of pregnancy, *Clin. Toxicol.,* 28, 467, 1990.

CHAPTER 7

Carbon Monoxide — From Tool to Neurotransmitter

Nanduri R. Prabhakar and Robert S. Fitzgerald

CONTENTS

1 INTRODUCTION

Stories of carbon monoxide (CO) poisoning can be read every day in newspapers across the world. But two other stories can be told about CO. In the past it has been used as a tool to assess the effects of hypoxia on the cardiorespiratory system. The peripheral chemoreceptors (i.e., the carotid and aortic bodies) are necessary for cardiorespiratory responses to low oxygen.

0-8493-4796-3/96/$0.00+$.50

Since CO has a greater effect on aortic than carotid bodies, it has been used as a tool to delineate the relative contribution of the two chemoreceptor systems to cardiorespiratory adaptations to hypoxia. Presently it is becoming evident not only that CO is formed in many different tissues but also that it may function as a transmitter in the nervous system. In fact, it is intriguing that CO may be a chemical messenger in one of the chemoreceptor organs, the carotid body, in which it was previously used as a tool to unravel the mechanisms of oxygen chemoreception.

2 CO AS A TOOL TO PROBE THE CARDIORESPIRATORY SYSTEMS

The toxic effects of inspired CO can be many, depending on the level and duration. But when carboxyhemoglobin saturation in the blood is raised to about 33% and maintained for some time, the principal toxic effect is due to the tissue hypoxia such a condition creates. Virtually all the inspired CO attaches to the hemoglobin, reducing its ability to carry oxygen (O_2) to the tissues; we shall call this CO hypoxia (COH).

What happens to the oxyhemoglobin dissociation curve in the presence of CO is both interesting and important. If an organ has a normal arterial blood PO_2 (PaO_2) flowing into it (~100 mmHg) and the arterio-venous difference in oxygen content across the organ is a normal 5 mls O_2/100 ml blood, the venous PO_2 (PvO_2), a good index of the PO_2 in the organ, is about 40 mmHg. If the PaO_2 is 100 mmHg but the O_2 content of the blood is reduced by the presence of 33% carboxyhemoglobin, then the PvO_2 is only 14 mmHg because CO displaces the oxyhemoglobin dissociation curve downward and to the left in this region of PO_2 values. By way of contrast, if the O_2 content is reduced by 33% by lowering the PaO_2 to about 38 mmHg (very severe, stressful hypoxic hypoxia [HH]) and the organ's O_2 consumption is the normal 5 ml O_2/100 ml blood, the PvO_2 is about 27 mmHg. Hence, COH leaves the organ with a lower PO_2 driving pressure than HH.

But, somewhat paradoxically, several years ago Montgomery and Rubin (1973), testing the effect of tissue hypoxia on the liver's ability to detoxify drugs, reported that at comparable levels of PvO_2, liver detoxification occurred much more rapidly under conditions of COH than under conditions of HH. One possibility was that blood flow to the liver during COH differed from that during HH. For example, greater blood flow would deliver more drug and oxygen; perhaps this explained the results during COH. Simultaneously with this report we were observing that the carotid body — a peripheral arterial chemoreceptor, long known to respond to HH — did not respond to COH. On the other hand, the aortic body — a second peripheral arterial chemoreceptor, also known to respond to HH — did respond to COH (Fitzgerald and Dehghani, 1990) (Figures 1A, 1B).

Inasmuch as one of the areas of cardiopulmonary research addresses the issue of how the cardiopulmonary system assists the organism in the capture of O_2 from the environment for distribution to the tissues, we developed an experimental model — using CO as a tool — to determine the effects of these two peripheral arterial chemoreceptors on blood flow to various organs during acute systemic hypoxia (Fitzgerald et al., 1991).

We used the anesthetized, artificially paralyzed cat instrumented with an aortic flow probe, and catheters in the femoral artery and vein, right auricle, pulmonary artery, left auricle and ventricle. In some of the experiments we injected radio labeled microspheres (^{141}Ce, ^{113}Sn, ^{85}Sc, ^{46}Sr) into the left ventricle, and measured blood flow using standard techniques. We calculated the vascular resistance in the organ from blood flow to the organ and the mean arterial and right atrial pressures (or alveolar pressure, if it was higher than right atrial pressure due to positive end expiratory pressure at the time of the injection).

Duplication of protective mechanisms and "backup" mechanisms are well known in physiology. Hence, we developed a design to test the *null hypothesis*: Systemic and organ vascular responses to systemic hypoxia with both chemoreceptors operating do not differ from the responses when only one set of chemoreceptors is operating. Because (1) the aortic bodies, but not the carotid bodies, increase their neural output during COH, while during HH both chemoreceptors increase their output, and (2) the aortic depressor nerves in the cat are quite easily separated from the vagus for transection, we challenged the cat with four hypoxic conditions in which the oxygen saturations were lowered to identical or very close levels (about 42%):

(A) HH/aortic nerves intact. Here both carotid and aortic bodies were sending impulses to the nucleus tractus solitarius for input into the vascular regulatory centers (Figures 1A, 1B, top panels).

(B) COH/aortic nerves intact. Here only the *aortic bodies* were sending impulses (Figure 1b, bottom panel). The carotid bodies are not stimulated by CO (Figure 1A, top panel).

(C) HH/aortic nerves transected. Here only the *carotid bodies* were sending impulses to the nucleus tractus solitarius; aortic bodies are responding to HH, but have been disconnected from the brain stem.

(D) COH/aortic nerves transected. neither carotid bodies nor aortic bodies were sending impulses; aortic bodies respond to CO, but are disconnected from the brain stem.

Mean arterial blood pressure rose only when both aortic and carotid bodies were operating (condition A), suggesting that to provoke sufficient reflex sympathetic output to the vasculature as a whole for overcoming the dilating effects of a massive local hypoxia, input from both sets of chemoreceptors was needed. Cardiac output also showed a significantly greater increase (153% of its control) when both sets of chemoreceptors were operating than

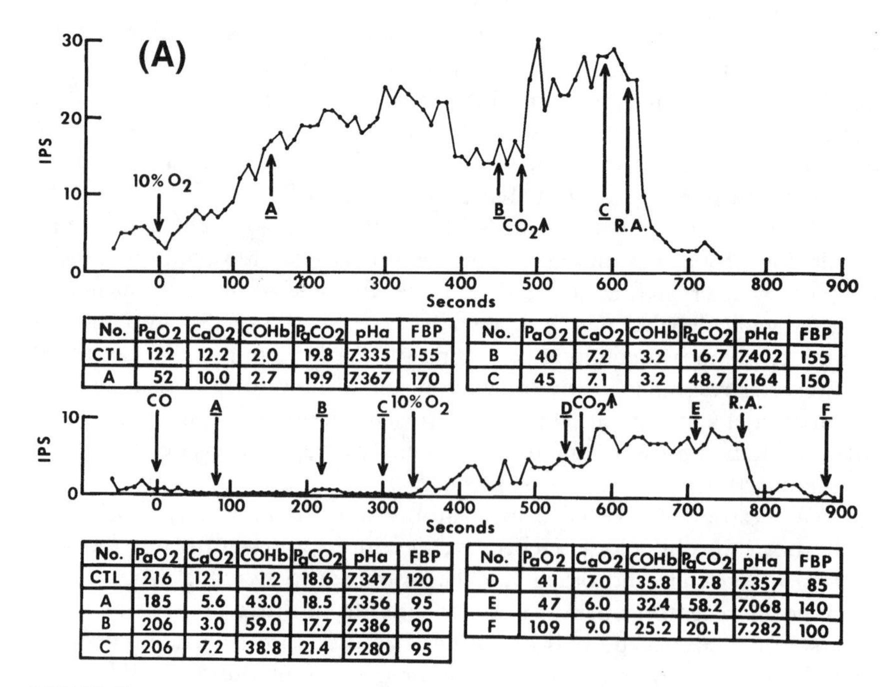

No.	P_aO_2	C_aO_2	COHb	P_aCO_2	pHa	FBP
CTL	122	12.2	2.0	19.8	7.335	155
A	52	10.0	2.7	19.9	7.367	170

No.	P_aO_2	C_aO_2	COHb	P_aCO_2	pHa	FBP
B	40	7.2	3.2	16.7	7.402	155
C	45	7.1	3.2	48.7	7.164	150

No.	P_aO_2	C_aO_2	COHb	P_aCO_2	pHa	FBP
CTL	216	12.1	1.2	18.6	7.347	120
A	185	5.6	43.0	18.5	7.356	95
B	206	3.0	59.0	17.7	7.386	90
C	206	7.2	38.8	21.4	7.280	95

No.	P_aO_2	C_aO_2	COHb	P_aCO_2	pHa	FBP
D	41	7.0	35.8	17.8	7.357	85
E	47	6.0	32.4	58.2	7.068	140
F	109	9.0	25.2	20.1	7.282	100

FIGURE 1A

Anesthetized cat. Response of a few fiber preparation of the carotid sinus nerve (from the carotid body) during the application of 10% O_2 in N_2 (top trace) and during the application of CO (bottom trace). IPS = impulses per second vs. time. Panels under trace give the blood gas values and blood pressure during control (CTL) and at various points (A, B, C, etc.) during the challenge. CO_2– = addition of CO_2 to inspirate; R.A. = application of room air. After 5 min application there was no response to CO; the carotid body was immediately challenged with 10% O_2 to determine if it was still viable after the CO, which had been terminated at 220 s.

when either the carotid body (137%) or the aortic body (142%) was acting alone. In all four challenges the calculated peripheral vascular resistance decreased; but the decrease was smallest when both sets of chemoreceptors were operating (to 87% of control vs. to 60%, 61%, and 72% of control for conditions B, C, and D above, respectively).

The effect of the hypoxias on the pulmonary vasculature produced some surprises. The pulmonary vasoconstrictor response to alveolar hypoxia is well known. This was seen during HH in both conditions A (Figure 2A) and C. The immediate rise in pulmonary artery pressure (Ppa) signals that effect. It is important to note that cardiac output continued to rise while Ppa remained constant (after the 2-minute mark). Because left atrial pressure remained constant, the constancy in Ppa in the face of the increase in flow means there had to be a decrease in pulmonary vascular resistance. We attributed this to a chemo-reflex effect. Such a decrease in pulmonary vascular resistance was also seen with CO during condition B when there was no alveolar hypoxia (Figure 2B). These data were consistent with the possibility

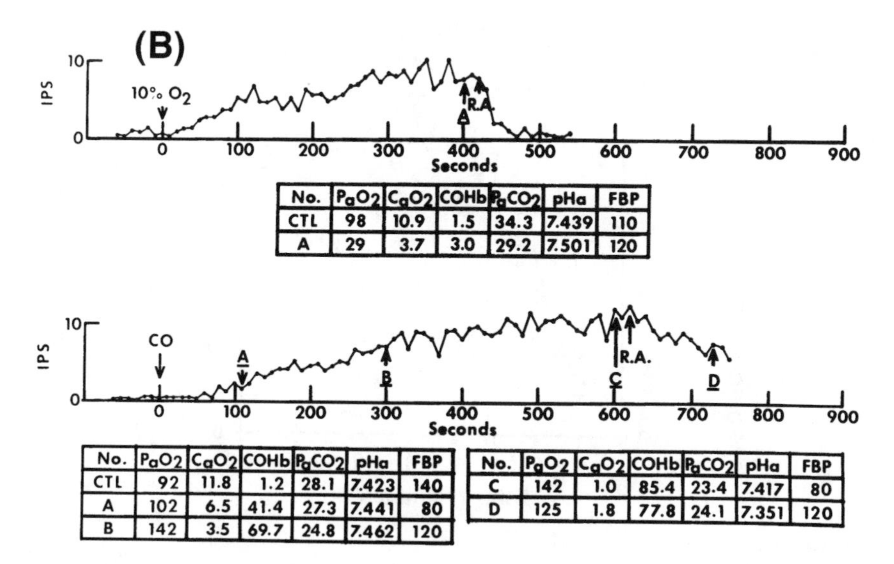

No.	P_aO_2	C_aO_2	COHb	P_aCO_2	pHa	FBP
CTL	98	10.9	1.5	34.3	7.439	110
A	29	3.7	3.0	29.2	7.501	120

No.	P_aO_2	C_aO_2	COHb	P_aCO_2	pHa	FBP
CTL	92	11.8	1.2	28.1	7.423	140
A	102	6.5	41.4	27.3	7.441	80
B	142	3.5	69.7	24.8	7.462	120

No.	P_aO_2	C_aO_2	COHb	P_aCO_2	pHa	FBP
C	142	1.0	85.4	23.4	7.417	80
D	125	1.8	77.8	24.1	7.351	120

FIGURE 1B

Anesthetized cat. Response of a few fiber preparation of the aortic depressor nerve (from the aortic body) during the application of 10% O_2 in N_2 (top trace) and during the application of CO (bottom trace). Explanation and abbreviations as in Figure 1A. Note the aortic body's response to CO.

that the carotid and aortic bodies provoked a reflex vasodilation in the pulmonary vasculature (Fitzgerald et al., 1992).

Among the various organs changes in the vascular resistances, not unexpectedly, varied depending on the type of hypoxia. However, the vascular resistances in the brain, heart, kidney, and liver of the anesthetized cat showed no influence attributable to the carotid or aortic bodies. Other organs did show such an influence (Table 1). Because in the stomach the decreases in vascular resistance in response to systemic hypoxia were significantly greater in the aortic body denervated animals, we concluded that the aortic body was responsible for maintaining a significant amount of tone in the vasculature of the stomach. The doubling of the vascular resistance in the spleen, when there was no input from the peripheral arterial chemoreceptors, could have been due to a local effect, perhaps similar to the well-known hypoxic pulmonary vasoconstriction. The 379% increase was significantly greater than the 300% and 265% increases (not different from each other, but different from the 209% increase) and suggested an interaction of the carotid and aortic bodies in controlling the vasculature of the spleen. Finally, the data suggested that the carotid body alone promoted a vasodilation in the eye in response to systemic hypoxia. Clearly, these data did not support our hypothesis.

Hence, CO turned out to be a valuable tool with which to probe the effects of a systemic hypoxic challenge on that system — the cardiopulmonary — whose principal function is to deliver O_2 to the organism.

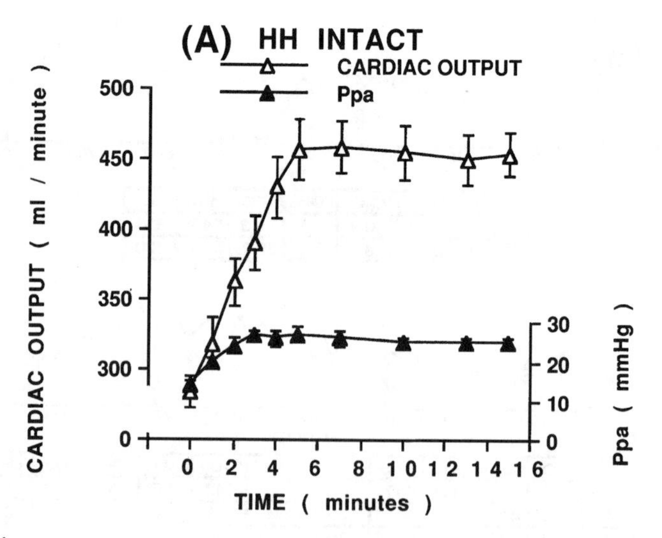

FIGURE 2A

Anesthetized cat. Response of the cardiac output and pulmonary artery pressure (Ppa) to acute systemic hypoxia generated by the application of 8% O_2 in N_2 to the ventilator (pH = 7.361; $PaCO_2$ = 36.5mmHg; PaO_2 = 28mmHg; HbO_2 = 43.7%). No further increase in Ppa occurs after 2 min while the cardiac output increases to twice its value at 2 min. Neither left atrial pressure nor alveolar pressure (not shown) changed. This necessitated a pulmonary vasodilatation.

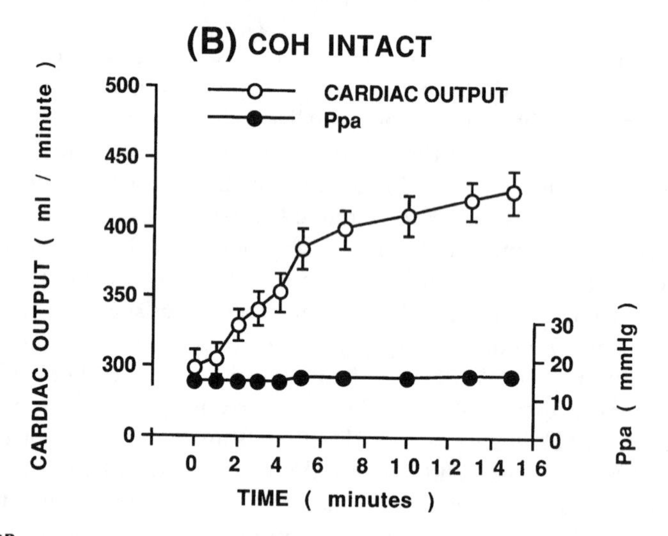

FIGURE 2B

Anesthetized cat. Response of the cardiac output and pulmonary artery pressure to acute systemic hypoxia generated by the application of CO to the ventilator (pH = 7.409; $PaCO_2$ = 34.3mmHg; PaO_2 = 130mmHg; HbO_2 = 44.8%). Ppa did not change, nor did either left atrial or alveolar pressures (not shown). Again the significant rise in cardiac output necessitated a pulmonary vasodilation.

TABLE 1
Vascular Resistance

Condition	Intact				Aortic Body Denervated			
Treatment	CTL	HH	CTL	COH	CTL	HH	CTL	COH
Stomach (% of control)	100	78*	100	68	100	58#	100	41
Spleen (% of control)	100	379*	100	300	100	265#	100	209
Eye (% of control)	100	48*	100	86	100	52#	100	84

Note: CTL = control value; HH = acute systemic hypoxia generated by applying 8% O_2 to the respirator; COH = acute systemic hypoxia generated by applying 1-2% CO to the respirator for ~90 s followed by 0.1% CO for the remainder of the exposure; * = significantly different from control; # = significantly different from corresponding intact value.

How curious and coincidental it is that CO now appears to be an important neurotransmitter or neuromodulator in one of the very organs that is so intimately involved in the cardiopulmonary response to systemic hypoxia, the carotid body. This is particularly so when elevating arterial carboxyhemoglobin to what seem to be extremely high levels does not stimulate an increase in neural activity (Figure 1B). This seems to emphasize the concept that it is not necessarily the level of CO, but rather the equipment and enzymes that are available to process CO that will determine whether or not it can generate an effect other than those resulting from its ability to occupy the site for the O_2 molecule on the hemoglobin molecule.

3 CO AS A NEUROTRANSMITTER

It is becoming increasingly apparent that gases such as nitric oxide (NO) and CO function as neurotransmitters (Moncada et al., 1991; Snyder, 1992). They seem to satisfy many of the criteria proposed for a transmitter: (1) they are synthesized in the nervous system; (2) they are released by the cells; (3) exogenous application influences neuronal activity; and (4) blockade of their synthesis affects many neuronal functions. However, NO and CO are unconventional transmitters in certain aspects. Unlike the classical transmitters, they are not stored in synaptic vesicles, nor do they act on membrane-bound receptors. Instead, they are produced according to need, and once formed, rapidly diffuse out of the cell. The only known "receptor" for NO and CO in the post-synaptic site is the iron in the heme moiety of proteins such as the enzyme soluble guanylate cyclase.

More than four decades ago, Sjorstrand (1949) provided the first evidence for the endogenous formation of CO. It is formed as a by-product of the catalytic oxidation of the heme molecule. The enzyme heme oxygenase (HO), in concert with microsomal NADPH-cytochrome P-450 reductase and molecular oxygen, catalyzes the physiological degradation of heme (Figure 3). One of the three molecules of oxygen used during the enzymatic cleavage of heme is utilized to generate CO. Purified NADPH-cytochrome P-450 reductase alone can catalyze oxidative degradation of the heme molecule (Docherty et al., 1984; Kim et al., 1987; Yoshinaga et al., 1982). However, only a small amount of CO is produced via this route. Therefore, heme oxygenases are the major enzyme system responsible for endogenous CO formation.

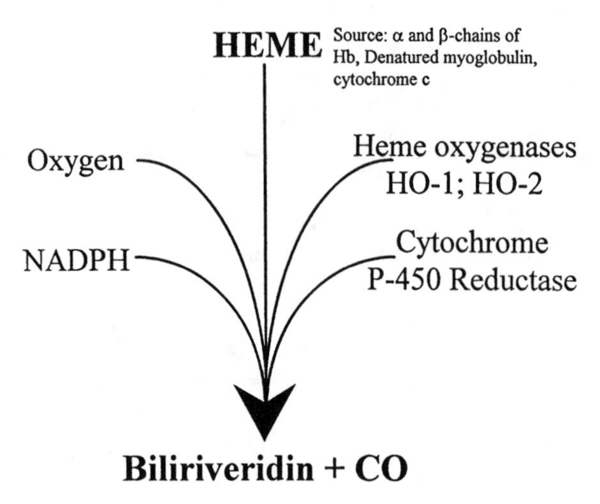

FIGURE 3
Schematic representation of the pathways involved in endogenous carbon monoxide formation. Abbreviations denote: CO, carbon monoxide; HO-1 and HO-2 heme oxygenase 1- and 2.

Virtually all tissues tested to date exhibit some degree of heme oxygenase activity where the enzyme is localized exclusively to the smooth endoplasmic reticulum (SER). However, the enzyme levels display considerable variations depending on age and sex. For instance, female organs have relatively more activity than males (Veltman and Maines, 1985). High levels of HO activity were reported in brain, spleen, and testis (approx. 11–12 nmol/mg/h), whereas in other tissues, HO activity ranged from 1–5 nmol/mg/h. The substrates for HO include the alpha and beta chains of hemoglobin, denatured myoglobin, methemoglobin, various heme c-derivatives, and proteolytic products of cytochrome c. Interestingly, oxyhemoglobin and intact myoglobin or cytochrome b5 are not substrates for HO (Kutty and Maines, 1981).

Molecular characterization revealed the existence of two distinct forms of heme oxygenase designated as HO-1 and HO-2 with differing characteristics (Kutty and Maines, 1989; Maines, 1988; Muller et al., 1987; Rotenberg and Maines, 1988). The most striking difference between the two forms is that HO-1, but not HO-2, can be induced by a variety of stimuli (Table 2). Metal ions like cobalt chloride were the first group of chemicals that were shown to induce HO-1 (Maines and Kappas, 1974). Other metals that affect

HO-1 include aluminum, tin, mercury, lead, and gold. In addition to the metals, a host of exogenous and endogenous chemicals as well as diverse factors such as X-ray irradiation, stress, fever, starvation, etc. can induce HO-1 activity. Some of the endogenous and exogenous chemicals that induce HO-1 activity include heme compounds, cemetidine (an histamine receptor antagonist), epinephrine, insulin, bacterial antigens, glucocorticoids, etc. (Maines, 1988). Hypoxia is another potent inducer of HO-1 in endothelial cells (Jornot and Junod, 1993).

TABLE 2

Similarities and Dissimilarities Between Heme Oxygenase-1 and Heme Oxygenase-2

Property	Heme Oxygenase-1	Heme Oxygenase-2
Molecular weight	30,000	36,000
Factors required for activity	Oxygen; NADPH; NADPH-cytochrome P-450 reductase	Oxygen; NADPH; NADPH-cytochrome P-450 reductase.
Km for hematin	0.24 uM	0.40 uM
Substrate	Hematin, Fe-hematoporphyrin, Fe-hematoporphyrin acetate, Hb, cytochrome P-450b, cytochrome P-420b, cytochrome P-420c.	Hematin, Fe-hematoporphyrin, Fe-hematoporphyrin acetate, Hb, cytochromE P-450b, cytochrome P-420b, cytochrome P-420c.
Product of heme degradation	Biliveridin IX, carbon monoxide.	Biliveridin IX, carbon monoxide.
Inducers	Heme, heat, metal ions, oxidative stress, fever, glucocorticoids etc.	Not known.
Morphological distribution	Spleen, liver, etc.	Neuronal tissues including brain, carotid body. Non-neuronal tissues include testis and liver.

The relative distribution of HO-1 and HO-2 also varies with tissues. High levels of HO-1 activity are seen in tissues with a high degree of heme turnover such as spleen and liver, whereas HO-2 is the major heme oxygenase in brain and testis (Trakshel et al., 1986a, 1986b). Although most tissues contain low levels of HO-1 activity, it can be increased hundreds of times by a variety of stimuli as described above.

Recent studies using *in situ* hybridization histochemistry and immunocytochemistry demonstrated the distribution of HO-2 in discrete neuronal population (Ewing and Maines, 1992; Verma et al., 1993). High levels of the enzymes were colocalized to olfactory epithelium, neuronal and granular layers of the olfactory bulb, piriform cortex, pyramidal cell layer of the hippocampus, dentate gyrus, granule and Purkinje cell layers of the cerebellum, pontine nucleus, tenia tecta, and islands of Caljae. Verma et al. (1993) have further shown that delta-aminolevulinic acid synthase, the rate-limiting enzyme for the synthesis of heme, is also colocalized in the same areas.

A good deal of information is available on the regulation of HO-1 activity. The nucleotide sequence of the HO-1 gene contains four introns and five exons. The ability of various stimuli to induce increases in enzyme levels can be explained by the presence of multiple promoters in the HO-1 gene. In eukaryotic cells, prolonged exposure to heat activates a specific set of genes resulting in induction of so-called heat-shock proteins. It has been shown that HO-1 can be induced by heat (Shibahara et al., 1987) and the biochemical properties of HO-1 suggest that it is similar to hsp-32, which is a heat shock-protein (Lindquist and Craig, 1988). Regulation of HO-1 activity resembles in many respects the control of metalothionen (Mt). Both the HO-1 and Mt gene can be induced by essentially the same type of stimuli, which include glucocorticoids, metals, hormones, and interferons that have a heat shock element in the 5′ flanking region (Shibahara et al., 1987). Compared to HO-1, relatively little is known about the regulation of HO-2. Chromosomal localization of human heme oxygenases has been recently reported (Kutty et al., 1994). HO-1 located on chromosome 22 q 12, and HO-2 to chromosome 16 p 13.3.

Several metalloporphyrins function as potent inhibitors of heme oxygenases by acting as false substrates. These include, zinc, tin, strontium, and cobalt protoporphyrin-9 (Kappas et al., 1984; Maines, 1981). The effects of various inhibitors on heme oxygenase seem to vary in different organs. For instance, Sn-PP-9 inhibits enzyme activity in the brain, liver, and spleen (Anderson et al., 1984) and markedly reduces CO production (Landau et al., 1987); whereas it has no effect on intestinal enzyme activity (Kim et al., 1987). Metals such as cadmium inhibit heme oxygenase activity, but this inhibition is secondary to reduced cytochrome reductase activity (Trakshel et al., 1986a). On the other hand, copper, nickel, and magnesium protoporphyrin-9 have little or no effect on heme oxygenase activity (see Dinerman et al., 1994). Therefore, the metalloporphyrins offer valuable tools to examine the role of endogenous CO in physiological systems.

Several lines of evidence suggest that CO increases cGMP levels by activating the enzyme soluble guanalyte cyclase. Maines et al. (1993) reported that newborn rats treated with buthionine-SR-sulfoximine, a substance that selectively depletes glutathione, showed marked induction of HO-1 mRNA as well as increased enzyme activity. cGMP levels of these animals were unaffected, despite marked reductions in nitric oxide synthase activity. These observations led to the suggestion that cGMP levels are maintained by increased generation of CO resulting from enhanced HO-1 activity. Further, CO was shown to increase cGMP levels in vascular smooth muscle cells (Lin and McGrath, 1989; Ramos et al., 1989). Using the partially purified soluble guanylate cyclase from platelets, it was shown that CO-induced enzyme activity is due to binding of CO to the iron of the soluble guanalyte cyclase (Brune et al., 1990). These observations suggest that CO regulates cGMP, an action analogous to that proposed for NO (Moncada et al., 1991; Snyder, 1992).

Verma et al. (1993) were the first to suggest that CO may function as a neurotransmitter. Using primary cultures of olfactory neurons, these authors demonstrated that odorants such as 1-isobutyl-3-methoxypyrazine (IBMP) markedly enhance cGMP levels. The odorant-induced increase, as well as basal levels of cGMP, could be attenuated by zinc protoporphrin-9, a potent inhibitor of HO-2. Similar attenuation of the response was also seen with hemoglobin, which traps CO. These observations, taken together with the colocalization of HO-2 with guanalyte cyclase in olfactory neurons, provided compelling evidence that endogenous CO regulates cGMP levels in olfactory neurons. CO production in olfactory neurons has been recently measured using radioactive [^{14}C] (Dinerman et al., 1994). CO production could be attenuated by the HO inhibitor zinc protoporpyrin-9, and the resulting changes in CO production paralleled the level of cGMP. These observations further support the notion that endogenous CO indeed regulates cGMP levels in central neurons.

Further evidence that CO is a transmitter in the central nervous system came from electrophysiological studies using inhibitors of HO and exogenous application of CO. Neurons in the nucleus of the solitary tract (nTS) integrate various afferent inputs arising from the cardiorespiratory systems. Glutamate is the principle excitatory transmitter (Leone and Gordon, 1989; McKitrick and Calaresu, 1989; Meely et al., 1989) and its actions are mediated by iono- and metabotropic glutamergic receptors. Glutamergic metaboreceptors have been implicated in integration of vagal sensory inputs in nTS and mediate baroreceptor reflexes (Glaum and Miller, 1992, 1993a). The synaptic effects of metaboreceptor activation can be mimicked by cGMP and are associated with activation of guanylate cyclase (Lewis et al., 1991). Because the effects of NO are coupled to cGMP, it was thought that NO is the mediator of metaboreceptor activation in nTS. However, the effects of activation of metaboreceptors was not mimicked by NO-generating substances (Glaum and Miller, 1993), and the distribution of NO-synthase did not coincide with guanalyte cyclase localization in nTS (Lin and McGrath, 1989). On the other hand, the distribution of HO-2 did coincide with the localization of guanalyte cyclase (Verma et al., 1993). Using the isolated brain slice preparation from rats (Glaum and Miller, 1993b), examined the role of CO in metaboreceptor activation in nTS. These authors tested the effects of (1S,3R)-1-aminocyclopentane-1,3-dicarboxylic acid (1S,3R)-ACPD), an agonist for glutamergic metaboreceptors, on synaptic activity in the dorsomedial subdivision of the nTS. ZnPP-9, an inhibitor of HO-2, but not the nitric oxide synthase inhibitor L-w-nitroarginine, attenuated the monosynaptic excitatory postsynaptic currents (EPSCs) evoked by (1S,3R)-ACPD. Moreover, the effects of ZnPP-9 seem to be selective because a relatively inactive analog, CuPP-9 (Drummond and Kappas, 1981), failed to attenuate the effects of (1s,3R)-ACPD. The effects of metaboreceptor activation could be mimicked by 8-bromo-cGMP, a membrane permeant analog of cGMP. However, ZnPP-9 had no effect on other types of glutamergic receptor activation. These results suggested that multiple metaboglutamergic receptors exist in the dorsomedial nTS and one of them is coupled to CO.

Long-term potentiation (LTP) is a mechanism that underlies certain forms of learning and memory. LTP requires activation of presynaptic sites via a retrograde messenger released from the postsynaptic site. It has been established that NO is one of the retrograde messengers mediating LTP (Bohme et al., 1991; Haley et al., 1992; Schuman and Madision, 1991; Zhuo et al., 1993; Zorumski and Izumi, 1993). Evidence that CO can serve as an intracellular messenger in the brain prompted several investigators to test its role in LTP. Stevens and Wang (1993) reported the involvement CO in LTP. Using hippocampal slices from rats and mice, these authors found that ZnPP-9, an inhibitor of heme oxygenase, prevented the induction of LTP without affecting long-term depression (LTD), another form of synpatic plasticity associated with memory. Heme oxygenase inhibitors also abolished LTP that was already established. Independent studies by other investigators reported similar results (Ikegaya et al., 1994; Zorumski and Izumi, 1993), and further demonstrated that exogenous CO enhances LTP. These observations support the idea that CO is a retrograde messenger in hippocampal pyramidal cells. Hippocampal pyramidal neurons contain not ony HO-2 (Verma et al., 1993), but also the endothelial type of NO synthase (Dinerman et al., 1994). The fact that inhibitors of both NO and CO synthesis abolish LTP suggest that both these gases may function in a coordinated fashion. In this context it is interesting to note that NO synthase structurally resembles cytochrome P-450 reductase (Bredt et al., 1991) .

Endogenous CO seems to play important roles in the peripheral nervous system as well. Evidence that CO is synthesized in neuronal tissues and that it might serve as an intracellular messenger in the nervous system prompted us to investigate its significance in carotid body chemoreception (Prabhakar et al., 1995). In collaboration with Dinerman and Snyder, we examined the distribution of HO-2 in rat and cat carotid bodies. Intense HO-2 immunoreactivity was found in many cells of the carotid body (Figure 4). Staining was localized exclusively to the cytoplasm. The cells that stained for HO-2 were also found to be positive for tyrosine hydroxylase, suggesting that they are glomus (type 1) cells. No immunoreactivity was evident either in nerve fibers or in the supporting cells. To assess the functional significance of HO-2, we monitored the effects of ZnPP-9, an inhibitor of HO, on carotid body sensory discharge *in vitro*. As little as 0.3 μM augmented the sensory discharge with maximal excitations at 3 μM. These doses are similar to those that inhibit HO activity. On the other hand, comparable concentrations of CuPP-9, a protoporphyrin that has negligible inhibitory influence on HO (Drummond and Kappas, 1981), had no effect on the carotid body activity (Figure 5). Furthermore, the excitatory effects of ZnPP-9 could be reversed by CO (Figure 6). These observations suggest that glomus cells, the putative transducers of the hypoxic stimulus produce CO, and that endogenous CO is inhibitory to the carotid body. The inhibitory actions of CO are consistent with the recent studies of Lahiri et al. (1993) who reported that low doses of CO inhibit the carotid body activity *in vitro*. However, exogenous CO seems to have little effect on carotid body sensory activity in anaesthetized animals (Figures 1A and 1B); whereas in blood-free *in vitro* preparations, CO inhibits the carotid body activity (Lahiri et al., 1993) at certain doses, which suggests that the lack of sensory response in intact animals could in part be due to its binding to hemoglobin, making it less available to the carotid body.

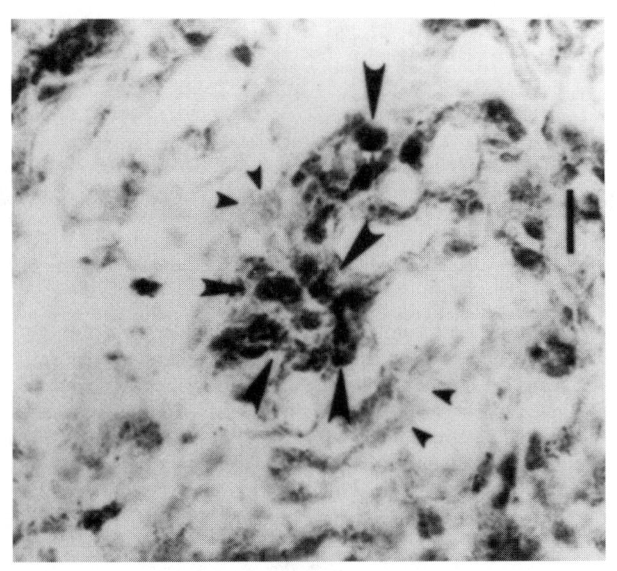

FIGURE 4
HO-2 immunoreactive cells in the cat carotid body. Glomeruli composed of individual glomus cells stain intensely for HO-2 (large arrows), whereas HO-2 immunoreactivity was not evident in supporting cells (small arrows). The bar length is 25 mm.

cGMP-like immunoreactivity has been demonstrated in type I and vascular endothelial cells of the carotid body (Wang et al., 1993). Given that HO-2 is also found in type I cells, one possibility is that the inhibitory actions of CO in the carotid body are coupled to the cGMP pathway. The fact that increases in cGMP levels are associated with inhibition of the carotid body (Wang et al., 1993), may support such a notion. Endogenous CO via cGMP mechanism can directly affect the systemic vascular resistence (Levere et al., 1990; Martasek, 1991; Sacerdoti et al., 1989). Therefore, another possibility is that CO, by dilating blood vessels in the carotid body, might improve tissue oxygenation resulting in inhibition of sensory discharge. It has been demonstrated that CO affects K^+ channel activity of glomus cells (Lopez-Lopez and Gonzalez, 1992). Alternatively, CO acting on nearby glomus cells in an autocrine or a paracrine fashion might influence the release of an excitatory transmitter necessary for carotid body excitation by interefering with K^+ channel activity.

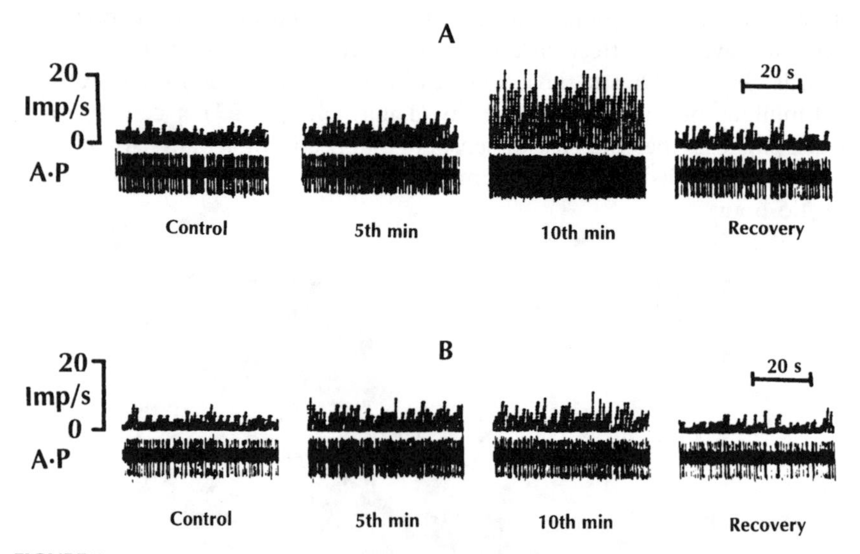

FIGURE 5
ZnPP-9 but not CuPP-9 increase carotid sinus nerve activity. (**A**) ZnPP-9 (3 μM) augmented the carotid body activity *in vitro*. The carotid body was perfused and superfused with Kreb's solution. Abbreviations are imp/s, integrated sensory discharge frequency impulses/s. A.P., Action potentials recorded from the carotid sinus nerve. Control, baseline activity during normoxia (pO_2 of the bathing medium = 140 mmHg; pCO_2 = 2 mmHg). 5th and 10th min, sensory activity at 5th and 10th min of perfusion with ZnPP-9 respectively. Recovery, sensory activity 15th min after terminating the perfusion with ZnPP-9. (**B**) Comparable doses of CuPP-9 (3 μM) failed to increase carotid body sensory activity . Abbreviations are the same as in A.

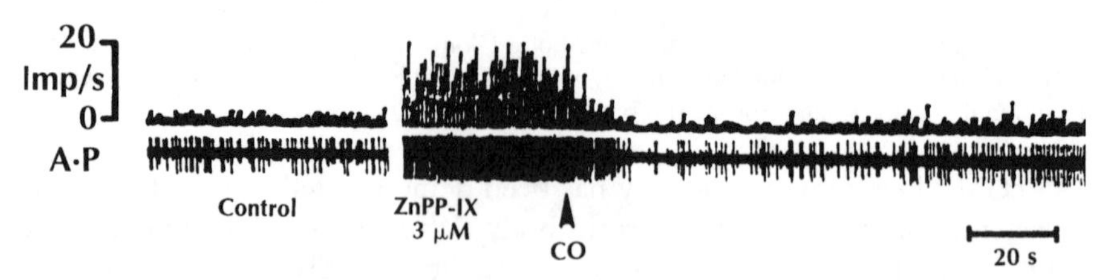

FIGURE 6
CO reverses ZnPP-9 induced augmentation of carotid body activity. Control, baseline sensory discharge during normoxia, ZnPP-9, sensory discharge during 10th min of perfusion with 3 μM of ZnPP-9. At arrow 0.3 ml of bathing medium bubbled with 100% CO (for 15 min) was given close to the carotid body. CO markedly attenuated the carotid excitation. Administration of same volume of medium without CO either had no effect or transiently diminished the activity for 10 s.

Nitric oxide synthase (NOS) has also been demonstrated in the carotid body (Prabhakar et al., 1993). NOS is found primarily in the nerve fibers innervating the chemoreceptor tissue that originate from the petrosal ganglion (Wang et al., 1993). Inhibitors of NOS activity augment carotid body activity, indicating that endogenous NO is an inhibitory chemical messenger as it is elsewhere in the peripheral nervous system. Interestingly, carotid body seems to synthesize both CO and NO albeit at different locations in the organ, and both seem to act as inhibitory chemical messengers. How might CO and NO contribute to oxygen chemoreception in the carotid body? It has been postulated that augmentation of carotid body sensory activity during hypoxia is due to decreased availability of an inhibitory neurotransmitter(s) (Kramer, 1978), whose identity remains elusive. The facts that the synthesis of both CO and NO has an absolute requirement for molecular oxygen, and that both are inhibitory to the sensory discharge led us to suggest that these gases may be the inhibitory neurotransmitters requisite for "disinhibition" (Prabhakar, 1994, 1995). Accordingly, during normoxia, constant production of CO and NO may maintain the carotid body activity at low levels. It follows that increased sensory discharge during hypoxia could be due to "disinhibition" resulting from decreased production of CO and NO. Such an idea seems partially justified by a recent demonstration that NOS activity in the carotid body could be inhibited by low levels of oxygen (Prabhakar et al., 1993).

Lloyd et al. (1968) postulated that a heme pigment is necessary for oxygen chemoreception at the carotid body. Because CO has high affinity for heme, they used it as a tool to test their hypothesis in awake human subjects. They observed that inhalation of CO (for few breaths) depressed ventilatory responses to hypoxia and attributed the attenuation of the response to carotid body inhibition. Accordingly, they concluded that heme-like pigment in the chemoreceptor tissue is essential for chemoreception. It is coincidental that CO, which was once used by respiratory physiologists as a probe to understand the oxygen chemoreception at the carotid body, has now been found to be produced by the chemoreceptor cells and may play an important role in transduction of the oxygen stimulus.

ACKNOWLEDGMENTS

Research in the authors' laboratories is supported by grants from National Institutes of Health (Heart, Lung and Blood Division) HL-45780; HL-52038(NRP) and a Research Career Development Award HL-02599 (NRP) and HL-50712 (RSF). NRP is grateful to Dr. Jeff Overholt for critical reading of the manuscript.

REFERENCES

Anderson, K.E., Simionatto, C.S., Drummond, G.S., and Kappas, A., Tissue distribution and disposition of tin-protoporphyrin, a potent competitive inhibitor of heme oxygenase, *J. Pharmacol. Exp. Therap.*, 228, 327, 1984.

Bohme, G.A., Bon, C., Stutzmann, J.M., Doble, A., and Blanchard, J.C., Possible involvement of nitric oxide in long-term potentiation, *Eur. J. Pharmacol.*, 199, 379, 1991.

Bredt, D.S., Hwang, P.H., Glatt, C., Lowenstein, C., Reed, R.R., and Snyder, S.H. Cloned and expressed nitric oxide synthase structurally resembles cytochrome P-450 reductase, *Nature,* 351, 714, 1991.

Brune, B., Schmidt, K., and Ullrich, V., Activation of soluble guanylate cyclase by carbon monoxide and inhibition by superoxide anion, *Eur. J. Biochem.,* 192, 683, 1990.

Dinerman, J.L., Dawson, T.M., Schell, M.J., Snowman, A., and Snyder, S.H., Endothelial nitric oxide synthase localized to hippocampal pyramidal cells: Implications for synaptic plasticity, *Proc. Natl. Acad. Sci. U.S.A.,* 91, 4214, 1994.

Dinerman, J.L., Prabhakar, N.R., and Snyder, S.H., Carbon monoxide: A neural messenger molecule, in *Pharmacology of Cerebral Ischemia,* Krieglstein, J. and Oberpichler-Schwenk, H., Eds., Wissenschfatliche Verlags Geselschaft mbH, Stuttgart, 1994, 1.

Docherty, J.C., Firneisz, G.D., and Schacter, B.A., Methene bridge carbon atom elimination in oxidative heme degradation catalyzed by heme oxygenase and NADPH-cytochrome P-450 reductase, *Arch. Biochem. Biophys.,* 235, 657, 1984.

Drummond, G.S. and Kappas, A., Prevention of neonatal hyperbilirubinemia by tin-protoporphyrin IX, a potent competitive inhibitor of heme oxidation, *Proc. Natl. Acad. Sci. U.S.A.,* 78, 6466, 1981.

Ewing, J.F. and Maines, M.D., In situ hybridization and immunohistochemical localization of heme oxygenase-2 mRNA and protein in normal rat brain: Differential distribution of isozyme 1 and 2 cell, *Mol. Cell Neurosci.,* 3, 559, 1992.

Fitzgerald, R.S. and Dehghani, G.A., Chemoreceptor control of organ vascular resistance during acute systemic hypoxia, in *Chemoreceptors and Chemoreceptor Reflexes,* Acker, H., Trzebski, A., and O'Regan, R.G., Eds., Plenum Press, New York, 1990, 217.

Fitzgerald, R.S., Dehghani, G.A., Sham, J.S.K., Shirahata, M., and Mitzner, W., Peripheral chemoreceptor modulation of the pulmonary vasculature in the cat, *J. Appl. Physiol.,* 73, 20, 1992.

Fitzgerald, R.S., Dehghani, G.A., Sham, J.S.K., and Shirahata, M., Pulmonary hypoxic vasoconstrictor response modulation by the peripheral arterial chemoreceptors, in *Response and Adaptation to Hypoxia, Organ to Organelle,* Lahiri, S., Cherniack, N.S., and Fitzgerald, R.S., Eds., Oxford University Press, New York, 1991, Chap. 19.

Glaum, S.R. and Miller, R.J., Metabolic glutamate receptors mediate excitatory transmission in the nucleus of the solitary tract, *J. Neurosci.,* 12, 2251, 1992.

Glaum, S.R. and Miller R.J., Zinc protoporphyrin-IX blocks the effects of metabotropic glutamate receptor activation in the rat nucleus tractus solitarii, *Mol. Pharmacol.,* 43, 965, 1993a.

Glaum, S.R. and Miller, R.J., Activation of metabotropic glutamate receptors produces reciprocal regulation of iontropic glutamate and GABA responses in the nucleus tractus solitarius of the rat, *J. Neurosci.,* 13, 1636, 1993b.

Haley, J.E., Wilcox, G.L., and Chapman P.F., The role of nitric oxide in hippocampal long-term potentiation, *Neuron,* 8, 211, 1992.

Ikegaya, Y., Saito, H., and Matsuki, N., Involvement of carbon monoxide in long-term potentiation in the dentata gyrus of anesthetized rats, *Jpn. J. Pharmacol.,* 64, 225, 1994.

Jornot, L. and Junod, A.F., Variable glutathione levels and expression of antioxidant enzymes in human endothelial cells, *Am. J. Physiol.,* 264, L482, 1993.

Kappas, A., Drummond, G.S., Simionatto, C.S., and Anderson, K.E., Control of heme oxygenase and plasma levels of bilirubin by a synthetic heme anaplasma levels of bilirubin by a synthetic heme analogue, tin-protoporphyrin, *Hepatology,* 4, 336, 1984.

Kim, C.B., Hintz, S.R., Vreman, H.J., Kwong, L.K., and Stevenson, D.K., Lack of intestinal heme oxygenase by antibiotics and tin protoporphyrin, *J. Pediatr. Gastroenterol. Nutr.,* 6, 302, 1987.

Kramer, E., Carotid body chemoreceptor function: Hypothesis based on a new circuit model, *Proc. Natl. Acad. Sci. U.S.A.,* 75, 2507, 1978.

Kutty, R.K. and Maines, M.D., Purification and characterization of bilveridin reductase from rat liver, *J. Biol. Chem.,* 256, 3956, 1981.

Kutty, R.K. and Maines, M.D., Oxidation of heme c derivatives by purified heme oxygenase. Evidence for the presence of one molecular species of heme oxygenase in rat liver, *J. Biol. Chem.*, 257, 9944, 1989.

Kutty, R.K., Kutty G., Rodriguez, I.R., Chader, G.J., and Wiggert, B., Chromosomal localization of the human heme oxygenase genes: Heme oxygenase -1 (HMOX-1) maps to chromosome 22q12 and heme oxygenase-2 (HMOX-2) maps to chromosome 16p13.3, *Genomics,* 20, 513, 1994.

Lahiri, S., Iturriaga, R., Mokashi, A., Ray, D.K., and Chug, D., CO reveals dual mechanism of O_2 chemoreception in the carotid body, *Resp. Physiol.*, 94, 227, 1993.

Landau, S.A., Sassa, S., Drummond, G.S., and Kappas, A., Proof that Sn-protoporphyrin inhibits the enzymatic catabolism of heme *in vivo*, *J. Exp. Med.*, 165, 1195, 1987.

Leone, C. and Gordon, F.J., Is L-glutamate a neurotransmitter of baroreceptor information in the nucleus of the tractus solitarius? *J. Pharmacol. Exp. Ther.*, 250, 953, 1989.

Levere, R.D., Martasek, P., Escalante, B., Schwartzman, M.L., and Abraham, N.G., Effect of heme arginate administration on blood pressure in spontaneously hypertensive rats, *J. Clin. Invest.*, 86, 213, 1990.

Lewis, S.L., Machado, B.H., Ohta H., and Talman W.T., Processing of cardiopulmonary afferent input with the nucleus tractus solitarii involves activation of soluble guanylate cyclase. *Eur. J. Pharmacol.*, 203, 327, 1991.

Lin, H. and McGrath, J.J., Carbon monoxide effects on calcium levels in vascular smooth muscle, *Biochem. Pharmacol.*, 38, 1368, 1989.

Lindquist, S. and Craig, E.A., The heat shock proteins, *Annu. Rev. Genet.*, 22, 631, 1988.

Lloyd, B.B., Cunningham, D.J.C., and Goode, R. C., Depression of hypoxic hyperventilation in man by sudden inspiration of carbon monoxide, in *Arterial Chemoreceptors,* Torrance, R. W., Ed., Blackwell Scientific, Oxford, 1968, 145.

Lopez-Lopez, J.R. and Gonzalez, C., Time course of K+ current inhibition by low oxygen in chemoreceptor cells of adult rabbit carotid body. Effects of carbon monoxide, *FEBS Lett.*, 299, 251, 1992.

Maines, M.D., Heme oxygenase: Function, multiplicity, regulatory mechanisms, and clinical applications, *FASEB J.*, 2, 2257, 1988.

Maines, M.D., Zinc-protoporphyrin is a selective inhibitor of heme oxygenase activity in the neonatal rat, *Biochem. Biophys. Acta,* 673, 339, 1981.

Maines, M.D. and Kappas, A., Cobalt induction of hepatic heme oxygenase; with evidence that cytochrome P-450 is not essential for this enzyme activity, *Proc. Natl. Acad. Sci. U.S.A.*, 71, 4293, 1974.

Maines, M.D., Mark, J.A., and Ewing, J.F., Heme oxygenase, a likely regulator of cGMP production in the brain: Induction *in vivo* of HO-1 compensates for depression in NO synthase activity, *Mol. Cell. Neurosci.*, 4, 398, 1993.

Martasek, P., Schwartzman, M.L., Goodman, A.I., Solingi, K.B., Levere, R.D., and Abraham, N.G., Hemin and L-arginine regulation of blood pressure in spontaneously hypertensive rats, *J. Am. Soc. Nephrol.*, 2, 1078, 1991.

McKitrick, D.J. and Calaresu, F.R., Baroreceptor activation of glutamate coinjection facilitates depressor responses to ANF microinjected into NTS, *Am. J. Physiol.*, 257, R405, 1989.

Meely, M.P., Underwood, M.D., Talman, W.T., and Reis, D.J., Content and *in vitro* release of endogenous amino acids in the area of the nucleus of the solitary tract of the rat, *J. Neurochem.*, 53, 1807, 1989.

Moncada, S., Palmer, R.M.J., and Higgs, E.A., Nitric oxide: Physiology, pathophysiology and pharmacology, *Pharmacol. Rev.*, 43, 109, 1991.

Montgomery, M.R. and Rubin, R.J., Oxygenation during inhibition of drug metabolism by carbon monoxide or hypoxic hypoxia, *J. Appl. Physiol.*, 35, 505, 1973.

Muller, R.M., Taguchi, H., and Shibahara, S., Nucleotide sequence and organization of the rat heme oxygenase gene, *J. Biol. Chem.*, 262, 7695, 1987.

Prabhakar, N.R., Gases as chemical Messengers in the carotid body: Role of nitric oxide and carbon monoxide in chemoreception, in *Control of Breathing: Modelling,* Adams, L.A. and Semple, S.J.C., Eds., Plenum Press, New York, 1995, chap. 58.

Prabhakar, N.R., Neurotransmitters in the carotid body, in *Arterial chemoreceptors: Cell to System,* O'Regan, R. et al., Eds., Plenum Press, New York, 1994, 57.

Prabhakar, N.R., Dinerman, J.L., Agani, F.A., and Snyder, S.H., Carbon monoxide: A role in carotid body chemoreception, *Proc. Natl. Acad. Sci. U.S.A.,* 92, 1994, 1995.

Prabhakar, N.R., Kumar, G.K., Chang, C.H., Agani, F.H., and Haxhiu, M.A., Nitric oxide in the sensory function of the carotid body, *Brain Res.,* 625, 16, 1993.

Ramos, K.S., Lin, H., and McGrath, J.J., Modulation of cyclic guanosine monophosphate levels in cultured aortic smooth muscle cells by carbon monoxide, *Biochem. Pharmacol.,* 38, 1368, 1989.

Rotenberg, M.O. and Maines, M.D., Isolation, characterization and expression in *Escherichia coli* of a cDNA encoding rat heme-oxygenase-2, *J. Biol. Chem.,* 263, 3348, 1988.

Sacerdoti, D., Escalante, B., Abraham, N.G., McGiff, J.C., Levere, R.D., and Schwartzman, M.L., Treatment with tin prevents the development of hypertension in spontaneously hypertensive rats, *Science,* 243, 388, 1989.

Schuman, E.M. and Madision, D.V., The intercellular messenger nitric oxide is required for long-term potentiation, *Science,* 254, 1503, 1991.

Shibahara, S., Muller, R., and Taguchi, M., Transcriptional control of rat heme oxygenase, *J. Biol. Chem.,* 262, 3348, 1987.

Sjorstrand, T., Endogenous formation of carbon monoxide in man under normal and pathological conditions, *J. Clin. Lab. Invest.,* 1, 201, 1949.

Snyder, S.H., Nitric oxide: First in a class of neurotransmitters, *Science,* 257, 494, 1992.

Stevens, C.F. and Wang, Y., Reversal of long-term potentiation by inhibitors of heme oxygenase, *Nature,* 364, 147, 1993.

Trakshel, G.M., Kutty, R.K., and Maines, M.D., Cadmium-mediated inhibition of testicular heme oxygenase activity: The role of NADPH-cytochrome (p-450) reductase, *Arch. Biochem. Biophys.,* 251, 175, 1986a.

Trakshel, G.M., Kutty, R.K., and Maines, M.D., Resolution of the rat brain heme oxygenase activity: Absence of a detectable amount of the inducible form (HO-1), *Arch. Biochem. Biophys.,* 260, 732, 1986b.

Veltman, J.C. and Maines, M.D., Sex difference in adrenal heme and cytochrome P-450 metabolism: Evidence for the repressive regulatory role of testosterone, *J. Pharmacol. Exp. Ther.,* 235, 71, 1985.

Verma, A., Hirsch, D.J., Glatt, C.E., Ronnett, G.V., and Snyder, S.H., Carbon monoxide: A putative neural messenger, *Science,* 259, 381, 1993.

Wang, W.-J., He, L., Chen, J., Dinger B., and Fidone S.J., Mechanisms underlying chemoreceptor inhibition induced by atrial natriuretic peptide in rabbit carotid body, *J. Physiol.,* 460, 427, 1993.

Wang, Z.-Z., Bredt, D.S., Fidone, S.J., and Stensas, L.J., Neurons synthesizing nitric oxide innervating the carotid body, *J. Comp. Neurol.,* 336, 419, 1993.

Yoshinaga, T., Sassa, S., and Kappas, A., A comparative study of heme degradation by NADPH-cytochrome c reductase alone and by the complete heme oxygenase system, *J. Biol. Chem.,* 257, 7794, 1982.

Zhuo, M., Small, S.A., Kandel, E.R., and Hawkins, R.D., Nitric oxide and carbon monoxide produce activity-dependent long-term synaptic enhancement in hippocampus, *Science,* 260, 1946, 1993.

Zorumski, C.F. and Izumi, Y., Nitric oxide and hippocampal synaptic plasticity, *Biochem. Pharmacol.,* 46, 777, 1993.

CHAPTER 8

Toxicity of Carbon Monoxide: Hemoglobin vs. Histotoxic Mechanisms

Claude A. Piantadosi

CONTENTS

0-8493-4796-3/96/$0.00+$.50

1 INTRODUCTION

The toxic mechanisms of action of carbon monoxide (CO) have held great scientific interest since Bernard (1857) first recognized that the gas produced tissue hypoxia by reversible combination with hemoglobin. Some years later, J.S. Haldane (1895) demonstrated that the interaction of CO with hemoglobin was antagonized by high oxygen partial pressures. Haldane also found that lethal CO poisoning in mice could be prevented with large quantities of dissolved oxygen at hyperbaric pressure. The physiological principles derived from early animal studies about the mechanism of toxicity of CO were simple: CO produced tissue hypoxia by binding tightly to hemoglobin and preventing it from transporting oxygen and releasing it to the tissues. Hence, tissues with high continuous requirements for O_2, such as the brain and the heart, would be most susceptible to the toxic effects of CO. These principles have served as the cornerstone of our understanding of the toxicity of the gas, and represent a longstanding paradigm about mechanisms of CO poisoning (Stewart, 1975) and its treatment (Kindwall, 1977). Certain aspects of CO poisoning, however, cannot be reconciled with simple cellular hypoxia derived from the presence of carboxyhemoglobin (COHb) (Piantadosi, 1987).

The failure of COHb to account for all of the pathophysiological features of CO exposure is illustrated by several interesting observations. These observations include poor correlation between neurological injury and COHb level (Savolainen et al., 1980), direct vasodilator effects of CO (Ramos et al., 1989), oxidation of CO to CO_2 in the body and by mitochondria (Young and Caughey, 1986), physiological effects of CO at low COHb levels (Horvath et al., 1975), and "remnant" effects of CO after COHb has been cleared from the circulation (Halperin et al., 1959). These curious but consistent findings have produced substantial efforts to understand mechanisms of intracellular CO toxicity (Coburn, 1979; Coburn and Forman, 1987) and have been used to justify hyperbaric oxygen therapy when blood COHb levels have returned to normal. More recently, it has become clear that intracellular uptake of the gas can alter neuronal energy provision *in vivo* (Brown and Piantadosi, 1992), and reoxygenation after CO hypoxia is associated with oxidative stress in animal models (Thom, 1990; Zhang and Piantadosi, 1992). Somewhat surprisingly, biochemical evidence also indicates that oxidative stress after CO hypoxia can be diminished by treatment with high inspired O_2 concentrations (Thom and Elbuken, 1991).

This chapter summarizes our present understanding of how CO interacts with hemoproteins in the body, beginning with the well-studied interactions of CO with hemoglobin and their effects on oxygen transport to tissue. This review will draw upon the results of a variety of experimental approaches in the literature to explain current concepts and controversies about uptake of the gas by intracellular hemoproteins and their potential roles in histotoxicity produced by CO poisoning. I will surmise from the literature that the toxic effects of CO result from the combination of tissue hypoxia due to COHb formation, and intracellular mechanisms of toxicity that arise during cellular hypoxia. Cellular hypoxia may be sufficient to promote significant CO uptake by tissues at COHb levels as low as 20 to 40%, and it is clearly sufficient to support these mechanisms at COHb levels equal to or greater than 50%.

2 EFFECTS OF CARBON MONOXIDE (CO) ON HEMOGLOBIN AND OXYGEN

2.1 Effects on Hemoglobin

Hemoglobin avidly binds CO at the iron (II) center of the heme moiety of the molecule. The heme-CO binding effect is competitive with O_2 binding, and the chemical affinity of hemoglobin for CO, known as the M value or Haldane constant, is 220 to 240 times that of molecular oxygen (Allen and Root, 1957; Coburn and Forman, 1987). Under steady state conditions, the ratio of COHb to O_2Hb is proportional to the ratio of their partial pressures. The proportional relationship between the affinity constant M and the partial pressures of CO and O_2 at saturation is described by the Haldane expression:

$$COHb/O_2Hb = M \cdot (PCO/PO_2)$$

Therefore, CO occupies the O_2-binding sites of hemoglobin at very low partial pressures of the gas, and because the rate of release of CO from heme is slow, the oxygen-carrying capacity of hemoglobin is diminished by an amount directly related to the concentration of COHb.

The oxygen-carrying capacity of hemoglobin, together with the hemoglobin concentration and its oxygen saturation, determine the oxygen content of blood. Hence, the arterial oxygen content (CaO_2) is diminished by the presence of COHb. CO binding to hemoglobin also alters the normal relationship between O_2Hb and PO_2, shifting the oxyhemoglobin dissociation curve to the left. The binding of CO to hemoglobin changes the shape of the sigmoidal oxyhemoglobin dissociation curve to that of a rectangular hyperbola (Figure 1). The shift of the oxyhemoglobin dissociation curve to the left interferes with release to the tissues of O_2, which remains bound to the hemoglobin molecule.

2.2 Effects on Oxygen Transport

Oxygen transport from the lungs to the body tissues is accomplished by two physical processes: convective or bulk O_2 transport from lungs to the

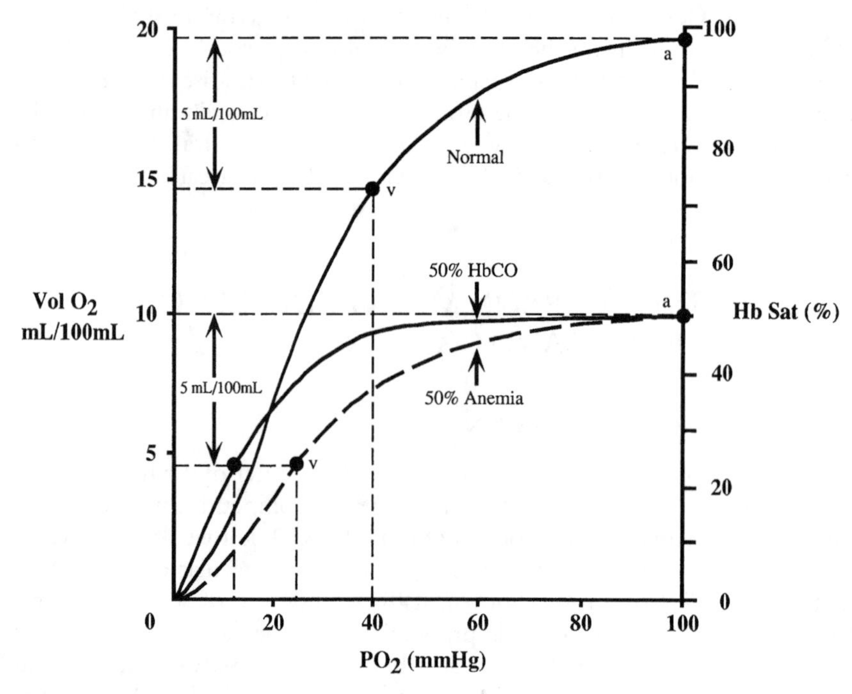

FIGURE 1
Interaction of CO with hemoglobin resulting in decreased O_2 carrying capacity and shift of the oxyhemoglobin dissociation curve to the left. a = arterial, v = venous values of PO_2, hemoglobin (Hb) saturation, and O_2 content.

capillaries, and diffusive O_2 transport from the capillaries to mitochondria. The effects of CO on hemoglobin produces a major effect on each of these two processes (Allen and Root, 1957). The convective O_2 transport is determined by the product of blood flow times CaO_2. Because the CaO_2 is decreased by COHb, this results in an anemia-like effect without the rheological benefit of decreased blood viscosity that is offered by anemia (Seisjo, 1978). The decrease in arterial oxygen content causes venous PO_2 to fall if tissue oxygen requirements remain constant and blood flow does not increase. The shift of the oxyhemoglobin dissociation curve to the left in the presence of COHb results in further lowering of venous PO_2 for any given arteriovenous content (A-VO_2) difference (Forster, 1970). This decreases the PO_2 available to the tissues by diffusion. The extent to which tissue PO_2 is decreased by COHb also may be much greater than that estimated by venous PO_2 (PvO_2) depending on the extent of AV shunting through the organ.

Although PvO_2 serves as an estimate of mean tissue oxygen tension (PtO_2) under normal conditions, the measurement is difficult to interpret in pathophysiological states (Tenney, 1977). Venous PO_2, and hence venous oxygen content (CvO_2), are weighted means of venous drainage from regions of variable oxygen demand, blood flow, and capillary density. If areas of the tissue that have low oxygen extraction ratios receive a high fractional blood flow, PvO_2 or CvO_2 can be maintained in the face of tissue oxygen insufficiency in other regions. As a result, despite "adequate" PvO_2, there may be inadequate concentrations of O_2 available to meet metabolic needs in specific tissues or regions of a tissue. Oxygen supply/demand mismatching could help explain regional differences in sensitivity to CO poisoning (e.g., within the brain) by enhancing tissue binding of CO in more hypoxia-sensitive areas. In experimental CO hypoxia *in vivo*, decreases in PvO_2 have been measured in the brain in response to toxic exposures (Doblar et al., 1977; Koehler et al., 1977; Paulson et al., 1973; Traystman et al., 1981). A fall in tissue PO_2 also has been measured using polarographic oxygen microelectrodes (Zorn, 1972). Increases in cerebral blood flow (CBF) in response to CO hypoxia and changes in cerebral tissue contents of phosphocreatine, adenine nucleotides, lactate, pyruvate, and energy charge suggest that significant tissue hypoxia occurs at COHb values of 20% or greater (MacMillan, 1975a, 1975b).

Several experimental studies have concluded that CO effects are due only to CO binding by hemoglobin resulting in decreased arterial oxygen content and shift of the oxyhemoglobin dissociation curve to the left (Doblar et al., 1977; Koehler et al., 1977; Paulson et al., 1973; Traystman et al., 1981) without advantageous reduction in blood viscosity (Paulson et al., 1973; Seisjo, 1978). This conclusion is based upon "equivalent" decreases in PvO_2 by CO hypoxia, hypoxic hypoxia, or anemia (Roth and Rubin, 1976a). It ignores differences in chemoreceptor and local microcirculatory responses to different ways of reducing tissue oxygen delivery and most importantly, differences in the recovery of oxidative metabolism post-hypoxia. Also, the oxyhemoglobin dissociation curve shift as a factor influencing tissue PO_2 is mitigated by the control of PtO_2 by local microcirculatory factors (Coburn, 1979). These considerations in CO hypoxia, along with difficulties explaining CO toxicity on the basis of COHb-related decreases in PtO_2 (Aronow and Isbell, 1973; Halperin et al., 1959; Horvath et al., 1975; Winston and Roberts, 1975), have left unanswered many important questions about the intracellular effects of the gas.

The earliest experimental study to suggest a role for intracellular uptake of the gas in CO poisoning were those of J.B.S. Haldane (1927). Later studies by Drabkin et al. (1943) and Goldbaum et al. (1975, 1977) found that increasing COHb by noninhalational routes did not produce the same lethal consequences as inhalational exposure to CO. Although the latter studies have been criticized because of flaws in experimental design, these investigators demonstrated that very high COHb levels were tolerated remarkably well when COHb was infused into the circulation, and suggested that the uptake of CO intracellularly during inhalation of the gas could contribute to its toxicity.

3 INTRACELLULAR EFFECTS OF CO

The high affinity of hemoglobin for CO and its relatively rapid equilibration with it have been used to suggest that uptake of CO by tissues contributes relatively little to the manifestations of CO poisoning. When considering the validity of this proposition, it is worth noting that an endogenous CO body store exists outside the vascular space, and biochemical mechanisms are available to manage it.

3.1 Endogenous Production of CO by the Body

When hemoglobin is degraded to bile pigments in the body, the a-carbon atom is separated from the porphyrin ring and converted by the microsomal enzyme, heme oxygenase, into CO. Heme oxygenase exists in at least two isoferous (HO-1 and HO-2) which are distributed differently in tissues. The major site of heme breakdown and thus the major organ producing CO endogenously is the liver (Berk et al., 1974, 1976). The spleen and the erythropoietic system are other important endogenous sources of CO. Heme oxygenase is also present within the brain (Maines, 1992). Catabolism of other hemoproteins such as myoglobin, cytochromes, and peroxidases contribute approximately one fourth of the total amount of CO produced by the body (Berk et al., 1976). Metabolic processes other than heme catabolism, e.g., lipid peroxidation, contribute only a very small amount of CO (Miyahara and Takahashi, 1971).

Processes that increase the destruction of the red cell and accelerate breakdown of other hemoproteins lead to increased production of CO. Hematomas, intravascular hemolysis, blood transfusions, and disordered erythropoiesis will increase the COHb level. Degradation of red blood cells under pathologic conditions such as hemolytic and congenital anemias and other blood disorders also will accelerate CO production (Berk et al., 1974; Solanki et al., 1988). Heme oxygenase activity can be induced by heme and by non-heme products including some hormones (Otterbein et al., 1995). In patients with hemolysis, CO production rates and COHb concentrations are several times higher than in healthy people (Coburn et al., 1966). Increased CO-production rates also have been reported after administration of certain drugs such as phenobarbital, diphenylhydantion (Coburn, 1970) and progesterone (Delivoria-Papadopoulos et al., 1970).

3.2 Intracellular Effects of Exogenous CO Exposure

The argument that hemoglobin acts as a "CO buffer" that prevents CO uptake by tissue would appear to be reasonable based on steady-state assumptions. Physiological data indicating uptake of the gas by tissue — e.g., muscle and brain — and the rapid shift of CO from intravascular to extravascular compartments during hypoxia, however, argue that kinetic factors are important in the distribution of the CO body burden during CO poisoning. That some aspects of CO toxicity are related to intracellular CO uptake is logical because other heme- and copper-containing proteins bind CO *in vitro*. Myoglobin (Mb)(Coburn and Mayers, 1971), cytochrome c oxidase (Keilin and Hartree, 1939), tryptophan dioxygenase and catalase (Forman and Feigelson, 1971), and cytochrome P450 (Omura et al., 1965) all bind sufficient CO *in vitro* to inhibit functions. Intracellular effects depend on tissue pressures of both CO and oxygen, since hemoproteins bind the gases competitively. This is the CO/O_2 ratio of the Warburg partition coefficient:

$$K = (n/1\text{-}n)\,(CO/O_2)$$

where n, the fraction of a compound bound to CO, is equal to 0.5. This gives K (ratio of CO to oxygen) for half-saturation with CO. CO/O_2 ratios have been determined for many hemoproteins from *in vitro* experiments, but these values are difficult to extrapolate to intracellular conditions *in vivo*. Intracellular PCO may be estimated in several ways, but PO_2 is difficult to measure and subject to variable tissue gradients. Intracellular PO_2 values measured by polarographic microelectrodes are limited by local tissue disturbance and consumption of O_2 by the electrode (Albanese, 1973). Estimates of tissue PCO obtained by measuring PCO in peritoneal cavity gas pockets created in small mammals provide values near 70% of the PCO in alveolar gas, but direct PCO measurements in tissue are not available (Gothert and Malorney, 1970). In view of these constraints, attempts to calculate CO/O_2 ratio in tissue are plagued with problems, however, *in vitro* CO/O_2 ratios can be used to prioritize candidate hemoproteins for probable CO effects. Approximate Warburg partition coefficients for three well-known candidate hemoproteins are 0.4 for myoglobin (Coburn and Forman, 1987), 1.0 for cytochrome P450 (Omura et al., 1965), and 5–15 for cytochrome a,a_3 (Coburn and Forman, 1987; Wohlrab and Ogunmola, 1971). On this basis alone, myoglobin would be the best site for intracellular CO uptake.

3.3 CO Binding by Myoglobin

The sarcoplasmic protein myoglobin is involved in O_2 transport from capillaries to mitochondria in cardiac and red skeletal muscles. Binding of CO by myoglobin has been demonstrated by extensive studies *in vitro* and *in vivo* (Clark and Coburn, 1975; Coburn, 1979; Coburn and Mayers, 1971; Coburn et al., 1973). CO binding to Mb in cardiac and skeletal muscle *in vivo* has been demonstrated at levels of COHb below 2% in heart and 1% in skeletal muscle (Coburn et al., 1973). The COMb/COHb ratio in canine heart muscle is near unity over a wide range of COHb values. In the presence of hypoxemia or hypoperfusion, the amount of CO uptake by myoglobin has been measured and shown to increase. A similar conclusion has been reached in humans during near maximal exercise where CO shifts from the intravascular to tissue compartments (Clark and Coburn, 1975).

The physiological significance of CO uptake by myoglobin is uncertain partly due to our incomplete understanding of its functional role in working muscle. At the partial pressures of oxygen found in working heart (2–5 torr), myoglobin is partially deoxygenated (Gayeski and Honig, 1991). Functional myoglobin facilitates O_2 diffusion from the sarcoplasm to the mitochondria so that the continuous O_2 demand of working muscle can be satisfied (Wittenberg et al., 1975; Wittenberg and Wittenberg, 1989). Myoglobin also may serve muscle function as an O_2 store, or by acting as an O_2 buffer to maintain uniform mitochondrial PO_2 during changes in O_2 supply. Functional myoglobin is necessary for maintenance of maximum O_2 uptake and mechanical tension in exercising skeletal muscle (Cole, 1982), and it facilitates oxidative phosphorylation (Wittenberg and Wittenberg, 1989). CO binding to Mb therefore would be expected to limit O_2 availability to mitochondria in working muscle. This possibility has been explored in model simulations by Hoofd and Kreuzer (1978) and Agostoni et al. (1980). The three-compartment (arterial and venous capillary blood, and Mb) computer model of Agostoni et al. (1980) predicted that COMb formation in low PO_2 regions of the heart (e.g., subendocardium) could be sufficient to impair intracellular O_2 transport to mitochondria at COHb saturations of 5 to 10%. The concentration of COMb also was predicted to increase during conditions of hypoxia, ischemia, and increased O_2 demand. It remains unknown, however, whether or not increases in COMb concentration in muscle can account for decreases in maximal O_2 uptake during exercise reported at COHb levels of 4 to 5%.

Some recent observations on the effects of CO on isolated cardiac myocytes are pertinent to the cellular mechanisms of toxicity in the heart. In cardiac myocytes, when carboxymyoglobin reaches 40% of the total cellular myoglobin pool, the rate of energy provision by oxidative phosphorylation, estimated from the ratio of phosphocreatine to ATP, has been reported to decrease significantly (Wittenberg and Wittenberg, 1993). This effect was found to occur independently of CO binding to cytochrome c oxidase, and was tested at physiologically relevant PO_2 values (≤5 torr). The authors predicted from their data that myoglobin-dependent oxidative phosphorylation would be affected when 20 to 40% of the arterial hemoglobin was in the form of COHb.

3.4 CO Binding by Cytochrome P450

Mixed-function oxidases (cytochrome P450) are involved in the metabolism of steroids and many drugs. Isoforms of these enzymes are distributed widely throughout mammalian tissues; the highest concentrations are found in the microsomes of liver, adrenal gland, and the lungs of some species (Estabrook et al., 1970). These oxidases also are present in low concentrations in kidney and brain tissues. Mixed-function oxidases catalyze hydroxylation reactions involving the uptake of a pair of electrons from NADPH with reduction of one atom of O_2 to H_2O and incorporation of the other into the substrate (White and Coon, 1980). These enzymes bind CO, and their Warburg binding coefficients range from 0.1 to 12 *in vitro* (see Coburn and Forman, 1987). The sensitivity of cytochrome P450 to CO is increased under conditions of rapid electron transport (Estabrook et al., 1970); however, calculations suggest that tissue PCO is too low to inhibit the function of these hemoproteins *in vivo* at less than 15 to 20% COHb (Coburn and Forman, 1987).

CO-binding coefficients for mixed function oxidases have been measured in several intact tissues. In isolated rabbit lung, the effects of CO on mixed-function oxidase are consistent with a Warburg coefficient of approximately 0.5 (Fisher and Dodi, 1981). In experimental animals *in vivo*, CO exposure decreases the rate of hepatic metabolism of hexobarbital and other drugs (Montgomery and Rubin, 1973; Roth and Rubin, 1976a, 1976b). These effects of CO on xenobiotic metabolism have been attributed to COHb-related tissue hypoxia because they are no greater than the effects of "equivalent" levels of hypoxic hypoxia. In studies of rat liver perfused *in situ* with hemoglobin-free buffers, optical measurements have demonstrated uptake of CO by cytochrome P450 systems at CO/O_2 ratios of 0.03 to 0.10 (Iyanagi et al., 1981; Sies and Brauser, 1970; Takano et al., 1985). In one study, significant inhibition of hexobarbital metabolism was found at a CO/O_2 of about 0.1 (Takano et al., 1985). This CO/O_2 ratio, if translated directly to COHb, would lead to COHb levels incompatible with life ($\approx$95%). In brief, scientific evidence is lacking for significant inhibition of the activity of cytochrome P450 systems *in vivo*, at COHb values of 20% or below, although further work is needed in this area.

3.5 CO Binding by Cytochrome c Oxidase

Cytochrome c oxidase, or cytochrome a,a_3, is the terminal enzyme in the mitochondrial electron transport chain that catalyzes the reduction of molecular O_2 to water. The effects of CO on mitochondrial respiration are important because of the fundamental relationship of the respiratory chain to cellular bioenergetics (Chance and Williams, 1956) (Figure 2). The Warburg constant for cytochrome oxidase is unfavorable for CO binding relative to the other hemoproteins mentioned so far, and the enzyme has an *in vitro* Michaelis-Menten constant (Km) for O_2 of less than 1 torr (Chance and Williams, 1956). Since intracellular PO_2 is normally greater than this, the oxidase should remain oxidized until significant impairment of O_2 delivery is present. Inhibition of cytochrome a,a_3 activity requires not only high CO/O_2 ratios, but only reduced cytochrome a_3 binds CO (Keilin and Hartree, 1939).

It has been known for many years, primarily through the work of Fenn (Fenn, 1970; Fenn and Cobb, 1932), that CO is slowly oxidized in the body

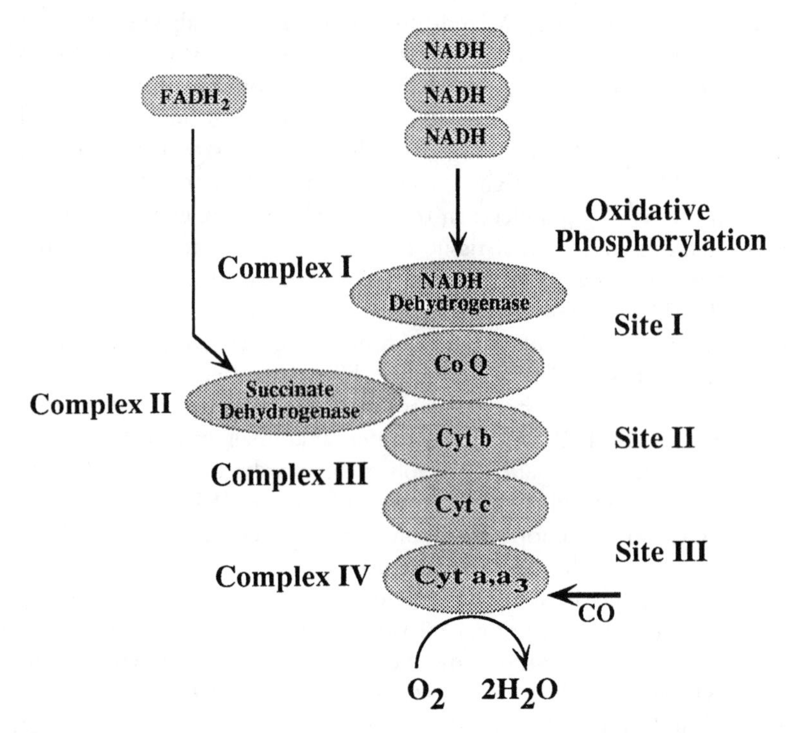

FIGURE 2
Schematic diagram of mitochondrial electron transport system. Electrons from the tricarboxylic acid cycle enter the system via NADH at Complex I and via FADH2 at Complex II. Molecular O_2 is reduced to water in a four-electron reaction by cytochrome *a,a*$_3$ in Complex IV. Adenosine triphosphate (ATP) is coupled to electron transport at three sites in the process of oxidative phosphorylation. CO inhibits respiration at Complex IV by binding to reduced cytochrome *a,a*$_3$. CoQ = Coenzyme Q, Cyt = cytochrome.

to CO_2. This process normally occurs at a much lower rate than the rate of endogenous CO production; however, the rate of CO oxidation increases in relation to the CO body burden (Luomanmaki and Coburn, 1969). That conversion of CO to CO_2 could be catalyzed by reduced cytochrome c oxidase was shown by Tzagoloff and Wharton (1965). Subsequently, Young et al. (1979) demonstrated that oxidized cytochrome oxidase promotes CO oxygenation, and that cytochrome oxidase in intact heart and brain mitochondria catalyzed the reaction at a CO/O_2 ratio of approximately 4 (Young and Caughey, 1986). Young and Caughey also have noted that the chemical form of the oxidase in the presence of CO will be determined by the CO/O_2 ratio (Young and Caughey, 1990). High CO/O_2 ratios favor the ferrous carbonyl form, while at intermediate CO/O_2 ratios, the CO-oxygenating form will be present, and at low CO/O_2 ratios the oxidized species will occur. In any event, electron flow ceases after CO binds to the oxidase whether or not CO oxygenation occurs (Young and Caughey, 1990).

Cytochrome a,a_3 in isolated mitochondria under resting conditions (State 4; low ADP concentrations) remains oxidized until oxygen concentration falls below 10^{-6} M and its reduction level only increases to 4% during rapid respiration (State 3) (Chance and Williams, 1956). At high turnover rates, the reduction level of the enzyme may approach 20% *in vitro* (Oshino et al., 1974). *In vivo*, the existence of such conditions would not favor CO interaction with cytochrome a,a_3. Chance et al. (1970), in studies of isolated pigeon heart mitochondria, found that CO/O_2 ratios as low as 0.2 markedly delayed the spectroscopic transient from anoxia to normoxia in uncoupled mitochondria. Other studies of respiring tissues, however, reported that CO/O_2 ratios of 12 to 20 were necessary for 50% inhibition of O_2 uptake (see Coburn and Forman, 1987). In this context, it is important to note that in a cell, the CO/O_2 ratio required to inhibit half of the O_2 uptake may require CO binding to more than one half of the oxidase molecules. This is because functional oxidase molecules may oxidize nearby respiratory complexes of inhibited oxidase molecules, thus causing the O_2 consumption to fall more slowly than predicted for a linear respiratory chain. The capacity of tissues to compensate for electron transport inhibition by this mechanism has not been investigated as a function of PO_2, cytosolic phosphorylation potential, or rate of electron transport during CO hypoxia *in vivo*.

The brain is a good organ to investigate the mitochondrial effects of CO because significant amounts of other CO-binding pigments, such as myoglobin, are absent. Tissues with high rates of oxygen utilization like the brain may have intracellular PO_2 gradients steep enough for mitochondrial CO uptake (Jones and Kennedy, 1982). Steep PO_2 gradients may explain why cerebral cytochrome a,a_3 *in vivo* is partly reduced under nonpathological circumstances (Jobsis, 1970; Jobsis and LaManna, 1978; Jobsis and Rosenthal, 1978; Kreisman et al., 1981). In fluorocarbon-circulated animals, where absorption changes by hemoglobin do not interfere, changes in the absorption spectrum of the brain cortex indicate CO binding to cytochrome a_3 at normal values ($\approx$100 torr) of arterial PO_2 (Piantadosi et al., 1985). In addition, oxidation–reduction (redox) responses of other mitochondrial cytochromes can be measured spectrally during CO binding to cytochrome a,a_3. These responses primarily involve energy-dependent redox changes in b-type cytochromes that occur after CO binding to the oxidase (Piantadosi, 1989; Young and Caughey, 1990). These redox responses occur at arterial CO/O_2 ratios of less than 0.01 (Piantadosi, 1989). In hemoglobin-circulated animals, evidence for CO uptake by cytochrome a,a_3 is found in the rat brain by spectroscopy at COHb levels of 50% and greater (Brown and Piantadosi, 1990) but has been difficult to detect at COHb levels less than 50%. Biochemical studies of these effects of CO in the brain have attempted to define the relative importance of COHb, intracellular CO binding and restoration of energy provision during reoxygenation after CO poisoning (Brown and Piantadosi, 1992).

Our understanding of the effects of CO hypoxia on rapidly metabolizing tissues has been assisted by the knowledge that CO hypoxia increases the reduction level of cytochrome c oxidase, which favors binding of the CO molecule to the enzyme. As noted above, this process occurs in the brain at COHb values of 50% or more, and it is reversed slowly after CO hypoxia (Brown and Piantadosi, 1990). It is not yet clear whether or not there is a lower limit for this effect; however, it is worth emphasizing that elevated levels of COHb can produce sufficient tissue hypoxia for CO to bind to the enzyme noncompetitively — that is, during local anoxia. In addition to the brain, this phenomenon has been demonstrated *in vivo* in the beating heart (Snow et al., 1988). The rate of recovery of cortical cytochrome a,a_3 reduction level after exposure also depends on inspired PO_2 during reoxygenation, indicating competition between CO and O_2 for the cytochrome a,a_3 binding sites. Studies of energy metabolism in the brain cortex of rats exposed to CO sufficiently to interfere with cytochrome a,a_3 function have indicated that the depletion of high energy stores (phosphocreatine) and intracellular acidosis during CO hypoxia can worsen after the CO exposure despite elimination of CO from hemoglobin (Brown and Piantadosi, 1992). The deterioration in energy metabolism during reoxygenation is related to CO uptake by cytochrome a,a_3 and can be prevented by hyperbaric oxygenation, again consistent with competition between CO and O_2 for the cellular binding site.

4 MECHANISM OF ACTION OF CO IN THE BRAIN

Clinical CO poisoning primarily affects the central nervous system (CNS), and the morphological features of CO toxicity involve a broad spectrum of both white and gray matter lesions of variable severity. CO-induced hypoxia and reoxygenation result in selective loss of regional neuronal viability, for example, cortex, deep cerebral white matter, pallidium, substantia nigra, and hippocampus (Ginsberg, 1985). The extent of tissue damage varies from coagulation necrosis with petechial bleeding or prominent congestion (Funata et al., 1982) to selective, delayed neuronal degeneration (Nabeshima et al., 1991). Regional variation in CO-induced brain damage is probably due to differences in time and concentration of CO exposure, severity of accompanying hypoxia, extent of CO uptake, and variations in circulatory anatomy (Funata et al., 1982). Delayed damage in hippocampus, reminiscent of that induced by brain ischemia–reperfusion (IR) (Ginsberg and Busto, 1989), is particularly interesting because the similarity may indicate shared mechanisms of pathogenesis. Mechanisms of neuronal damage after CO hypoxia are probably just as complex as they are after IR, and may involve some of the same factors such as excitotoxicity, catecholamine release, NO· production and reactive oxygen species (ROS) generation. Moreover, evidence from other experimental models indicates that delayed degeneration may be mediated by programmed cell death, mechanisms such as apoptosis (Dragunow et al., 1993; McManus et al., 1992). These factors are discussed below and illustrated in Figure 3.

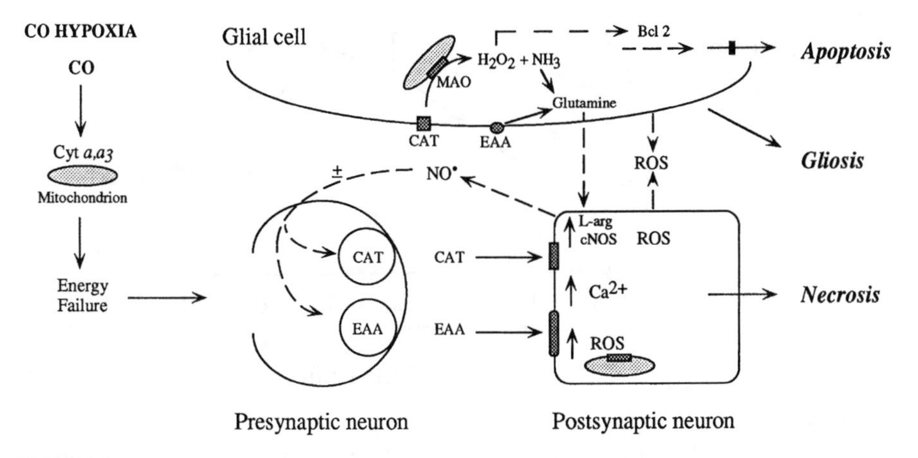

FIGURE 3
Hypothesis diagram of possible mechanisms of neuronal injury after CO hypoxia. Dashed lines indicate untested possibilities. EAA = excitatory amino acids, CAT = catecholamines, Cyt a,a_3 = cytochrome c oxidase, MAO = monoamine oxidase, L-arg = L-arginine, c NOS = constitutive nitric oxide synthase, ROS = reactive oxygen species. Chemical abbreviations given in the text.

4.1 Excitotoxicity and CO Hypoxia

One major hypothesis to explain delayed degeneration after brain ischemia is neuronal excitation. Excitatory amino acids (EAA), such as glutamate, accumulate in synaptic clefts during neuronal depolarization, probably due to both excessive presynaptic release and failure in ATP-dependent re-uptake mechanisms (Amara, 1992). EAA activate at least three glutamate receptors including N-methyl-D-aspartic acid (NMDA), D-amino-3-hydroxy-5-methyl-4-isoxazolepropionic acid (AMPA), and kainic acid (KA) (Choi, 1988). Increased receptor activities upon EAA binding, coupled with intracellular calcium influx (Ca^{2+}), are significant factors in cell death after IR (Takano et al., 1985). At present, little is known about the effects of EAA on tissue damage by CO except that NMDA antagonism attenuates delayed neuronal degeneration in hippocampus after CO poisoning (Ishimaru et al., 1992), indicating that EAA and subsequent increase in the intracellular Ca^{2+} concentration may be involved. This hypothesis is supported by preliminary data that interstitial glutamate concentration increases in the hippocampus during and after CO exposure (Figure 4).

Interstitial EAA accumulation and subsequent neuronal damage may be modulated by retrograde nitric oxide (NO·) production during IR. NO· is

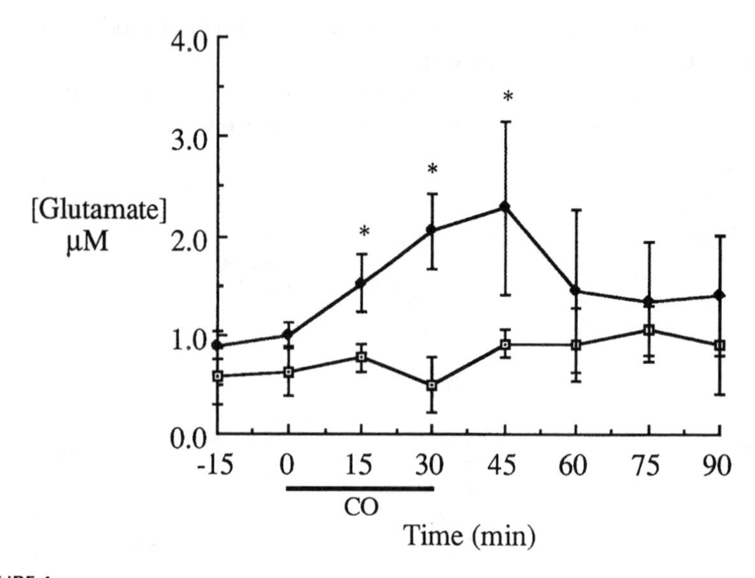

FIGURE 4
Glutamate concentration in the brain interstitium of the rat during and after CO hypoxia. Measurements were made *in vivo* by intracerebral microdialysis of the hippocampus of the anesthetized rat. Values are mean ±SE for 4 animals in each group. Final COHb levels after 30 min of exposure were 60 to 70%. (J. Zhang and C.A. Piantadosi, unpublished observations).

synthesized in the brain by the NO· synthases (NOS), which require O_2, calcium-calmodulin, and tetrahydrobiopterin. In the presence of these cofactors, NOS utilizes L-arginine in a five-electron oxidation of its guanidino nitrogen to produce NO· and citrulline. In animals experiencing stroke, inhibition of NOS can protect the brain against ischemic damage (Nagafuji et al., 1992, Nowicki et al., 1991). In transient forebrain IR, a model producing characteristic delayed degeneration, NOS inhibition tends to aggravate the hippocampal lesions (Sancesario et al., 1992) and enhance delayed death (McManus et al., 1992) which suggests a protective role of NO· in this injury.

The role of NO· in CO poisoning has not been well defined. CO poisoning causes profound increases in cerebral blood flow, and consequently, levels of tissue oxygen and glucose delivery and lactate clearance are higher compared to brain ischemia. These conditions, accompanied by interstitial EAA accumulation, may facilitate NO· production after CO exposure. In addition, both NO· and CO activate guanylyl cyclase causing increases in cGMP (Verma et al., 1993). This cellular effect of CO may be responsible for vasodilation during CO poisoning (Ramos et al., 1989). In preliminary experiments, however, NOS inhibitors appear to interfere with the vascular effects of CO in the rat. The exact biological role of NO· remains unclear because its fate is complex; it is susceptible to both oxidation and reduction, and interactions occur with biological metals and thiols (Stamler et al., 1992). Some redox forms of NO retain NO·-like activity as demonstrated by the potent bioactivity of S-nitrosothiols (RS-NO) (Mohr et al., 1994). To understand the role of NO· production in CO hypoxia, it will be necessary to examine the extent to which NO· production increases, and determine how NOS inhibition alters interstitial EAA accumulation and changes in RS-NO pools during and after CO exposure. It is also important to determine how much CO alters the activities of guanylyl cyclase and NOS *in vivo*. NO· overproduction also may lead to direct cytotoxicity in the presence of superoxide due to the formation of peroxynitrite anion (ONOO-), which is a strong oxidant with reactive hydroxyl radical-like activity (Beckman et al., 1990).

4.2 Catecholamines and CO Hypoxia

Excessive accumulation of monoamine neurotransmitters — for example, norepinephrine (NE) and dopamine (DA) — may also influence development of neuronal death after reoxygenation of ischemic brain (Globus et al., 1987, 1989). Catecholamines activate voltage-dependent NMDA channels, thereby contributing to calcium influx and neuronal damage after cerebral ischemia (Globus et al., 1988). Chemical depletion of catecholamines or nigral lesioning before ischemia appear to be protective (Globus et al., 1987). Interestingly, increased release of catecholamines also has been indicated after CO hypoxia, and both DA and NE concentrations can be elevated in brain homogenates for weeks after CO exposure (Newby at al., 1978). Increases in DA also correlate with behavioral changes after CO exposure (Newby at al., 1978). Enhanced synaptic accumulation of catecholamines during early CO exposure is highly likely because blockade of their metabolism significantly attenuates production of reactive oxygen species after CO poisoning (Piantadosi et al., 1995).

The mechanisms of catecholamine accumulation in the interstitium are not understood fully; however, EAA may play an important role in this process. Activation of NMDA receptors causes release of NE and DA in synaptosomal preparations from guinea pig cerebral cortex and cultured neurons (Hanbauer et al., 1992; Montague et al., 1994). More remarkably, these NMDA-induced DA and NE surges appear to be related to NO· production because both events can be prevented by NOS inhibition (Hanbauer et al., 1992; Montague et al., 1994). The involvement of NO· in DA and NE release is supported by the observation that cGMP increases the activity of tyrosine hydroxylase, the first and rate-limiting enzyme in catecholamine synthesis (Roskoski and Roskoski, 1987). Hence, it is likely that EAA release, catecholamine metabolism and NO· production are closely related *in vivo*, and can be influenced by CO hypoxia and, theoretically, by intracellular uptake of CO.

4.3 Reactive Oxygen Species (ROS) and CO Hypoxia

Oxygen, when reduced incompletely, produces ROS such as superoxide anion (O_2^-), hydrogen peroxide (H_2O_2) and hydroxyl radical ($OH^{\cdot}$) (Halliwell, 1992). ROS production increases in both forebrain IR (Carney and Floyd, 1991) and CO hypoxia (Thom, 1990) with the highest oxidative stress occurring in the most vulnerable brain regions (Hall et al., 1993; Zhang and Piantadosi, 1992). This stress may result from lower antioxidant capacity (Matsuyama et al., 1993) and/or higher tissue concentrations of iron (Zecca et al., 1994) in these regions. ROS production follows increased EAA and intracellular calcium and can be attributed to several sources, including mitochondria (Halestrap et al., 1993), cycloxygenase (Hall et al., 1993), xanthine oxidase (Lindsay et al., 1991) and neutrophils (Barone et al., 1992). Data for ROS production during excitotoxicity are supported by studies showing exogenously supplied antioxidants attenuate NMDA neurotoxicity *in vitro* (Lafon-Cazal et al., 1993) and cerebral ischemia *in vivo* (Carney and Floyd, 1991).

Intracellular generation of H_2O_2 has been demonstrated in the rat forebrain after CO hypoxia and reoxygenation. Subsequent studies of forebrain mitochondria isolated during and after CO hypoxia have implicated them as the source of H_2O_2 (Zhang and Piantadosi, 1992). This is consistent with the inhibition of electron transport by CO binding to cytochrome a,a_3. In the mitochondria, decreases in the ratio of reduced to oxidized glutathione (GSH/GSSG) were found immediately and 2 h after CO exposure. Glutathione is a major defense mechanism against oxidative stress in mitochondria, serving as a substrate for glutathione peroxidase, which catalyzes the conversion of H_2O_2 to water resulting in formation of GSSG. GSSG is then reduced to GSH by glutathione reductase. GSH depletion data during and after CO hypoxia have been supported by studies of $OH^{\cdot}$ generation in the brain using sodium salicylate as a probe. Hydroxylation of salicylate by nonenzymatic mechanisms was increased 3–4-fold in mitochondria isolated from the forebrain during and after CO hypoxia. These changes were not found in the cytosol, again implicating mitochondria as a primary source of oxidative stress (Zhang and Piantadosi, 1992).

ROS produced from mitochondria after CO hypoxia can be blunted by HBO therapy at 1.5 ATA and enhanced at 2.5 ATA (Piantadosi et al., 1995). This increased ROS production at 2.5 ATA in the CO-exposed brain appears to be related in part to monoamine oxidase (MAO) activity and/or catecholamine release since it can be ameliorated by the MAO inhibitor, pargyline. MAO (isoforms A and B) located on the outer mitochondrial membrane catalyzes oxidative deamination of catecholamines and produces H_2O_2 and NH_3, and utilizes molecular O_2 in the process. The Km for O_2 of the enzyme is approximately 100 μM; therefore, high tissue PO_2 values can increase its activity. Inhibition of MAO activity with pargyline decreases H_2O_2 generation in the brain during hyperoxia and protects against CNS O_2 toxicity (Zhang and Piantadosi, 1991). This mechanism may be responsible for the increased ROS production after CO hypoxia at the high tissue PO_2 values produced by hyperbaric oxygen therapy at 2.5 ATA.

In addition to EAA-mediated oxidative stress, autooxidation or oxidative deamination of catecholamines during reperfusion may also contribute to ROS production (Simonson et al., 1993). ROS production after CO exposure can be inhibited by partially blocking of type B monoamine oxidase (MAO-B) (Piantadosi et al., 1995), which is located predominantly in glia. Glial responses, such as gliosis, previously believed to be a response to tissue damage, may play a role in cellular damage after cerebral hypoxia. Persistent gliosis can precede delayed cell death in vulnerable brain regions after ischemia (Petito et al., 1992). Increased gliosis and MAO-B activity also have been implicated in the pathogenesis of Alzheimer's disease (Jossan et al., 1991), amyotrophic lateral sclerosis (Ekblom et al., 1993), and Parkinson's disease (Forno et al., 1992). It should be determined whether gliosis after CO exposure enhances MAO-B activity, and when combined with increased catecholamine release, augments H_2O_2 production and contributes to delayed degeneration in vulnerable brain regions.

4.4 Programmed Neuronal Cell Death and Cerebral Hypoxia

ROS, regardless of their source, may exacerbate EAA toxicity by potentiating synaptic EAA accumulation (Gilman et al., 1994; Pellegrini-Giampietro et al., 1990). In addition, direct damage is produced by oxidizing essential cellular constituents, and ROS may mediate programmed cell death — that is, apoptosis. Apoptotic cell death requires activation and/or expression of specific cellular processes, some of which may act through oxidant pathways (Hockenbery et al., 1993). For example, in developing neuronal cell lines, a mitochondrially targeted protein, Bcl-2, inhibits apoptosis. In adult animals, Bcl-2 is expressed in microglia, and may play a role in inhibiting neuronal death by ischemia, b-amyloid peptide and glutamate. Overexpression of Bcl-2 suppresses lipid peroxidation (Hockenbery et al., 1993), and exogenously supplied antioxidants prevent both glutamate-induced oxidative stress and apoptosis (Ratan et al., 1994). Although precise mechanisms by which glutamate/ROS induces cell death are unclear, activation of poly-ADP synthetase may be involved (Zhang et al., 1994). It is reasonable to hypothesize that impaired mitochondrial energy provision in CO hypoxia leads to neuronal depolarization, EAA and catecholamine release, and failure of re-uptake until energy metabolism is restored during reoxygenation. These processes, normally modulated by NO· production, could contribute to degeneration of neurons in vulnerable regions, possibly by enhancing mitochondrial ROS generation which can initiate apoptosis (Bredesen, 1994). A role for CO uptake by cytochrome oxidase in a process such as this will require some new tools and a great deal of research to investigate.

4.5 Putative Physiological Role for CO in the Brain

A new concept in the biology of CO is the possibility that endogenous CO production in the brain plays a neurotransmitter role similar to NO· (Maines, 1992; Schmidt, 1992; Zhuo et al., 1993). Support for this hypothesis comes from observations showing inhibition of heme oxygenase in the brain, which produces CO will decrease depolarization-induced, calcium-dependent glutamate release (Shinomura et al., 1994) and reverses established, long-term potentiation (Stevens and Wang, 1993). If effects of inhibiting heme oxygenase are related to CO production, like NO·, they may be achieved in part by activating soluble guanylyl cyclase (Verma et al., 1993). Notably, CO affects guanylyl cyclase activity 100 times less potently than NO· hence, the physiological activity of CO in the brain may be considerably less than that of NO· in modulating cGMP dependent processes except in regions where NO· production is absent. In any event, toxic exposures to CO may interfere with these signaling processes in some cases.

5 CONCLUSIONS

The primary toxic effects of exposure to CO result from cellular hypoxia due to COHb formation. Cellular hypoxia itself can be responsible for significant tissue injury, but it does not explain all of the pathophysiological features of CO hypoxia. Cellular hypoxia favors movement of CO from the blood to intracellular compartments where it binds to myoglobin, cytochrome c oxidase, guanylylcyclase, and perhaps other hemoproteins. The cellular uptake of CO by brain mitochondria can alter energy provision during, and in particular, after CO poisoning. In addition, CO hypoxia leads to significant intracellular oxidative stress in the brain, which may persist after COHb has been cleared from the circulation. These principles may explain the development of some of the unusual, prolonged, and delayed pathophysiological manifestations of exposure to toxic concentrations of CO.

REFERENCES

Agostoni, A., Stabilini, R., Viggiano, G., Luzzsana, M., and Samaja, M., Influence of capillary and tissue PO_2 on carbon monoxide binding to myoglobin: A theoretical evaluation, *Microvasc. Res.,* 20, 81, 1980.

Albanese, R.B., On microelectrode distortion of tissue oxygen tensions, *J. Theor. Biol.,* 38, 143, 1973.

Allen, T.A. and Root, W.S., Partition of carbon monoxide and oxygen between air and whole blood of rats, dogs and men as affected by plasma pH, *J. Appl. Physiol. Respir. Environ. Exercise Physiol.,* 10, 186, 1957.

Amara, S.G., Neurotransmitter transporters. A tale of two families [news; comment], Nature, 360, 420, 1992.
Aronow, W.S. and Isbell, M.W., Carbon monoxide effect of exercise-induced angina pectoris, *Ann. Int. Med.*, 79, 392, 1973.
Barone, F.C., Schmidt, D.B., Hillegass, L.M., Price, W.J., White, R.F., Feuerstein, G.Z., Clark, R.K., Lee, E.V., Griswold, D.E., and Sarau, H.M., Reperfusion increases neutrophils and leukotriene B4 receptor binding in rat focal ischemia, *Stroke,* 23, 1337, 1992.
Beckman, J.S., Beckman, T.W., Chen, J., Marshall, P.A., and Freeman, B.A., Apparent hydroxyl radical production by peroxynitrite: Implications for endothelial injury from nitric oxide and superoxide, *Proc. Natl. Acad. Sci., U.S.A.,* 87, 1620, 1990.
Berk, P.D., Blaschke, T.F., Scharschmidt, B.F., Waggoner, J.G., and Berlin, N.I., A new approach to quantitation of the various sources of bilirubin in man, *J. Lab. Clin. Med.*, 87, 767, 1976.
Berk, P.D., Rodkey, F.L., Blaschke, T.F., Collison, H.A., and Waggoner, J.G., Comparision of plasma bilirubin turnover and carbon monoxide production in man, *J. Lab. Clin. Med.*, 83, 29, 1974.
Bernard, C., Lecons sur les effects des substances toxiques et medicamenteuses, Bailliere, Paris, 1857.
Bredesen, D.E., Neuronal apoptosis: Genetic and biochemical modulation, in *Current Communications in Cell and Molecular Biology,* Tomei, L.D. and Cope, F.O., Eds., Cold Spring Harbor Laboratory Press, Maine, 1994.
Brown, S.D. and Piantadosi, C.A., *In vivo* binding of carbon monoxide to cytochrome c oxidase in rat brain, *J. Appl. Physiol.*, 68, 604, 1990.
Brown, S.D. and Piantadosi, C.A., Recovery of energy metabolism in rat brain after carbon monoxide hypoxia, *J. Clin. Invest.*, 89, 666, 1992.
Carney, J.M. and Floyd, R.A., Protection against oxidative damage to CNS by alpha-phenyl-tert-butyl nitrone (pbn) and other spin-trapping agents: A novel series of nonlipid free radical scavengers, *J. Mol. Neurosci.*, 3, 47, 1991.
Chance, B., Erecinska, M., and Wagner M., Mitochondrial responses to carbon monoxide toxicity, *Ann. N.Y. Acad. Sci.*, 174, 193, 1970.
Chance, B. and Williams, G.R., The respiratory chain and oxidative phosphorylation, *Adv. Enzymol.*, 17, 65, 1956.
Choi, D.W., Calcium-mediated neurotoxicity: Relationship to specific channel types and role in ischemic damage [review], *Trends Neurosci.*, 11, 465, 1988.
Clark, B.J. and Coburn, R.F., Mean myoglobin oxygen tension during exercise at maximal oxygen uptake, *J. Appl. Physiol. Respir. Environ. Exercise Physiol.*, 39, 135, 1975.
Coburn, R.F., Enhancement by phenobarbital and diphenylhydantoin of carbon monoxide production in normal man, *N. Engl. J. Med.*, 283, 512, 1970.
Coburn, R.F., Mechanisms of carbon monoxide toxicity, *Prev. Med.*, 8, 310, 1979.
Coburn, R.F. and Forman, H.J., Carbon monoxide toxicity, *Handb. Physiol.*, 4, 439, 1987.
Coburn, R.F. and Mayers, L.B., Myoglobin O_2 tension determined from measurements of carboxymyoglobin in skeletal muscle, *Am. J. Physiol.*, 220, 66, 1971.
Coburn, R.F., Ploegmakers, F., Gondrie, P., and Abboud, R., Myocardial myoglobin oxygen tension, *Am. J. Physiol.*, 224, 870, 1973.
Coburn, R.F., Williams, W.J., and Kahn, S.B., Endogenous carbon monoxide production in patients with hemolytic anemia, *J. Clin. Invest.*, 45, 460, 1966.
Cole, R.P., Myoglobin function in exercising skeletal muscle, *Science,* 216, 523, 1982.
Delivoria-Papadopoulos, M., Coburn, R.F., and Forster, R.E., Cyclical variation of rate of heme destruction and carbon monoxide production (V˙CO) in normal women, *Physiologist,* 13, 178, 1970.
Doblar, D.D., Santiago, T.V., and Edelman, N.J., Correlation between ventilatory and cerebrovascular responses to inhalation of CO, *J. Appl. Physiol. Respir. Environ. Exercise Physiol.*, 43, 455, 1977.
Drabkin, D.L., Lewey, F.H., Bellet, S., and Ehrich, W.H., The effect of replacement of normal blood by erythymocytis saturated with carbon monoxide, *Am. J. Med. Sci.*, 205, 755, 1943.

Dragunow, M., Young, D., Hughes, P., Macgibbon, G., Lawlor, P., Singleton, K., Sirimanne, E., Beilharz, E., and Gluckman, P., Is c-jun involved in nerve cell death following status epilepticus and hypoxic-ischaemia brain injury? [published erratum appears in *Brain Res. Mol. Brain Res.,* 1993, Oct. 20(1–2), 179], *Brain Res. Mol. Brain Res.,* 18, 347, 1993.

Ekblom, J., Jossan, S.S., Bergstrom, M., Oreland, L., Walum, E., and Aquilonius, S.M., Monoamine oxidase-b in astrocytes, *Glia,* 8, 122, 1993.

Estabrook, R.W., Franklin, M.R., and Hildebrandt, A.G., Factors influencing the inhibitory effect of carbon monoxide on cytochrome P-450-catalyzed mixed function oxidation reactions, in *Biological Effects of Carbon Monoxide,* Coburn, R.F., Ed., *Ann. N.Y. Acad. Sci.,* 1970, 218.

Fenn, W.O. and Cobb, D.M., The burning of carbon monoxide by heart and skeletal muscle, *Am. J. Physiol.,* 102, 393, 1932.

Fenn, W.O., The burning of CO in the tissues, in *Biological Effects of Carbon Monoxide,* Coburn, R.F., Ed., *Ann. N.Y. Acad. Sci.,* 1970, 64.

Fisher, A.B. and Dodi, C., Lung as a model for evaluation of critical intracellular PO_2 and PCO, *Am. J. Physiol.,* 241, E47, 1981.

Forman, H.J. and Feigelson, P., Kinetic evidence indicating the absence during catalysis of an unbound ferro-protoporphyrin form of tryptophan oxidase, *Biochemistry,* 10, 760, 1971.

Forno, L.S., Delanney, L.E., Irwin, I., Di Monte, D., and Langston, J.W., Astrocytes and Parkinson's disease, [review], *Prog. Brain Res.,* 94, 429, 1992.

Forster, R.E., Carbon monoxide and the partial pressure of oxygen in tissue, *Ann. N.Y. Acad. Sci.,* 174, 233, 1970.

Funata, N., Okeda, R., Takano, T., Miyazaki, Y., Higashino, F., Yokoyama, K., and Manabe, M., Electron microscopic observations of experimental carbon monoxide encephalopathy in the acute phase, *Acta Pathol. Jpn.,* 32, 219, 1982.

Gayeski, T.E. and Honig, C.R., Intracellular PO_2 in individual cardiac myocytes in dogs, cats, rabbits, ferrets and rats, *Am. J. Physiol.,* 260, H522, 1991.

Gilman, S.C., Bonner, M.J., and Pellmar, T.C., Free radicals enhance basal release of d-[3h] aspartate from cerebral cortical synaptosomes, *J. Neurochem.,* 62, 1757, 1994.

Ginsberg, M.D., Carbon monoxide intoxication: Clinical features, neuropathology and mechanisms of injury, *J. Toxicol. Clin. Toxicol.,* 23, 281, 1985.

Ginsberg, M.D. and Busto, R., Rodent models of cerebral ischemia [review]. *Stroke,* 20, 1627, 1989.

Globus, M.Y., Busto, R., Dietrich, W.D., Martinez, E., Valdes, I., and Ginsberg, M.D., Direct evidence for acute and massive norepinephrine release in the hippocampus during transient ischemia, *J. Cereb. Blood Flow Metab.,* 9, 892, 1989.

Globus, M.Y., Busto, R., Dietrich, W.D., Martinez, E., Valdes, I., and Ginsberg, M.D., Effect of ischemia on the *in vivo* release of striatal dopamine, glutamate, and gamma-aminobutyric acid studied by intracerebral microdialysis, *J. Neurochem.,* 51, 1455, 1988.

Globus, M.Y., Ginsberg, M.D., Harik, S.I., Busto, R., and Dietrich, W.D., Role of dopamine in ischemic striatal injury: Metabolic evidence, *Neurology,* 37, 1712, 1987.

Goldbaum, L.R., Orellano, T., and Dergal, E., Studies on the relation between carboxyhemoglobin concentration and toxicity, *Aviat. Space Environ. Med.,* 48, 969, 1977.

Goldbaum, L.R., Ramirez, R.G., and Absalon, K.B., What is the mechanism of carbon monoxide toxicity? *Aviat. Space Environ. Med.,* 46, 1289, 1975.

Gothert, M., Lutz, F., and Malorney, G., Carbon monoxide partial pressure in tissue of different animals, *Environ. Res.,* 3, 303, 1970.

Haldane, J.B.S., Carbon monoxide as a tissue poison, *Biochem. J.,* 21, 1068, 1927.

Haldane, J.S., The relation of the action of carbonic oxide to oxygen tension, *J. Physiol. (London),* 18, 201, 1895.

Halestrap, A.P., Griffiths, E.J., and Connern, C.P., Mitochondrial calcium handling and oxidative stress [review], *Biochem. Soc. Trans.,* 21, 353, 1993.

Hall, E.D., Andrus, P.K., Althaus, J.S., and Vonvoigtlander, P.F., Hydroxyl radical production and lipid peroxidation parallels selective post-ischemic vulnerability in gerbil brain, *J. Neurosci. Res.,* 34, 107, 1993.

Halliwell, B., Reactive oxygen species and the central nervous system [review], *J. Neurochem.*, 59, 1609, 1992.
Halperin, M.H., McFarland, R.A., Niven, J.I., and Roughton, F.J.W., The time course of the effects of carbon monoxide on visual thresholds, *J. Physiol.*, 146, 583, 1959.
Hanbauer, I., Wink, D., Osawa, Y., Edelman, G.M., and Gally, J.A., Role of nitric oxide in NMDA-evoked release of [3h]-dopamine from striatal slices, *Neuroreport,* 3, 409, 1992.
Hockenbery, D.M., Oltvai, Z.N., Yin, X.M., Milliman, C.L., and Korsmeyer, S.J., Bcl-2 functions in an antioxidant pathway to prevent apoptosis, *Cell,* 75, 241, 1993.
Hoofd, L. and Kreuzer, F., Calculation of the facilitation of O_2 or CO transport by Hb or Mb by means of a new method for solving the carrier-diffusion problem, in *Oxygen Transport to Tissue,* Vol. 3, Silver, I.A., Erecinska, M., and Bicher, H.I., Eds., Plenum Press, New York, 1978.
Horvath, S.M., Raven, P.B., Dahms, T.E., and Gray, D.J., Maximal aerobic capacity at different levels of carboxyhemoglobin, *J. Appl. Physiol. Respir. Environ. Exercise Physiol.*, 38, 300, 1975.
Ishimaru, H., Katoh, A., Suzuki, H., Fukuta, T., Kameyama, T., and Nabeshima, T., Effects of n-methyl-d-aspartate receptor antagonists on carbon monoxide-induced brain damage in mice, *J. Pharmacol. Exp. Ther.*, 261, 349, 1992.
Iyanagi, T., Suzaki, T., and Kobaysashi, S., Oxidation-reduction states of pyridine nucleotide and cytochrome P450 during mixed-function oxidation in perfused rat liver, *J. Biol. Chem.*, 256, 12933, 1981.
Jobsis, F.F., Oxidative metabolic effects of cerebral hypoxia, in *Advances in Neurology,* Fahn, S., et al., Eds., Raven Press, New York, 1970, 219.
Jobsis, F.F. and LaManna, J.C., Kinetic aspects of intracellular redox reactions, *in vivo* effects during and after hypoxia and ischemia, in *Extra Pulmonary Manifestations of Respiratory Disease,* Robin, E.D., Ed., Marcel Dekker, New York, 1978, 63.
Jobsis, F.F. and Rosenthal, M., Behavior of the mitochondrial respiratory chain *in vivo*, in *Cerebral Vascular Smooth Muscle and Its Control,* Ciba Foundation Symposium 56, Elsevier/Exerpta Medica/North-Holland, Amsterdam, 1978, 429.
Jones, D.P. and Kennedy, F.G., Intracellular oxygen supply during hypoxia, *Am. J. Physiol.*, 243, C247, 1982.
Jossan, S.S., Gillberg, P.G., Gottfries, C.G., Karlsson, I., and Oreland, L., Monoamine oxidase b in brains from patients with Alzheimer's disease: A biochemical and autoradiographical study, *Neuroscience,* 45, 1, 1991.
Keilin, D. and Hartree, E.F., Cytochrome and cytochrome oxidase, *Proc. R. Soc. (London) Ser. B*(1), 27, 167, 1939.
Kindwall, E.P., Carbon monoxide and cyanide poisoning, in *Hyperbaric Oxygen Therapy,* Davis, J.C., Ed., Undersea Medical Society, Rockville, MD, 1977, 179.
Koehler, R.C., Jones, M.D., and Traystman, R.J., Cerebral circulatory response to carbon monoxide and hypoxic hypoxia in the lamb, *Am. J. Physiol.*, 243, H27, 1977.
Kreisman, N.R., Sick, T.J., LaManna, J.C., and Rosenthal, M., Local tissue oxygen tension — cytochrome a,a_3 redox relationships in rat cerebral cortex *in vivo*, Brain Res., 218, 161, 1981.
Lafon-Cazal, M., Pietri, S., Culcasi, M., and Bockaert, J., NMDA-dependent superoxide production and neurotoxicity, *Nature,* 364, 535, 1993.
Lindsay, S., Liu, T.H., Xu, J.A., Marshall, P.A., Thompson, J.K., Parks, D.A., Freeman, B.A., Hsu, C.T., and Beckman, J.S., Role of xanthine dehydrogenase and oxidase in focal cerebral ischemic injury to rat, *Am. J. Physiol.*, 261, H2051, 1991.
Luomanmaki, K. and Coburn, R.F., Effects of metabolism and distribution of carbon monoxide on blood and body stores, *Am. J. Physiol.*, 217, 354, 1969.
MacMillan, V., The effects of acute carbon monoxide intoxication on the cerebral energy metabolism of the rat, *Can. J. Physiol. Pharmacol.*, 52, 354, 1975a.
MacMillan, V., Regional cerebral blood flow of the rat in acute carbon monoxide intoxication, *Can. J. Physiol. Pharmacol.*, 53, 644, 1975b.

Maines, M.D., Review. Carbon monoxide: An emerging regulator of cGMP in the brain, *Neurosci. Lett.*, 4, 389, 1992.
Matsuyama, T., Michishita, H., Nakamura, H., Tsuchiyama, M., Shimizu, S., Watanabe, K., and Sugita, M., Induction of copper-zinc superoxide dismutase in gerbil hippocampus after ischemia, *J. Cereb. Blood Flow Metab.*, 13, 135, 1993.
McManus, J.P., Buchan, A.M., Hill, I.E., Rasquinha, I., Preston, E., Weissman, B.A., Kadar, T., Brandeis, R., and Shapira, S., Global ischemia can cause DNA fragmentation indicative of apoptosis in rat brain, *Neurosci. Lett.*, 146, 139, 1992.
Miyahara, S. and Takahashi, H., Biological CO evolution: Carbon monoxide evolution during auto- and enzymatic oxidation of phenols, *J. Biochem., (Tokyo)*, 69, 231, 1971.
Mohr, S., Stamler, J.S., and Brüne, B., Mechanism of covalent modification of glyceraldehyde-3-phosphate dehydrogenase at its active site thiol by nitric oxide, peroxynitrite and related nitrosating agents, *FEBS Lett.*, 348, 223, 1994.
Montague, P.R., Gancayco, C.D., Winn, M.J., Marchase, R.B., and Friedlander, M.J., Role of no production in NMDA receptor-mediated neurotransmitter release in cerebral cortex, *Science*, 263, 973, 1994.
Montgomery, M.R. and Rubin, R.J., Oxygenation during inhibition of drug metabolism by carbon monoxide or hypoxic hypoxia, *J. Appl. Physiol.*, 35, 505, 1973.
Nabeshima, T., Katoh, A., Ishimaru, H., Yoneda, Y., Ogita, K., Murase, K., Ohtsuka, H., Inari, K., Fukuta, T., and Kameyama, T., Carbon monoxide-induced delayed amnesia, delayed neuronal death and change in acetylcholine concentration in mice, *J. Pharmacol. Exp. Ther.*, 256, 378, 1991.
Nagafuji, T., Matsui, T., Koide, T., and Asano, T., Blockade of nitric oxide formation by n omega-nitro-l-arginine mitigates ischemic brain edema and subsequent cerebral infarction in rats, *Neurosci. Lett.*, 147, 159, 1992.
Newby, M.B., Roberts, R.J., and Bhatnagar, R.K., Carbon monoxide and hypoxia induced effects on catacholamines in the mature and developing rat brain, *J. Pharm. Exp. Therapeut.*, 206, 61, 1978.
Nowicki, J.P., Duval, D., Poignet, H., and Scatton, B., Nitric oxide mediates neuronal death after focal cerebral ischemia in the mouse, *Eur. J. Pharmacol.*, 204, 339, 1991.
Omura, T., Sato, R., Cooper, D.Y., Rosenthal, O., and Estabrook, R.W., Function of cytochrome P450 of microsomes, *Fed. Proc.*, 24, 1181, 1965.
Oshino, N., Sugano, T., Oshino, R., and Chance, B., Mitochondrial function under hypoxic conditions: The steady states of cytochrome $a + a_3$ and their relation to mitochondrial energy states, *Biochim. Biophys. Acta*, 368, 298, 1974.
Otterbein, L., Sylvester, S.L., and Choi, A.M.K., Hemoglobin provides protection against lethal endotoxemia in rats: the role of heme oxygenase-1., *Am. J. Resp. Cell and Mol. Biol.*, 13, 595, 1995.
Paulson, O.B., Parving, H.H., Olesen, J., and Skinhoj, E., Influence of carbon monoxide and of hemodilution on cerebral blood flow and blood gases in man, *J. Appl. Physiol. Respir. Environ. Exercise Physiol.*, 35, 111, 1973.
Pellegrini-Giampietro, D.E., Cherici, G., Alesiani, M., Carla, V., and Moroni, F., Excitatory amino acid release and free radical formation may cooperate in the genesis of ischemia-induced neuronal damage, *J. Neurosci.*, 10, 1035, 1990.
Petito, C.K., Chung, M., Halaby, I.A., and Cooper, A.J., Influence of the neuronal environment on the pattern of reactive astrocytosis following cerebral ischemia [review], *Prog. Brain Res.*, 94, 381, 1992.
Piantadosi, C.A., Carbon monoxide, oxygen transport, and oxygen metabolism, *J. Hyperbaric Med.*, 2, 27, 1987.
Piantadosi, C.A., Spectrophotometry of cytochrome b in rat brain *in vivo* and *in vitro*, *Am. J. Physiol.*, 256, C840, 1989.
Piantadosi, C.A., Sylvia, A.L., Saltzman, H.A., and Jobsis-VanderVliet, F.F., Carbon monoxide-cytochrome interactions in the brain of the fluorocarbon-perfused rat, *J. Appl. Physiol.*, 58, 665, 1985.

Piantadosi, C.A., Tatro, L., and Zhang, J., Hydroxyl radical production in the brain after CO hypoxia in rats, *Free Radical Biol. Med.*, 18, 603, 1995.

Ramos, K.S., McGrath, H.A., and McGrath, J.J., Modulation of cyclic guanosine monophosphate levels in cultured aortic smooth muscle cells by carbon monoxide, *Biochem. Pharmacol.*, 38, 1368, 1989.

Ratan, R.R., Murphy, T.H., and Baraban, J.M., Oxidative stress induces apoptosis in embryonic cortical neurons, *J. Neurochem.*, 62, 376, 1994.

Roskoski, R., Jr. and Roskoski, L.M., Activation of tyrosine hydroxylase in pc12 cells by the cyclic GMP and cyclic AMP second messenger systems, *J. Neurochem.*, 48, 236, 1987.

Roth, R.A, Jr. and Rubin, R.J., Comparison of the effect of carbon monoxide and of hypoxic hypoxia. I. *In vivo* metabolism, distribution and action of hexobarbital, *J. Pharmacol. Exp. Ther.*, 199, 53, 1976a.

Roth, R.A., Jr. and Rubin, R.J., Role of blood flow in carbon monoxide- and hypoxic-induced alterations in hexobarbital metabolism in rats, *Drug Metab. Dispos.*, 4, 460, 1976b.

Sancesario, G., Iannone, M., D'Angelo, V., Nistico, G., and Bernardi, G., N omega-nitro-l-arginine-methyl ester inhibits electrocortical recovery subsequent to transient global brain ischemia in Mongolian gerbil, *Funct. Neurol.*, 7, 123, 1992.

Savolainen, H., Kurppa, K., Tenhunen, R., and Kivisto, H., Biochemical effects of carbon monoxide poisoning in rat brain with special reference to blood carboxyhemoglobin and cerebral cytochrome oxidase activity, *Neurosci. Lett.*, 19, 319, 1980.

Schmidt, H.H.H.W., NO, CO and OH endogenous soluble guanylyl cyclase-activating factors, *FEBS Lett.*, 307, 102, 1992.

Seisjo, B.K., *Brain Energy Metabolism*, John Wiley & Sons, New York, 1978.

Shinomura, T., Nakao, S., and Mori, K., Reduction of depolarization-induced glutamate release by heme oxygenase inhibitor: Possible role of carbon monoxide in synaptic transmission, *Neurosci. Lett.*, 166, 131, 1994.

Sies, H. and Brauser, B., Interaction of mixed function oxidase with its substrates and associated redox transitions of cytochrome P450 and pyridine necleotides in perfused rat liver, *Eur. J. Biochem.*, 15, 531, 1970.

Simonson, S.G., Zhang, J., Canada, A.T., Jr., Su, Y.F., Benveniste, H., and Piantadosi, C.A., Hydrogen peroxide production by monoamine oxidase during ischemia-reperfusion in the rat brain, *J. Cereb. Blood Flow Metab.*, 13, 125, 1993.

Snow, T.R., Vanoli, E., De Ferrari, G., Stramba-Badiale, M., and Dickey, D.T., Response of cytochrome a,a_3 to carbon monoxide in canine hearts with prior infarctions, *Life Sci.*, 42, 927, 1988.

Solanki, D.L., McCurdy, P.R., Cuttitta, F.F., and Schechter, G.P., Hemolysis in sickle cell disease as measured by endogenous carbon monoxide production: A preliminary report, *Am. J. Clin. Pathol.*, 89, 221, 1988.

Stamler, J.S., Singel, D.J., and Loscalzo, J., Biochemistry of nitric oxide and its redox-activated forms, *Science*, 258, 1898, 1992.

Stevens, C.F. and Wang, Y., Reversal of long-term potentiation by inhibitors of heme oxygenase, *Nature*, 364, 147, 1993.

Stewart, R.D., The effect of carbon monoxide on humans, *Annu. Rev. Pharmacol.*, 17, 409, 1975.

Takano, T., Motohashi, Y., Miyazaki, Y., and Okeda, R., Direct effect of carbon monoxide on hexobarbital metabolism in the isolated perfused liver in the absence of hemoglobin, *J. Toxicol. Environ. Health*, 15, 847, 1985.

Tenney, S.M., A theoretical analysis of the relationships between venous blood and mean tissue oxygen pressure, *Resp. Physiol.*, 20, 283, 1977.

Thom, S.R., Carbon monoxide mediated brain lipid peroxidation in the rat, *J. Appl. Physiol.*, 68, 997, 1990.

Thom, S.R. and Elbuken, M.E., Oxygen-dependent antagonism of lipid peroxidation, *Free Rad. Biol. Med.*, 10, 413, 1991.

Traystman, R.J., Fitzgerald, R.S., and Loscutoff, S.C., Cerebral circulatory responses to arterial hypoxia in normal and chemodenervated dogs, *Circ. Res.*, 42, 649, 1981.

Tzagoloff, A. and Wharton, D.C., Studies on the electron transfer system: LXII. The reaction of cytochrome oxidase with carbon monoxide, *J. Biol. Chem.,* 240, 2628, 1965.

Verma, A., Hirsch, D.J., Glatt, C.E., Ronnett, G.V., and Snyder, S.H., Carbon monoxide: A putative neural messenger [see comments] [published erratum appears in *Science,* 1994, Jan. 7; 263 (5143), 15], *Science,* 259, 381, 1993.

White, R.E. and Coon, M.J., Oxygen activation by cytochrome P450, *Annu. Rev. Biochem.,* 49, 325, 1980.

Winston, J.M. and Roberts, R.J., Influence of carbon monoxide, hypoxic hypoxia or potassium cyanide pretreatment on acute carbon monoxide and hypoxic hypoxia lethality, *J. Pharm. Exp. Therapeut.,* 193, 713, 1975.

Wittenberg, B.A. and Wittenberg, J.B., Effects of carbon monoxide on isolated heart muscle cells, Health Effects Institute Research Report 62, Cambridge, MA, 1993.

Wittenberg, B.A. and Wittenberg, J.B., Transport of oxygen in muscle, *Ann. Rev. Physiol.,* 51, 857, 1989.

Wittenberg, B.A., Wittenberg, J.B., and Caldwell, P.R.B., Role of myoglobin in the oxygen supply to red skeletal muscle, *J. Biol. Chem.,* 250, 9038, 1975.

Wohlrab, H. and Ogunmola, B.G., Carbon monoxide binding studies of cytochrome a,a_3 hemes in intact rat liver mitochondria, *Biochemistry,* 10, 1103, 1971.

Young, L.J. and Caughey, W.S., Mitochondrial oxygenation of carbon monoxide, *Biochem. J.,* 239, 225, 1986.

Young, L.J. and Caughey, W.S., Pathobiochemistry of CO poisoning, *FEBS Lett.,* 272, 1, 1990.

Young, L.J., Choc, M.G., and Caughey, W.S., Role of oxygen and cytochrome c oxidase in the detoxification of CO by oxidation to CO_2, in *Biochemical and Clinical Aspects of Oxygen: Proceedings of a Symposium,* September, 1975, Fort Collins, CO, Caughey, W.S. and Caughey, H., Eds., Academic Press, New York, 1979, 355.

Zecca, L., Pietra, R., Goj, C., Mecacci, C., Radice, D., and Sabbioni, E., Iron and other metals in neuromelanin, substantia nigra, and putamen of human brain, *J. Neurochem.,* 62, 1097, 1994.

Zhang, J. and Piantadosi, C.A., Mitochondrial oxidative stress after carbon monoxide hypoxia in the rat brain, *J. Clin. Invest.,* 90, 1193, 1992.

Zhang, J. and Piantadosi, C.A., Prevention of H_2O_2 generation by monoamine oxidase protects against CNS O_2 toxicity, *J. Appl. Physiol.,* 71, 1057, 1991.

Zhang, J., Dawson, V.L., Dawson, T.M., and Snyder, S.H., Nitric oxide activation of poly(adp-ribose) synthetase in neurotoxicity, *Science,* 263, 687, 1994.

Zhuo, M., Small, S.A., Kandel, E.R., and Hawkins, R.D., Nitric oxide and carbon monoxide produce activity-dependent long-term synaptic enhancement in hippocampus, *Science,* 260, 1946, 1993.

Zorn, H., The partial oxygen pressure in the brain and liver at subtoxic concentrations of carbon monoxide, *Staub-Reinhalt. Lug. Engl. Ed.,* 32, 24, 1972.

CHAPTER 9

Carbon Monoxide-Induced Impairment of Learning, Memory, and Neuronal Dysfunction

Masayuki Hiramatsu, Tsutomu Kameyama, and Toshitaka Nabeshima

CONTENTS

0-8493-4796-3/96/$0.00+$.50

1 INTRODUCTION

Carbon monoxide (CO) poisoning has been a well-known cause of a wide variety of neurologic and psychiatric problems. Following acute CO poisoning, 25 to 40% of patients died on initial exposure, while 10 to 30% of patients developed neuropsychiatric problems 1 to 3 weeks after exposure (Ginsberg, 1979). Usually delayed neuropsychiatric problems occur after a clear period of apparent recovery and show one or more symptoms of dementia, such as loss of intellectual ability of sufficient severity to interfere with social or occupational functioning, memory impairment, impairment of abstract thinking or judgment, disturbance of higher cortical functioning (aphasia, apraxia, agnosia, or personality change) (Min, 1986) (Table 1). Necroses of the cerebral cortex, the hippocampus, the substantia nigra, and the globus pallidus have been discovered through anatomical study (Lapresle and Fardeau, 1967), computed tomography (Sawada et al., 1980; 1983), and magnetic resonance scanning (Horowitz et al., 1987). Nevertheless, the etiology of this phenomenon remains unclear.

TABLE 1

Characteristics of CO-poisoning

Acute Stage	Chronic Stage
Hot flush	Intellectual disturbance
Cardiopalmus	Amnesia
Vertigo (dizziness)	Psychosis
Vomiting (emesis)	Paralysis
Headache	Chorea
Hypotension	Cortical blindness
Atonia of leg	Aparaxia, agonia
Psychiatric symptoms: excitation, negativism, mannerism, depression, hysteria	Incontinence
Disturbance of consciousness	
Derangement of capacity to register, disturbance of memory, disorientation, fabrication	

Multiinfarct dementia as well as dementia of the Alzheimer's type is an increasing problem in a senescent society and the vascular type of dementia has been reported to be more frequent in Japan as compared to Europe or the United States. Cerebral ischemia that produces deficiency of oxygen and glucose in the brain results in a selective pattern of neuronal degeneration within the central nervous system in a delayed manner in both humans and experimental animals (Brierly and Graham, 1984; Kirino, 1982; Levine and Payan, 1966; Pulsinelli and Brierly, 1979). Although links between circulatory failure and the subsequent development of dementia are not established, brain tissue hypoxia suffered during CO poisoning and ischemia is assumed to damage some vulnerable neurons in the brain such as the hippocampal CA1 subfield in experimental animals.

Animal models and techniques for inducing hypoxic brain injury for detailed study depend on several factors that elude complete control. First, the experimental insult should parallel as closely as possible the clinical one. Second, test procedures should elicit similar responses in each individual tested. Third, the experimental animal pathology should be similar to that in humans. Although each species has different anatomical and physiological variability, it is possible to obtain more uniform responses to experimental hypoxic injury. In fact, using experimental animals, hypoxia produces several dysfunctions in the central nervous system. Memory deficiency occurs 24 h after ischemia in rats in a passive avoidance task (Yamazaki et al., 1984; Yasumatsu et al., 1987) and in gerbil in the performance of habituation and the passive avoidance (Amano et al., 1993). The function of the acetylcholinergic neurons in ischemic rats was lower than in control rats (Take et al., 1984a). In addition, "delayed neuronal death" can occur even after recovery from changes in biochemical and electrophysiological parameters induced by ischemic insult (Take et al., 1984b). Delayed neuronal damage and deficiencies in learning and memory were also produced after CO exposure in mice as described below. This memory deficiency also develops in a delayed manner, more than 3 d after CO exposure. Thus, changes in the experimental animal model may be similar to the memory deficit that occurs after hypoxic insults in humans.

In this chapter, we summarize some of data on CO-induced impairment of learning, memory, and neuronal dysfunction in mice and rats.

2 METHODS FOR CARBON MONOXIDE (CO) EXPOSURE

Mice (ddY strain, 7 weeks old) were put into a transparent plastic vessel (diameter 6 cm, height 10 cm) with a pipe feeding into it and two holes at the bottom to remove air. They were exposed to pure CO gas once or 3 times at 1-h intervals at the rate of 35 ml/min or 10 ml/min (Hiramatsu et al., 1992b; Yoshida et al., 1992). The mice were exposed to CO each time until they began gasping, or chronic convulsions were observed, and maintained in that state for 5–7 s. As a result, CO exposure lasted for 30 and 55 s. Mortality rate ranged from 10 to 20%. The mice were kept on a hot plate for 2 h to maintain their body temperature at 38–39°C (Hiramatsu et al., 1994a).

To examine learning and memory deficiency after CO exposure, we used two types of amnesia: (1) delayed amnesia 7 d after CO exposure, and (2) impairment of spatial working memory 5–7 d after CO exposure. A step-down type passive avoidance task was used to test for the former, as we previously described (Nabeshima et al., 1988), and a Y-maze test (Sarter et al., 1988) was used for the latter.

3 BIOCHEMICAL CHANGES

3.1 CO–Hemoglobin

CO readily combines with hemoglobin to form carboxyhemoglobin (COHb), preventing the transfer of oxygen to tissues. Tissues such as the brain depend mainly on aerobic metabolism, so a deficiency of oxygen supply damages normal cell functions. The oxygen-carrying capacity of hemoglobin is progressively reduced with rising COHb saturation (Penney, 1990). Because blood COHb concentration was kept at a high level at least for 2 h after a single CO exposure (Kinbara et al., 1989), large amounts of COHb will be maintained in blood for a long time by successive CO exposures.

3.2 Acetylcholine and Choline Levels

A decrease in extracellular cortical acetylcholine levels has been reported following occlusion of the middle cerebral artery in rats (Scremin and Jenden, 1989). In agreement with this finding, acetylcholine levels in the frontal cortex were significantly lower than those of the control group 1 and 7 d after CO exposure (Nabeshima et al., 1991b). In the striatum, acetylcholine levels in the CO-treated group were significantly lower than those in the control group 7 d after CO exposure. In contrast, acetylcholine levels in the hippocampus tended to be lower 7 d after CO exposure, but choline levels were significantly higher 1 and 7 d after CO exposure. In other regions, neither acetylcholine nor choline was changed by CO exposure.

3.3 Biogenic Amine Levels

It is reported that not only the massive increase in excitatory amino acids but also catecholamines play a very important role in the pathogenesis of hypoxic cell damage, because depletion of catecholamine by the administration of α-methyl-*p*-tyrosine protects dopamine and glutamate nerve terminals from ischemia-induced injury (Weinberger et al., 1985). A few studies have examined the possible effects of CO exposure on cerebral biogenic amines (Akiyama et al., 1991; Hiramatsu et al., 1994d). It has been reported that exposure to hypoxia (8–15% O_2 in N_2) increased dopamine levels in striatal dialysate in rats (Akiyama et al., 1991). Similarly, during CO exposure, dopamine levels rise dramatically in the extracellular fluid of the rat striatum, whereas its metabolites, 3,4-dihydroxyphenylacetic acid (DOPAC) and homovanillic acid decreased about 20–25% (Hiramatsu et al., 1994d). In contrast, the increase of serotonin release after CO exposure was much less compared to that of dopamine release and there was no significant difference in 5-hydroxyindoleacetic acid levels in rats (Hiramatsu et al., 1994d). Newby et al. (1978) found that striatal dopamine levels in rats are increased after 2 h of exposure to 0.15% CO, but not following 2 h of low oxygen. Dopamine levels in the striatum in the rats exposed to CO and low oxygen remained higher than in the control rats after treatment with a catecholamine synthesis inhibitor, indicating a decreased dopamine turnover. Akiyama et al. (1991) have suggested that exocytotic release of dopamine from nerve terminals is unaffected by hypoxia and that the hypoxia-induced increase in the extracellular dopamine levels is mainly the result of inhibition of the dopamine re-uptake mechanism by hypoxia. The effect of CO on dopamine turnover persisted for several weeks after a single 5 h CO exposure of 8-day-old rats (Newby et al., 1978).

Dopamine and DOPAC levels in the frontal cortex in mice 7 d after CO exposure were higher compared to the control group. However, there was no significant difference in the levels of dopamine and its metabolites in the striatum in mice (Hiramatsu et al., 1994c).

Krebs et al. (1991a, 1991b) have indicated that striatal dopaminergic nerve terminals possess NMDA receptors and that these presynaptic receptors are involved in a facilitative control of dopamine release. However, systemic administration of MK-801 did not antagonize CO-induced dopamine release (Hiramatsu et al., 1994d).

3.4 [^{3}H] Quinuclidinyl Benzilate, [^{3}H] Glutamate, [^{3}H] Glycine, and [^{3}H]TCP Binding

In *in vitro* autoradiography experiments, CO exposure slightly increased the binding of [^{3}H]quinuclidinyl benzilate (QNB), a ligand for muscarinic acetylcholine receptors, in the striatum and the cortex, but not in the hippocampus 7 d after CO exposure (Nabeshima et al., 1991b). Further, because excitotoxicity involves primarily NMDA- and kainate-type receptors, glutamatergic neurons may be affected by neurodegeneration, as has been reported in neuronal cultures and ischemic models (Jorgensen and Diemer, 1982; Simon et al., 1984). With regard to [^{3}H]glutamate and [^{3}H]glycine binding, no effects were observed in any of the brain regions tested up to 7 d after CO exposure (Nabeshima et al., 1991b). However, specific binding of [^{3}H]1-[1-(2-thienyl) cychohexyl] piperidine ([^{3}H]TCP) in the hippocampal CA1 was increased 7 d after CO exposure (Hiramatsu et al., 1992a).

The lack of change in glutamate and glycine binding sites observed previously (Nabeshima et al., 1991b) may be due to the experimental procedures, since whole brain homogenates were used. Histological studies have revealed only discrete and moderate modification, affecting only the CA1 layer of the hippocampus (Ishimaru et al. 1992; Nabeshima et al., 1991b). Further functional studies may be required to precisely quantify the level of dysfunction in glutamatergic systems after CO exposure. However, at present, the findings do not exclude the possibility that damaged glutamatergic neurotransmission could also be implicated in the resulting amnesia.

4 HISTOCHEMISTRY

Investigations into the mechanism of ischemia-induced neurodegeneration and its possible therapeutic amelioration have been pursued for many years. Cerebral ischemia in experimental animals results in a selective pattern of neuronal degeneration within the central nervous system, and neuronal death of cells in the hippocampal CA1 area occurs gradually over a period of a few days (Kirino, 1982; Pulsinelli et al., 1982). This phenomenon has been termed delayed neuronal death. Delayed neuronal damage also observed after CO exposure, but its extent was more mild than that induced by ischemia (Ishimaru et al., 1991).

It has been reported that neuronal damage is potentiated by repeated cerebral ischemia in the rat and gerbil (Kato et al., 1989, Nakano et al., 1989). Neuronal damage induced by successive CO exposures was more extensive and severe than that by single CO exposure (Hiramatsu et al., 1992a; Ishimaru et al., 1991). Those observations are consistent with the previous finding that repeated cerebral ischemia has a cumulative effect because the neuronal damage is greater following 3 times 5-min ischemia at 1-h intervals in the gerbil than 15-min ischemia produced as a single insult (Tomida et al., 1987).

4.1 Survival of Cells in the Hippocampal CA1 Subfield

CO exposure affected the survival of pyramidal cells in delayed manner in the hippocampal CA1 subfield. One day after CO exposure, the number of pyramidal cells in the hippocampal CA1 subfield was not significantly lower than that in the control group, but 3, 5, and 7 d after CO exposure, it was significantly lower (Nabeshima et al., 1991b). Seventeen percent of the pyramidal cells were damaged 7 d after CO exposure. Seven days after CO exposure, the layer of pyramidal cell in the hippocampal CA1 subfield was thinner than that in the control group. In addition, the disappearance of the lateral hippocampal CA1 pyramidal cells and the atrophy of the medial hippocampal CA1 pyramidal cells were observed (Nabeshima et al., 1991b). Although the layer of the pyramidal cells of both hemispheres in the CO-treated group was thinner, the degree of neuronal damage was different in the two hemispheres. The number of pyramidal cells in both hemispheres in the CO-treated group was significantly smaller than that in the control group. However, damage of CA1 pyramidal cells was asymmetric in the two hemispheres of mouse brain exposed to CO (Nabeshima et al., 1991b).

4.2 Survival of Cells in the Parietal Cortex

Although necrosis of the cerebral cortex in humans takes place, there were no significant changes in the number and size of neuronal cells in the parietal cortex in mice at 7 d after CO exposure compared to those in the control group (Nabeshima et al., 1991b).

5 BEHAVIORAL CHANGES

Acute severe CO exposure produces its major toxic effects by binding to hemoglobin and interfering with oxygen delivery. It induces hypothermia, hypotension, bradycardia hemoconcentration, and hyperglycemia before death supervenes (Penney et al., 1989, 1991). Sutariya et al. (1989) found that whole-body cooling or natural induction of hypothermia during CO poisoning increased survival and improved neurologic outcome. However, hypothermia continued for several hours following CO poisoning decreases survival, while rapid rewarming to normal body temperature increased survival (Sutariya et al., 1990).

5.1 Influence of Body Temperature Changes on Cell Damage and Memory

When body temperature was uncontrolled, rectal temperature dropped about 3°C 15 min after each CO exposure (Ishimaru et al., 1991). Kaltwasser et al. (1977) reported that CO toxicity was reduced in rats with chlorpromazine-induced hypothermia. It has also been reported that post-ischemic hypothermia acts protectively against ischemic brain damage in the rat and gerbil (Boris-Möller et al., 1989; Busto et al., 1987; Clifton et al., 1989). Dialysate levels of glutamate in the hippocampus are decreased by post-ischemic hypothermia (Globus et al., 1988a, 1988b; Mitani and Kataoka, 1991). In fact, the maintenance of body temperature in normal range during CO exposures significantly increased the degree of delayed neuronal damage (Ishimaru et al., 1991) and memory deficiency (Maurice et al., 1994b) induced by successive CO exposure at 1-h intervals.

5.2 CO-Induced Acute Memory Impairment

When CO exposure was carried out immediately after training on passive avoidance response, step-down latency in the retention testing, which was conducted 24 h after the exposure, was shorter compared to the control group (acute amnesia) (Nabeshima et al., 1990a). This is the point of acquisition of learning — that is, before memory was perfectly consolidated. However, CO exposure was carried out more than 4 h after training, when memory was consolidated; memory was not impaired in the retention testing of passive avoidance until at least 48–72 h after training (Kinbara et al., 1989; Nabeshima et al., 1991b). These results suggest that CO-induced acute amnesia may be a model of consolidation impairment, but not of retrieval impairment. Using this model, we showed that pre-training administration of (+)MK-801 potentiated CO-induced acute amnesia, but nefiracetam

(Hiramatsu et al., 1992b), NIK-247 (9-amino-2,3,5,6,7,8-hexahydro-1H-cyclopenta(b)-quinoline monohydrate hydrochloride) and tacrine (Yoshida et al., 1992) ameliorated it.

5.3 CO-Induced Retrograde Memory Impairment

When mice were exposed to CO 24 h after training for passive avoidance response, step-down latency in the retention testing was shorter compared to the control group in delayed manner (Nabeshima et al., 1991b). For example, when the retention test was carried out 5 and 7 d after CO exposure, the step-down latency in the CO-treated group was significantly shorter than in the control group, but not 1 and 3 d after CO exposure (Nabeshima et al., 1991b). These results suggest that memory is damaged in a retrograde manner even after memory is consolidated.

5.4 CO-Induced Learning and Memory Deficiency

5.4.1 Passive Avoidance Performance

When mice were exposed to CO before training, step-down latency in the retention test which was conducted 1 d after training was affected in delayed manner. For example, when mice were exposed to CO 5 and 7 d before training, the step-down latency in the CO-treated group in the retention testing was significantly shorter than that in the control group, but not when the mice were exposed to CO 1 and 3 d before training (Hiramatsu et al., 1994a; Nabeshima et al., 1991b) (Figure 1A). These results indicate that amnesia was produced by pretraining CO exposure after a delay of more than 5 d ("delayed amnesia"). The delayed amnesia appears to be developed in parallel with "delayed neuronal damage" in the hippocampal CA1 subfield (Nabeshima et al., 1991b).

Successive CO exposure seems to be more effective to produce delayed amnesia compared to single CO exposure (Ishimaru et al., 1991; Maurice et al., 1994b). When CO exposure was performed at room temperature and body temperature was not regulated, the decrease in step-down latency was less, but less significant than in mice exposed 3 times to CO and with maintained body temperature at 38°C (Maurice et al., 1994b).

5.4.2 Spontaneous Alternation Performance

In the Y-maze test, successive CO exposures decreased the spontaneous alternation performance in a time-related manner, a significant difference from control levels being observed 3 d after CO exposure and lasting for at least 7 d after the exposure (Hiramatsu et al., 1994a; Maurice et al., 1994b) (Figure 1B). When mice were exposed to CO 3 times, the alternation in

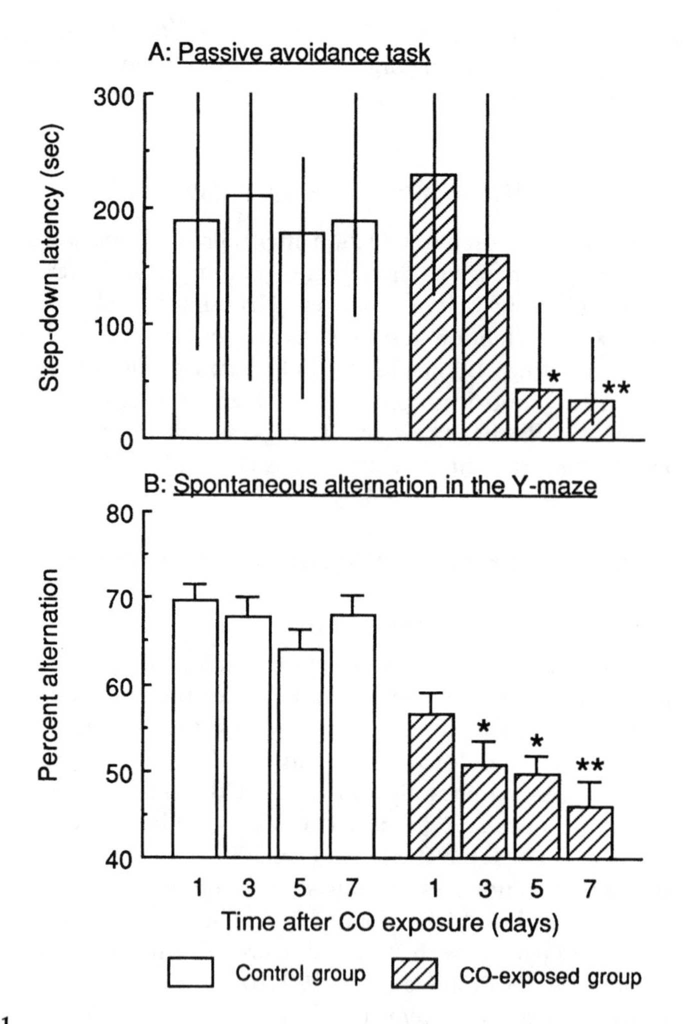

FIGURE 1
Delayed amnesia induced by pretraining CO exposure in mice in the passive avoidance task and spontaneous alternation in the Y-maze. (A) Mice were exposed to CO 1, 3, 5 and 7 d before training. Retention test was carried out 24 h after training. Each value shows median and interquartile ranges. (B) Mice were exposed to CO 1, 3, 5, and 7 d before the Y-maze test. Each value shows mean and S.E. * $p<0.05$, ** $p<0.01$ vs. non-CO-treated control group.

performance 5 d after exposure decreased significantly; however, differences between control and CO-exposed groups in the total number of arm entries during the 8-min session, one of the parameters of motor activity, were not significant (Maurice et al., 1994b).

Total arm entries were not affected by exposure to CO 5 d before the test. Although it is possible that several other behaviors could modify the alternation in the Y-maze test, no side-effect behaviors, such as modification of locomotion, hypothermia, etc., were observed 5 d after CO exposure.

We therefore assumed that the decrease in spontaneous alternation directly reflected the impairment of working memory, as defined by Sarter et al. (1988) and Parada-Turska and Turski (1990).

6 MECHANISMS UNDERLYING NEURONAL DYSFUNCTION AFTER CO EXPOSURE

6.1 Involvement of Excitatory Amino Acids in Memory Impairment and Cell Damage

Recent evidence implicates the endogenous excitatory amino acids, such as glutamate and aspartate, in ischemic neurodegeneration (Beneviste et al., 1984; Jorgensen et al., 1982; Simon et al., 1984). The accumulation of glutamate and aspartate has been implicated in the selective vulnerability of some neurons in the hippocampus and striatum that receive input using these amino acids (Meldrum and Garthwaite, 1990). The high density of NMDA receptors in the hippocampus may also explain the vulnerability of this structure to excitotoxicity (Gill et al., 1987; Monaghan and Cotman, 1985). Both excitatory amino acids are known to cause edema and cell death at high concentration, and several experiments have shown that anoxia also enhances glutamate release and that large amounts of glutamate are accumulated in the brain during ischemia (Globus et al., 1988a). It is known that glutamate induces a rapid accumulation of Na^+, Cl^-, and H_2O into cells after acute exposure; in the presence of Ca^{2+}, cells exposed to glutamate for a brief period exhibit increased Ca^{2+} influx and delayed cell damage or death (Flynn et al., 1989).

The neuronal network that modulates learning and memory processes appears to be located in the hippocampus, and this may explain the severe impairment of learning and delayed amnesia observed in ischemic models (Davis et al., 1987; Volpe et al., 1984). Blockade of NMDA receptors by a noncompetitive antagonist (+)MK-801, prevents CO-induced neuronal damage in mice (Ishimaru et al., 1991) and delayed memory deficiency in mice (Maurice et al., 1994b; Nabeshima et al., 1990a) as well as ischemia-induced neurodegeneration in the gerbil (Gill et al., 1987; Rothman and Olney, 1986). CPP (3-[(±)-carboxypiperazin-4-yl]-propyl-1-phosphonic acid), a competitive NMDA receptor blocker, and 7-chlorokynurenic acid, an antagonist of glycine binding sites, also significantly reduced the CO-induced neurodegeneration (Ishimaru et al., 1992). Ifenprodil (an antagonist of polyamine binding sites), however, failed to protect it (Ishimaru et al., 1992). Furthermore, (+)MK-801 also effectively protected CO-induced delayed amnesia in passive avoidance performance, and in alternation performance with no effect being shown on the total number of arm entries in the Y-maze test (Maurice et al., 1994b). From these results, it is clear that

NMDA receptor/ion channel complex is involved in the mechanism of CO-induced delayed neurodegeneration and memory impairment.

6.2 Other Mechanisms Underlying Learning, Memory, and Neuronal Dysfunction

Complete ischemia and hypoxia decreases protein synthesis and degradation (Hossmann, 1985). During the recirculation period protein synthesis remains depressed, probably because of a chain-initiating defect. Recovery depends on the duration of the hypoxic period. Selectively vulnerable areas of brain are reported to undergo a delayed suppression of protein synthesis after initial recovery; this interesting finding may be related to defects in Ca^{2+}-calmodulin-dependent phosphorylation, the effects of which require time to appear because the cell may have stores of proteins that maintain short-term function (Hossmann, 1985).

7 COMPOUNDS ACTING AS NEUROPROTECTANTS AGAINST CO-INDUCED DELAYED AMNESIA

7.1 MK-801, CPP

(+)MK-801, a selective noncompetitive NMDA receptor antagonist, effectively protected successive CO-induced delayed amnesia in passive avoidance performance (Maurice et al., 1994b; Nabeshima et al., 1990a) and in spontaneous alternation performance, with no effect being shown on the total number of arm entries in the Y-maze test when (+)MK-801 was administered subcutaneously before CO exposure (Maurice et al., 1994b). Treatment with MK-801, before successive CO exposures also significantly reduced the delayed neuronal damage in mice with no effect on body temperature (Ishimaru et al., 1992) as well as ischemia-induced neurodegeneration in the gerbil (Gill et al., 1987; Rothman and Olney, 1986). Furthermore, a competitive NMDA receptor antagonist (CPP) also significantly reduced the CO-induced neurodegeneration (Ishimaru et al., 1992). These results indicate that NMDA receptor antagonists have neuroprotective properties against CO-induced hypoxia.

7.2 Chlorokynurenic Acid, Ifenprodil

The NMDA receptor/ion channel complex also includes strychnine-insensitive glycine binding sites and polyamine binding sites, which are closely associated with the receptor/ion channel complex (Bristow et al., 1986; Carter et al., 1989). As described above, an antagonist of glycine binding sites (7-chlorokynurenic acid) also significantly reduced the CO-induced neurodegeneration (Ishimaru et al., 1992). Ifenprodil (an antagonist of polyamine binding sites), however, failed to protect it (Ishimaru et al., 1992). From these results, compounds antagonizing NMDA receptor/ion channel complex will be a good protectant against hypoxia-induced delayed neurodegeneration or neurological disorders associated with overactivation of NMDA receptor/ion channel complex. However, none of these are available in clinical application yet.

7.3 Cholecystokinin-Related Peptides (CCK, Sulfated CCK-8, Ceruletide)

Systemic and central administration of ceruletide, CCK, and sulfated CCK-8 prevented CO-induced delayed amnesia (Maurice et al., 1994b). Under these conditions, the mnemonic capacity, alternation performance in the Y-maze test, and step-down latency in the passive avoidance test were in the same range as values for the saline-treated non-CO-exposed mice (the control group). L-365,260, a selective CCK-B antagonist (Lotti and Chang, 1989), completely blocked the antiamnesic properties of centrally administered ceruletide. Conversely, L-364,718, a selective CCK-A receptor antagonist (Chang and Lotti, 1986; Lotti et al., 1987), had no effect (Maurice et al., 1994b). Therefore, the neuroprotective effect of centrally administered CCK peptides involves brain CCK-B receptors.

CCK-8 and ceruletide have been reported to decrease brain and body temperature after systemic administration (Eigyo et al., 1992; Katsuura and Itoh, 1981); this could help to explain their neuroprotective action. The effect of ceruletide was greater when the body temperature of the mice was maintained at 38°C during successive CO exposures than when the mice were kept at room temperature (Maurice et al., 1994b). Therefore, the hypothermia induced by CCK peptides does not account for the neuroprotective effect of the CCK peptides. Recent studies have examined the mechanism underlying the neuroprotective effects that CCK-related peptides exert on glutamate neurotoxicity in the central nervous system. Shinohara et al. (1992) have reported that CCK-8 and ceruletide suppressed the increases in intracellular free calcium concentrations induced by glutamate in rat neuronal cultures. This effect appeared to be mediated via CCK-B type receptors. Since, CCK-8 effectively suppressed the increase in intracellular calcium induced by kainate, but not that induced by NMDA (Shinohara et al., 1992), the neuroprotective actions of CCK-related peptides and MK-801 may involve different types of glutamate receptors. Tamura et al. (1992) have shown that CCK-8 is effective in reducing both kainate- and NMDA-induced neurotoxicity in rat cortical neuronal cultures. They showed that CCK-8 did not affect the NMDA-induced calcium influx, but did inhibit nitric oxide formation triggered by NMDA receptor activation.

8 COMPOUNDS AMELIORATING CO-INDUCED DELAYED AMNESIA

8.1 NEFIRACETAM (DM-9384)

Nefiracetam, a pyrrolidone derivative, is a novel compound that has recently been proposed to treat cognitive disorders of senility (Nabeshima, 1993; Sarter, 1991). Memory deficiency is induced when mice are exposed to CO 30 min after a training period with passive avoidance performance, and before memory is totally consolidated (acute amnesia) (Nabeshima et al., 1990a, 1992a). When retention testing was carried out 24 h after CO exposure, pretraining administration of this compound significantly ameliorated CO-induced acute amnesia (Hiramatsu et al., 1992b; Nabeshima et al., 1992a).

When training was carried out 7 d after CO exposure, learning impairment was developed in a delayed manner (delayed amnesia). Pretraining administration of nefiracetam significantly ameliorated the CO-induced delayed amnesia when retention testing was carried out one day after training. When mice were exposed to CO 24 h after a training period, delayed amnesia was also induced 7 d after the exposure. Nefiracetam also ameliorated CO-induced delayed amnesia when it was given before the training period and the effect of nefiracetam was blocked by scopolamine (Hiramatsu et al., 1992b).

It has been reported that nefiracetam ameliorates chlordiazepoxide- and cycloheximide-induced amnesia with dysfunction in the cholinergic neuronal system in mice (Nabeshima et al., 1990b, 1991a), and facilitates the acquisition of shuttle avoidance and discrimination learning in rats (Sakurai et al., 1989). Nefiracetam also stimulates acetylcholine release in the frontal cortex (Kawajiri et al., 1990). It is possible that this compound ameliorates the dysfunction of cholinergic neuronal function and modifies impairment of acquisition and/or consolidation of learning and memory induced by CO.

8.2 TACRINE AND NIK-247

It has been reported that cerebral hypoxia induced by KCN, $NaNO_2$ or CO impairs the synthesis of brain acetylcholine (Gibson et al., 1978; Shimada, 1981) and decreases the content of acetylcholine in the frontal cortex (Nabeshima et al., 1991b). Physostigmine, a typical acetylcholinesterase inhibitor, decreases the lethality of hypoxia induced by KCN, $NaNO_2$, or N_2 with 5% oxygen in mice and improves the synthesis of acetylcholine impaired by the hypoxia (Gibson and Blass, 1976; Scremin and Scremin, 1979). It has been known that 4-aminopyridine derivatives, such as tacrine and NIK-247 are potent, centrally acting acetylcholinesterase inhibitors (Sarter, 1991). When tacrine or NIK-247 was administered immediately after training or before the retention, the CO-induced acute amnesia was attenuated (Yoshida et al., 1992). When mice were exposed to CO 7 d before the retention, NIK-247 and tacrine also administered before the retention, improved memory retrieval in CO-exposed mice (Yoshida et al., 1992). These results suggest that hypoxia induced by CO exposure develops dysfunction of cholinergic neurotransmission, and NIK-247 and tacrine may counteract this deficiency by acting as acetylcholinesterase inhibitors.

8.3 Nicotine

Pretraining administration of (–)-nicotine ameliorated CO-induced memory dysfunction (Hiramatsu et al., 1994b) in agreement with previous findings showing facilitation of learning in both active (Barlow et al., 1970; Sansone et al., 1991) and passive (Nordberg and Bergh, 1985; Sansone et al., 1991) avoidance tasks. The doses indicating antiamnesic effects were far below that required to alter either the shock sensitivity or locomotor activity (Hiramatsu et al., 1994b). Administration of a low dose of mecamylamine, a nicotinic acetylcholine receptor antagonist, prior to injection of (–)-nicotine, blocked the antiamnesic effect of (–)-nicotine on delayed amnesia induced by pre-training CO exposure (Hiramatsu et al., 1994b). It has been reported that (–)-nicotine may increase acetylcholine release in brain (Loiacono and Mitchelson, 1990; Terry et al., 1993). These results suggest that (–)-nicotine acts directly on nicotinic acetylcholine receptors and/or indirectly by facilitating acetylcholine release. This may suggest that nicotinic cholinergic systems are also impaired by CO exposure in mice.

8.4 Naftidrofuryl Oxalate

Naftidrofuryl oxalate has an antiamnesic action on the scopolamine-, cycloheximide-, and basal-forebrain lesion-induced amnesia models in rats (Kameyama et al., 1988) and bicuculline- and picrotoxin-induced amnesia models in mice (Kameyama et al., 1987). It has been shown that this compound also improves cerebral-circulation and cerebral metabolism (Avéret et al., 1983; Meynaud et al., 1973) and has weak serotonin-2-receptor blocking action (Nabeshima et al., 1992b). When naftidrofuryl oxalate was administered before training and immediately after mice were exposed to CO, CO-induced acute amnesia was prevented (Nabeshima et al., 1989). Naftidrofuryl oxalate also ameliorated, but less effective compared to acute amnesia, CO-induced delayed amnesia (Nabeshima et al., 1989). There remains no clear explanation for the antiamnesic action of this compound in CO exposure.

8.5 Sigma Receptor-Related Compounds

Sigma ligands such as 1,3-di-(2-tolyl)guanidine (DTG), N-allylnormetazocine ((+)-SKF 10,047) (Weber et al., 1986) reversed the delayed amnesia induced by successive CO-exposure in mice when they were given before behavioral testing in Y-maze and training period in the passive avoidance (Maurice et al., 1994a). This effect seems to be mediated through sigma sites, because a putative sigma receptor antagonist, α-(4-fluorophenyl-2-pyrimidinyl)-1-piperazine butanol (BMY 14802) (Taylor and Dekleva, 1987), exhibited no effect by itself, but blocked the effect of DTG on both short- and long-term memory impairment (Maurice et al., 1994a).

The NMDA system regulates acetylcholine release from septal, cortical, striatal, and nucleus accumbens slices (Jones et al., 1987; Lodge et al., 1988; Nishimura et al., 1990; Snell and Johnson, 1986). Furthermore, various *in vitro* and *in vivo* findings have suggested that hypofunctioning of NMDA receptor-mediated neurotransmission could be reversed by sigma ligands; the potentiating effect of sigma ligands on NMDA responses in the rat has been shown in superfusion studies (Monnet et al., 1991; Roman et al., 1991) and electrophysiologically (Monnet et al., 1990, 1992). The modulatory effect of sigma ligands on cholinergic hypofunction induced by CO exposure cannot be excluded. This modulation could be direct or indirect, since (+)-SKF 10,047 has been reported to increase acetylcholine release, both in guinea-pig cortical slices (Siniscalchi et al., 1987), and in rat frontal cortex *in vivo* at high doses (Matsuno et al., 1992; 1993).

8.6 Dynorphin A (1–13)

Administration of dynorphin A (1–13), a kappa opioid receptor agonist, before the training session, immediately after the training session and before the retention test, prolonged the decreased step-down latency in the successive CO-exposed mice (Hiramatsu et al., 1995). Dynorphin A (1–13) also ameliorated CO-induced impairment of spontaneous alternation performance (Hiramatsu et al., 1996), in agreement with the previous findings indicating improvement of scopolamine-induced impairment of alternation performance in mice (Itoh et al., 1993). These ameliorating effects of dynorphin were almost completely antagonized by nor-binaltorphimine, a kappa opioid receptor antagonist (Hiramatsu et al., 1995) (Table 2). However, dynorphin A (1–13) did not facilitate the acquisition of memory in normal mice (Hiramatsu et al., 1995). These results suggest that dynorphin A (1–13) may be capable of ameliorating cholinergic dysfunction via the kappa opioidergic system. This hypothesis may be supported by the previous finding that dynorphin potentiates learning in basal forebrain-lesioned rats in a step-through-type passive avoidance task (Ukai et al., 1993).

9 CONCLUSIONS

Evidence for the involvement of cholinergic neurotransmission in learning and memory are substantial. Systemic administration of muscarinic cholinergic

TABLE 2

Antagonism by Nor-Binaltorphimine of the Effects of Dynorphin A (1-13) on CO-Induced Delayed Amnesia in Mice Using Passive Avoidance and Spontaneous Alternation in the Y-maze

Treatments (nmol/mouse, i.c.v.)		N	Median	Range
Passive Avoidance				
Control				
Saline	Saline	16	178.5	(138.5–299.3)
CO exposure				
Saline	Saline	16	40.50	(17.3–91.3)*
Saline	Dynorphin A (1-13) 1.5	14	193.5	(169.0–269.8)**
nBNI 5.4	Dynorphin A (1-13) 1.5	15	69.0	(28.5–118.5)***
nBNI 5.4	Saline	15	28.0	(11.5–61.5)
Spontaneous Alternation in the Y-Maze				
Control				
Saline	Saline	27	70.6%	(60.7–77.1%)
CO exposure				
Saline	Saline	25	61.1%	(57.7–64.7%)*
Saline	Dynorphin A (1–13) 1.5	27	70.0%	(61.6–75.6%)**
nBNI 5.4	Dynorphin A (1–13) 1.5	16	60.0%	(48.1–67.3%)***
nBNI 5.4	Saline	14	55.9%	(45.8–60.8%)

Note: Training was carried out 7 d after CO exposure. Mice were treated intracerebroventricularly with nor-binaltorphimine (nBNI; 5.4 nmol/mouse) and dynorphin A (1–13) 30 and 15 min, respectively, before the first training session, and the retention test was carried out 24 h after training. Each value shows the median and ranges. *N* shows the number of mice used. *$p<0.05$ vs. normal control (Mann-Whitney U-test), **$p<0.05$ vs. CO + saline + saline, ***$p<0.05$ vs. CO + saline + dynorphin (Bonferroni's test).

antagonists, such as scopolamine, disrupts the performance of experimental animals in a wide variety of learning and memory tasks, including inhibitory (passive) avoidance (Kameyama et al., 1986) and spatial maze tasks (Buresova et al., 1986; Wirsching et al., 1984). However, recent findings have indicated that independent manipulation of the cholinergic receptor blockade in experimental animals may be an inadequate model of cognitive dysfunction (Nabeshima et al., 1993). Consistent with this hypothesis, cholinergic dysfunction observed in aging and Alzheimer's disease is accompanied by changes in other neurotransmitter systems, such as peptidergic (Hiller et al., 1987; Jiang et al., 1989) and noradrenergic (Scarpace and Abrass, 1988) systems, which may be important in memory modulation.

The hypoxia induced by CO exposure has been shown to involve the neurotoxicity of excitatory amino acids (Nabeshima et al., 1991b). Moderate neuronal damage has been observed in the hippocampal CA1 subfield

(Ishimaru et al., 1991, 1992), and appeared to parallel the onset of delayed amnesia in a passive avoidance test (Nabeshima et al., 1991b) and the impairment of spontaneous alternation behavior (Maurice et al., 1994b). As mentioned above, animals exhibit dysfunction in not only the cholinergic but also monoaminergic neurons in the frontal cortex, striatum, and hippocampus, which are important brain regions in learning and memory. Inasmuch as several neurotransmitter systems are also implicated in learning and memory processes in disease states, a CO-induced amnesia model could be a useful model for the investigation of memory deterioration. Although the exact mechanisms underlying CO-induced amnesia remain to be examined, we hope that this animal model could be used to develop therapeutic drugs for dementia.

ACKNOWLEDGMENTS

The authors thank Dr. Kiyofumi Yamada for his helpful discussions. This work was supported in part by a grant from the Uehara Memorial Foundation to M.H. and T.N. and grants-in-aid for scientific research from the Ministry of Education, Science and Culture, Japan (No. 05780588 for M.H., No. 05454582 for T.K. and Nos. 03304036, 05454148, 00092093 for T.N.).

REFERENCES

Akiyama, Y., Koshimura, K., Ohue, Y., Lee, K., Miwa, S., Yamagata, S., and Kikuchi, H., Effects of hypoxia on the activity of the dopaminergic neuron system in the rat striatum as studied by *in vivo* brain microdialysis, *J. Neurochem.*, 57, 997, 1991.

Amano, M., Hasegawa, M., Hasegawa, T., and Nabeshima, T., Characteristics of transient cerebral ischemia-induced deficits on various learning and memory tasks in male mongolian gerbils, *Jpn. J. Pharmacol.*, 63, 469, 1993.

Avéret, N., Rigoulet, M., and Cohadon, F., Effect du naftidrofuryl sur lóedème cérébral vasogénique chez le lapin, *Circ. Métab. Cerveau*, 2, 145, 1983.

Barlow, R.B., Oliverio, A., Satta, M., and Thompson, G.M., Some central effects in mice of compounds related to nicotine, *Br. J. Pharmacol.*, 39, 647, 1970.

Beneviste, H., Drejer, J., Schousboe, A., and Diemer, N.H., Elevation of extracellular concentrations of glutamate and aspartate in rat hippocampus during transient cerebral ischaemia monitored by intracerebral microdialysis, *J. Neurochem.*, 43, 1369, 1984.

Boris-Möller, F., Smith, M.-L., and Siesjö, B.K., Effects of hypothermia on ischemic brain damage: A comparison between preischemic and postischemic cooling, *Neurosci. Res. Commun.*, 5, 87, 1989.

Brierly, J.B. and Graham, D.I., Hypoxia and vascular disorders of central nervous system, in *Greenfield's Neuropathology*, Adams, J.H., Corsellis, J.A.N., and Duchen, L.W., Eds., Edward Arnold, London, 1984, 125.

Bristow, D.R., Bowery, N.G., and Woodruff, G.N., Light microscopic autoradiographic localization of [^{3}H]glycine and [^{3}H]strychnine binding sites in rat brain, *Eur. J. Pharmacol.*, 126, 303, 1986.

Buresova, O., Bolhuis, J.J., and Bures, J., Differential effects of cholinergic blockade on performance of rats in the water tank navigation task and in a radial water maze, *Behav. Neurosci.*, 100, 476, 1986.

Busto, R., Dietrich, W.D., Globus, M.Y., Valdes, I., Scheeinberg, P., and Ginsberg, M.D., Small differences in intraischemic brain temperature critically determine the extent of ischemic neuronal injury, *J. Cereb. Blood Flow Metab.*, 7, 729, 1987.

Carter, C., Rivy, J.-P., and Scatton, B., Ifenprodil and SL 82.0715 are antagonists at the polyamine site of the N-methyl-D-aspartate (NMDA) receptor, *Eur. J. Pharmacol.*, 164, 611, 1989.

Chang, R.S.L. and Lotti, V.J., Biochemical and pharmacological characterization of an extremely potent and selective nonpeptide cholecystokinin antagonist, *Proc. Natl. Acad. Sci. U.S.A.*, 83, 4923, 1986.

Clifton, G.L., Taft, W.C., Blair, R.E., Choi, S.C., and DeLorenzo, R.J., Conditions for pharmacological evaluation in the gerbil model of forebrain ischemia, *Stroke*, 20, 1545, 1989.

Davis, H.P., Baranowski, J.R., Pulsinelli, W. A., and Volpe, B.T., Retention of reference memory following ischemic hippocampal damage, *Physiol. Behav.*, 39, 783, 1987.

Eigyo, M., Katsuura, G., Shinaku, H., Shinohara, S., Katoh, A., Shiomi, T., and Matsushita, A., Systemic administration of a cholecystokinin analogue, ceruletide, protects against ischemia-induced neurodegeneration in gerbils, *Eur. J. Pharmacol.*, 214, 149, 1992.

Flynn, C.J., Farooqui, A.A., and Horrocks, L.A., Ischemia and hypoxia, in *Basic Neurochemistry: Molecular, Cellular, and Medical Aspects,* 4th ed., Siegel, G.J., Agranoff, B.W., Albers, R.W., and Molinoff, P.B., Eds., Raven Press, New York, 1989, 783.

Gibson, G.E. and Blass, J.P., Impaired synthesis of acetylcholine in brain accompanying mild hypoxia and hypoglycemia, *J. Neurochem.*, 27, 37, 1976.

Gibson, G.E., Shimada, M., and Blass, J.P., Alterations in acetylcholine synthesis and cyclic nucleotides in mild cerebral hypoxia, *J. Neurochem.*, 31, 757, 1978.

Gill, R., Foster, A.C., and Woodruff, G.N., Systemic administration of MK-801 protects against ischemia-induced hippocampal neurodegeneration in the gerbil, *J. Neurosci.*, 7, 3343, 1987.

Ginsberg, M.D., Delayed neurological deterioration following hypoxia, *Adv. Neurol.*, 26, 21, 1979.

Globus, M.Y.-T., Busto, R., Dietrich, W.D., Martinez, E., Valdes, I., and Ginsberg, M.D., Effect of ischemia on the *in vivo* release of striatal dopamine, glutamate, and γ-aminobutyric acid studied by intracerebral microdialysis, *J. Neurochem.*, 51, 1455, 1988a.

Globus, M.Y.-T., Busto, R., Dietrich, W.D., Martinez, E., Valdes, I., Scheinberg, P., and Ginsberg, M.D., Intra-ischemic extracellular release of dopamine and glutamate is associated with striatal vulnerability to ischemia, *Neurosci. Lett.*, 91, 36, 1988b.

Hiller, J.M., Itzhak, Y., and Simon, E.J., Selective changes in mu, delta and kappa opiod receptor binding in certain limbic regions of the brain in Alzheimer's disease patients, Brain Res., 406, 17, 1987.

Hiramatsu, M., Kameyama, T., and Nabeshima, T., Experimental techniques for developing new drugs acting on dementia (6) — Carbon monoxide-induced amnesia model in experimental animals, *Jpn. J. Psychopharmacol.*, 14, 305, 1994a.

Hiramatsu, M., Koide, T., Isihara, S., Kameyama, T., and Nabeshima, T., Changes in brain receptors and their contents of acetylcholine in carbon monoxide-induced amnesia in mice, in *Proc. 112th Annual Meeting Pharmaceutical Society of Japan,* 3, 36, 1992a.

Hiramatsu, M., Koide, T., Ishihara, S., Shiotani, T., Kameyama, T., and Nabeshima, T., Involvement of the cholinergic system in the effects of nefiracetam (DM-9384) on carbon monoxide (CO)-induced acute and delayed amnesia, *Eur. J. Pharmacol.*, 216, 279, 1992b.

Hiramatsu, M., Sasaki, M., and Kameyama, T., Effects of dynorphin A (1-13) on carbon monoxide (CO)-induced delayed amnesia in mice studied in a step-down type passive avoidance task, *Eur. J. Pharmacol.*, 282, 185, 1995.

Hiramatsu, M., Sasaki, M., Nabeshima, T., and Kameyama, T., Effects of dynorphin A (1-13) on carbon monoxide-induced delayed amnesia in mice studied in spontaneous alternation performance, *Pharmacol. Biochem. Behav.*, in press, 1996.

Hiramatsu, M., Satoh, H., Kameyama, T., and Nabeshima, T., Nootropic effect of nicotine on carbon monoxide (CO)-induced delayed amnesia in mice, *Psychopharmacology,* in press, 1994b.

Hiramatsu, M., Satoh, H., Murai, M., Kameyama, T., and Nabeshima, T., Studies of (–)-nicotine on carbon monoxide-induced learning impairment in mice, in *Proc. Int. Symp. on Nicotine, The Effects of Nicotine on Biological Systems II,* 1994c.

Hiramatsu, M., Yokoyama, S., Nabeshima, T., and Kameyama, T., Changes in concentrations of dopamine, serotonin, and their metabolites induced by carbon monoxide (CO) in the rat striatum as determined by *in vivo* microdialysis, *Pharmacol. Biochem. Behav.*, 48, 9, 1994d.

Horowitz, A.L., Kaplan, R., and Sarpel, G., Carbon monoxide toxicity: MR imaging in brain, *Radiology,* 162, 787, 1987.

Hossmann, K.-A., Post-ischemic resuscitation of the brain: Selective vulnerability versus global resistance, in *Progress in Brain Research, Vol. 63,* Kogure, K., Hossmann, K.-A., Siesjö, B.K., and Welsh, F.A., Eds., Elsevier, New York, 1985.

Ishimaru, H., Nabeshima, T., Katoh, A., Suzuki, H., Fukuta, T., and Kameyama, T., Effects of successive carbon monoxide exposures on delayed neuronal death in mice under the maintenance of normal body temperature, *Biochem. Biophys. Res. Commun.*, 179, 836, 1991.

Ishimaru, H., Kotah, A., Suzuki, H., Fukuta, T., Kameyama, T., and Nabeshima, T., Effects of N-methyl-D-aspartate receptor antagonists on carbon monoxide-induced brain damage in mice, *J. Pharmacol. Exp. Ther.*, 261, 349, 1992.

Itoh, J., Ukai, M., and Kameyama, T., Dynorphin A-(1-13) markedly improves scopolamine-induced impairment of spontaneous alternation performance in mice, *Eur. J. Pharmacol.,* 236, 341, 1993.

Jiang, H.-K., Owyang, V., Hong, J.-S., and Gallagher, M., Elevated dynorphin in the hippocampal formation of aged rats: Relation to cognitive impairment on the spatial learning task. *Proc. Natl. Acad. Sci. U.S.A.,* 86, 2948, 1989.

Jones, S.M., Snell, L.D., and Johnson, K.M., Inhibition by phencyclidine of excitatory amino acid-stimulated release of neurotransmitter in the nucleus accumbens, *Neuropharmacology,* 26, 173, 1987.

Jorgensen, M.B. and Diemer, N.H., Selective neuron loss after cerebral ischemia in the rat: Possible role of transmitter glutamate, *Acta Neurol. Scand.*, 66, 536, 1982.

Kaltwasser, K., Kleinau, H., Pankow, D., Ponsold, W., and Strahl, U., Hypothermie und kohlenmonoxid-toxizitat, *Zeitschr. Exp. Chirurg.*, 10, 45, 1977.

Kameyama, T., Nabeshima, T., and Noda, Y., Cholinergic modulation of memory for step-down type passive avoidance task in mice, *Res. Commun. Psychol. Psychiat. Behav.*, 11, 193, 1986.

Kameyama, T., Nabeshima, T., and Noda, Y., The effect of naftidrofuryl oxalate (LS-121) on learning and memory, *Pharmacometrics,* 33, 843, 1987.

Kameyama, T., Nabeshima, T., Katoh, A., and Ogawa, S., Effect of naftidrofuryl oxalate (LS-121) on experimental amnesia model in rats, *Folia Pharmacol. Jpn.*, 92, 241, 1988.

Kato, H., Kogure, K., and Nakano, S., Neuronal damage following repeated brief ischemia in the gerbil, *Brain Res.*, 479, 366, 1989.

Katsuura, G. and Itoh, S., Effect of cholecystokinin octapeptide on body temperature in the rat, *Jpn. J. Physiol.*, 31, 849, 1981.

Kawajiri, S., Taniguchi, K., Sakurai, T., Kojima, H., and Yamasaki, T., Effects of DM-9384 on extracellular acetylcholine in the rat frontal cortex measured with microdialysis, *Eur. J. Pharmacol.*, 183, 928, 1990.

Kinbara, K., Nabeshima, T., Yoshida, S., and Kameyama, T., Characterization of carbon monoxide-induced amnesia: Effects of carbon monoxide exposure on the step-down passive avoidance response in mice, *Jpn. J. Pharmacol.*, 49 (Suppl.), 270, 1989.

Kirino, T., Delayed neuronal death in the gerbil hippocampus following ischemia, *Brain Res.*, 239, 57, 1982.

Krebs, M.O., Desce, J.M., Kemel, M.L., Gauchy, C., Godeheu, G., Cheramy, A., and Glowinski, J., Glutamatergic control of dopamine release in the rat striatum: Evidence for presynaptic N-methyl-D-aspartate receptors on dopaminergic nerve terminals, *J. Neurochem.*, 56, 81, 1991a.

Krebs, M.O., Trovero, F., Desban, M., Gauchy, C., Glowinski, J., and Kemel, M.L., Distinct presynaptic regulation of dopamine release through NMDA receptors in striosome- and matrix-enriched areas of the rat striatum, *J. Neurosci.*, 11, 1256, 1991b.

Lapresle, J. and Fardeau, M., The central nervous system and carbon monoxide poisoning. Anatomical study of brain lesions following intoxication with carbon monoxide (22 cases), *Prog. Brain Res.*, 24, 31, 1967.

Levine, S. and Payan, H., Effects of ischemia and other procedures on the brain and retina of the gerbil (*Meriones unguiculatus*), *Exp. Neurol.* 16, 255, 1966.

Lodge, D., Aram, J.A., Church, J., Davies, S.N., Fletcher, E., and Martin, D., Electrophysiological studies of the interaction between phencyclidine/sigma receptor agonists and excitatory amino acid neurotransmission on central mammalian neurons, in *Sigma and Phencyclidine Like Compounds as Molecular Probes in Biology,* Domino, E.F. and Kamenka, J.M., Eds., NPP Books, Ann Arbor, MI, 1988, 239.

Loiacono, R.E. and Mitchelson, F.J., Effect of nicotine and tacrine on acetylcholine release from rat cerebral cortical slices, *Naunyn-Schmiedeb. Arch. Pharmacol.,* 342, 31, 1990.

Lotti, V.J. and Chang, R.S.L., A new potent and selective non-peptide gastrin antagonist and brain cholecystokinin receptor (CCK-B) ligand: L-365,260, *Eur. J. Pharmacol.*, 162, 273, 1989.

Lotti, V.J., Pendleton, R.G., Gould, R.J., Hanson, H.M., Chang, R.S.L., and Clineschmidt, B.V., *In vivo* pharmacology of L-364,718, a new potent nonpeptide peripheral cholecystokinin antagonist, *J. Pharmacol. Exp. Ther.*, 241, 103, 1987.

Matsuno, K., Matsunaga, K., and Mita, S., Increase of extracellular acetylcholine level in rat frontal cortex induced by (+) N-allylnormetazocine as measured by brain microdialysis, *Brain Res.*, 575, 315, 1992.

Matsuno, K., Matsunaga, K., Senda, T., and Mita, S., Increase in extracellular acetylcholine level by sigma ligands in rat frontal cortex, *J. Pharmacol. Exp. Ther.*, 265, 851, 1993.

Maurice, T., Hiramatsu, M., Kameyama, T., Hasegawa, T., and Nabeshima, T., Behavioral evidence for a modulating role of σ ligands in memory processes: II. Reversion of carbon monoxide (CO)-induced amnesia, *Brain Res.,* 647, 57, 1994a.

Maurice, T., Hiramatsu, M., Kameyama, T., Hasegawa, T., and Nabeshima, T., Cholecystokinin-related peptides, after systemic or central administration, prevent carbon monoxide-induced amnesia in mice, *J. Pharmacol. Exp. Ther.*, 269, 665, 1994b.

Meldrum, B. and Garthwaite, J., Excitatory amino acid neurotoxicity and neurodegenerative disease, in *Trends in Pharmacological Science, A TiPS Special Report 1991, The Pharmacology of Excitatory Amino Acids*, Lodge, D. and Collingridge, G., Eds., Elsevier Trends Journals, U.K., 1990, 54.

Meynaud, A., Grand, M., and Fontaine, L., Effect of naftidrofuryl upon energy metabolism of the brain, *Arzneim.-Forsch. (Drug Res.)*, 23, 1431, 1973.

Min, S.K., A brain syndrome associated with delayed neuropsychiatric sequelae following acute carbon monoxide intoxication, *Acta Psychiatr. Scand.*, 73, 80, 1986.

Mitani, A. and Kataoka, K., Critical levels of extracellular glutamate mediating gerbil hippocampal delayed neuronal death during hypothermia: Brain microdialysis study, *Neuroscience*, 42, 661, 1991.

Monaghan, D.T. and Cotman, C.W., Distribution of N-methyl-D-aspartate-sensitive l-[^{3}H]glutamate binding sites in rat brain, *J. Neurosci.*, 5, 2909, 1985.

Monnet, F.P., Debonnel, G., Junien, J.L., and De Montigny, C., N-Methyl-D-aspartate-induced neuronal activation is selectively modulated by sigma receptors, *Eur. J. Pharmacol.*, 179, 441, 1990.

Monnet, F.P., Blier, P., Debonnel, G., and De Montigny, C., The modulation by sigma ligands of NMDA-evoked [^{3}H]noradrenaline release involves $G_{i/o}$ proteins, *Soc. Neurosci. Abstr.*, 17, 1340, 1991.

Monnet, F.P., Debonnel, G., and De Montigny, C., *In vivo* electrophysiological evidence for a selective modulation of N-methyl-D-aspartate-induced neuronal activation in rat CA3 dorsal hippocampus by sigma ligands, *J. Pharmacol. Exp. Ther.*, 261, 123, 1992.

Nabeshima, T., Ameliorating effects of nefiracetum (DM-9384) on brain dysfunction, *Jpn. J. Neuropsychopharmacol.*, 16, 309, 1994.

Nabeshima, T., Behavioral aspects of cholinergic transmission: Role of basal forebrain cholinergic system in learning and memory, *Prog. Brain Res.*, 98, 405, 1993a.

Nabeshima, T., Hiramatsu, M., Koide, T., Ishihara, S., Katoh, A., Ishimru, H., Ichihashi, H., Shiotani, T., and Kameyama, T., Involvement of the cholinergic system in the effects of DM-9384 on carbon monoxide (CO)-induced acute and delayed amnesia, in *The Role of Neurotransmitters in Brain Injury,* Globus, M. and Dietrich, W. D., Eds., Plenum Press, New York, 1992a, 219.

Nabeshima, T., Hiramatsu, M., Niwa, K., Fuji, K., and Kameyama, T., Effect of naftidrofuryl oxalate on 5-HT2 receptors in the mouse brain: Evaluation by using quantitative autoradiography and head-twitch response, *Eur. J. Pharmacol.*, 223, 109, 1992b.

Nabeshima, T., Katoh, A., Ishimaru, H., Yoneda, Y., Ogita, K., Murase, K., Ohtsuka, H., Inari, K., Fukuta, T., and Kemeyama, T., Carbon monoxide-induced delayed amnesia, delayed neuronal death and change in acetylcholine concentration in mice, *J. Pharmacol. Exp. Ther.*, 256, 378, 1991a.

Nabeshima, T., Morinaka, H., Kinbara, K., and Kameyama, T., Effects of naftidrofuryl oxalate on the CO-induced acute and delayed amnesia, *Res. Commun. Psychol. Psychiat. Behav.*, 14, 225, 1989.

Nabeshima, T., Tohyama, K., and Kameyama, T., Effects of DM-9384, a pyrrolidone derivative, on alcohol- and chlordiazepoxide-induced amnesia in mice, *Pharmacol. Biochem. Behav.*, 36, 233, 1990a.

Nabeshima, T., Tohyama, K., Murase, K., Ishihara, S., Kameyama, T., Yamasaki, T., Hatanaka, S., Kojima, H., Sakurai, T., Takasu, Y., and Shiotani, T., Effects of DM-9384, a cyclic derivative of GABA, on amnesia and decreases in GABAA and muscarinic receptors induced by cycloheximide, *J. Pharmacol. Exp. Ther.*, 257, 271, 1991b.

Nabeshima, T., Yoshida, S., and Kameyama, T., Effects of the novel compound NIK 247 on impairment of passive avoidance response in mice, *Eur. J. Pharmacol.*, 154, 263, 1988.

Nabeshima, T., Yoshida, S., Morinaka, H., Kameyama, T., Thurkauf, A., Rice, K.C., Jacobson, A.E., Monn, J.A., and Cho, A.K., MK-801 ameliorates delayed amnesia, but potentiates acute amnesia induced by CO, *Neurosci. Lett.*, 108, 321, 1990b.

Nakano, S., Kato, H., and Kogure, K., Neuronal damage in the rat hippocampus in a new model of repeated reversal transient cerebral ischemia, *Brain Res.*, 490, 178, 1989.

Newby, M.B., Roberts, R.J., and Bhatnagar, R.K., Carbon monoxide- and hypoxia-induced effects on catecholamines in the mature and developing rat brain, *J. Pharmacol. Exp. Ther.*, 206, 61, 1978.

Nishimura, L.M. and Boegman, R.J., N-methyl-D-aspartate-evoked release of acetylcholine from the septal/diagonal band of the rat brain, *Neurosci. Lett.*, 115, 259, 1990.

Nordberg, A. and Bergh, C., Effect of nicotine on passive avoidance behavior and motor activity in mice, *Acta Pharmacol. Toxicol.*, 56, 337, 1985.

Parada-Turska, J. and Turski, W.A., Excitatory amino acid antagonists and memory: Effect of drugs acting at N-methyl-D-Aspartate receptors in learning and memory tasks, *Neuropharmacology,* 29, 1111, 1990.

Penney, D.G., Acute carbon monoxide poisoning: Animal models. A review, *Toxicology,* 62, 123, 1990.

Penney, D.G., Helfman, C.C., Dunbar, J.C., Jr., and McCoy, L.E., Acute severe carbon monoxide exposure in the rat: Effects of hyperglycemia and hypoglycemia on mortality, recovery, and neurologic deficit, *Can. J. Physiol. Pharmacol.*, 69, 1168, 1991.

Penney, D.G., Verma, K., and Hull, J.A., Cardiovascular, metabolic and neurologic effects of acute carbon monoxide poisoning in the rat, *Toxicol. Lett.*, 45, 207, 1989.

Pulsinelli, W.A. and Brierly, J.B., A new model of bilateral hemispheric ischemia in the unanesthetized rat, *Stroke*, 10, 267, 1979.

Pulsinelli, W.A., Brierly, J.B., and Plum, F., Temporal profile of neuronal damage in a model of transient forebrain ischemia, *Ann. Neurol.*, 11, 491, 1982.
Roman, F.J., Pascaud, X., Duffy, O., and Junien, J.L., Modulation by neuropeptide Y and peptide YY of NMDA effects in hippocampal slices: Role of sigma receptors, in *NMDA Related Agents: Biochemistry, Pharmacology and Behavior,* Kameyama, T., Nabeshima, T., and Domino, E.F., Eds., NPP Books, Ann Arbor, MI, 1991, 211.
Rothman, S.M. and Olney, J.W., Glutamate and pathophysiology of hypoxic-ischemic brain damage, *Ann. Neurol.*, 19, 105, 1986.
Sakurai, T., Ojima, H., Yamasaki, T., Kojima, H., and Akashi, A., Effects of N-(2,6-dimethylphenyl)-2-(2-oxo-1-pyrrolidinyl) acetamide (DM-9384) on learning and memory in rats, *Jpn. J. Pharmacol.*, 50, 47, 1989.
Sansone, M., Castellano, C., Battaglia, M., and Ammassari-Teule, M., Effects of oxiracetam-nicotine combinations on active and passive avoidance learning in mice, *Pharmacol. Biochem. Behav.*, 39, 197, 1991.
Sarter, M., Taking stock of cognition enhancers, *Trends Pharmacol. Sci.*, 12, 456, 1991.
Sarter, M., Bodewitz, G., and Stephens, D.N., Attenuation of scopolamine-induced impairment of spontaneous alternation behavior by antagonist but not inverse agonist and agonist β-carbolines, *Psychopharmacology,* 94, 491, 1988.
Sawada, Y., Ohashi, N., and Maemura, K., Computerized tomography as an indication of long-term outcome after acute carbon monoxide poisoning, *Lancet,* 1, 783, 1980.
Sawada, Y., Sakamoto, T., and Nishide, K., Correlation of pathological findings with computed tomographic findings after acute carbon monoxide poisoning, *N. Engl. J. Med.*, 308, 1296, 1983.
Scarpace, P.J. and Abrass, I.B., Alpha- and beta-adrenergic receptor function in the brain during senescence, *Neurobiol. Aging,* 9, 53, 1988.
Scremin, A.M.E. and Scremin, O.V., Physostigmine-induced cerebral protection against hypoxia, *Stroke,* 10, 142, 1979.
Scremin, O.U. and Jenden, D.J., Focal ischemia enhances choline output and decreases acetylcholine output from rat cerebral cortex, *Stroke,* 20, 92, 1989.
Shimada, M., Alteration of acetylcholine synthesis in mouse brain cortex in mild hypoxic hypoxia, *J. Neural Transm.*, 50, 233, 1981.
Shinohara, S., Katsuura, G., Eigyo, M., Shintaku, H., Ibii, N., and Matsushita, A., Inhibitory effect of CCK-8 and ceruletide on glutamate-induced rises in intracellular free calcium concentrations in rat neurons cultures, *Brain Res.*, 588, 223, 1992.
Simon, R.P., Swan, J.H., Griffith, T., and Meldrum, B.S., Blockade of N-methyl-D-aspartate receptors may protect against ischemic damage in the brain, *Science,* 226, 850, 1984.
Siniscalchi, A., Cristofori, P., and Veratti, E., Influence of N-allylnormetazocine on acetylcholine release from brain slices: Involvement of muscarinic receptors, *Arch. Pharmacol.*, 336, 425, 1987.
Snell, L.D. and Johnson, K.M., Characterization of the inhibition of excitatory amino acid-induced neurotransmitter release in the rat striatum by phencyclidine-like drugs, *J. Pharmacol. Exp. Ther.*, 238, 938, 1986.
Sutariya, B., Penney, D.G., Barnes, J., and Helfman, C., Hypothermia protects brain function in acute carbon monoxide poisoning, *Vet. Hum. Toxicol.*, 31, 436, 1989.
Sutariya, B.B., Penney, D.G., and Nallamothu, B.G., Hypothermia following acute carbon monoxide poisoning increases mortality, *Toxicol. Lett.*, 52, 201, 1990.
Take, Y., Narumi, S., Nagai, Y., Kurihara, E., Saji, Y., and Nagawa, Y., Neurochemical study of the temporary cerebral ischemic rats produced by bilateral vertebral and carotid artery occlusion, *Folia Pharmacol. Jpn.*, 84, 485, 1984a.
Take, Y., Yamazaki, N., Fukuda, N., Saji, Y., and Nagawa, Y., Physiological and biochemical study of temporary cerebral ischemic rats produced by bilateral vertebral and carotid artery occlusion, *Folia Pharmacol. Jpn.*, 84, 471, 1984b.
Tamura, Y., Sato, Y., Akaike, A., and Shiomi, H., Mechanisms of cholecystokinin-induced protection of cultured cortical neurons against N-methyl-D-aspartate receptor-mediated glutamate cytotoxicity, *Brain Res.*, 592, 317, 1992.

Taylor, D.P. and Dekleva, J., Potential antipsychotic BMY 14802 selectively binds to sigma sites, *Drug Dev. Res.*, 11, 65, 1987.

Terry, A.V., Jr., Buccafusco, J.J., and Jackson, W.J. Scopolamine reversal of nicotine enhanced delayed matching-to-sample performance in monkeys, *Pharmacol. Biochem. Behav.*, 45, 925, 1993.

Tomida, S., Nowak, T.S., Jr., Vass, K., Lohr, J.M., and Klatzo, I., Experimental model for repetitive ischemic attacks in the gerbil: The cumulative effect of repeated ischemia insult, *J. Cereb. Blood Flow Metab.*, 7, 773, 1987.

Ukai, M., Kobayashi, T., and Kameyama, T., Dynorphin A-(1-13) attenuates basal forebrain-lesion-induced amnesia in rats, *Brain Res.*, 625, 355, 1993.

Volpe, B.T., Pulsinelli, W.A., Tribuna, J., and Davis, H.P., Behavioral performance of rats following transient forebrain ischemia, *Stroke*, 15, 558, 1984.

Weber, E., Sonders, M., Quarum, M., McLean, S., Pou, S., and Keana, J.F.W., 1,3-Di-(2-[5-^{3}H]tolyl)guanidine: A selective ligand that label sigma-type receptors for psychotomimetic opiates and antipsychotic drugs, *Proc. Natl. Acad. Sci. U.S.A.*, 83, 8784, 1986.

Weinberger, J., Nieves-Rosa, J., and Cohen, G., Nerve terminal damage in cerebral ischemia: Protective effects of alpha-methyl-para-tyrosine, *Stroke*, 16, 874, 1985.

Wirsching, B.A., Beninger, R.J., Jhamandas, K., Boegman, R.J., and El-Defrawy, S.R., Differential effects of scopolamine on working and reference memory of rats in the radial maze, *Pharmacol. Biochem. Behav.*, 20, 659, 1984.

Yamazaki, N., Take, Y., Nagaoka, A., and Nagawa, Y., Beneficial effect of idebenon (CV-2619) on cerebral ischemia-induced amnesia in rats, *Jpn. J. Pharmacol.*, 36, 349, 1984.

Yasumatsu, H., Yamamoto, Y., Takamuku, H., Anami, K., Takehara, S., Setoguchi, M., and Maruyama, Y., Pharmacological studies on Y-8894 (VII). Effects on transient cerebral ischemia-induced amnesia in rats, *Folia Pharmacol. Jpn.*, 90, 321, 1987.

Yoshida, S., Nabeshima, T., Kinbara, K., and Kameyama, T., Effects of NIK-247 on CO-induced impairment of passive avoidance in mice, *Eur. J. Pharmacol.*, 214, 247, 1992.

CHAPTER 10

Behavioral Effects of Carbon Monoxide Exposure: Results and Mechanisms

Vernon A. Benignus

CONTENTS

0-8493-4796-3/96/$0.00+$.50

1 INTRODUCTION

Short-term exposure to carbon monoxide (CO) will produce frank behavioral effects (unconsciousness) if carboxyhemoglobin (COHb) rises to sufficiently high levels. Despite the fact that the above effect has been known for many years, and despite the fact that many experiments have been conducted, the portion of the human dose-effects curve between zero effects and unconsciousness is empirically unknown. Although there are many reports of effects below the level of COHb needed to produce unconsciousness, there are always equally persuasive reports of no effects at the same or greater COHb. The uncertainty is so great as to dissallow conclusions about the dose-effects curve. This may be due to a number of factors, including (a) the inherent (and previously unappreciated) complexity of the problem, (b) the possibility of a specially sensitive group of people, and (c) methodological problems that are prevalent in the extant research.

The lack of data about less-than-frank behavioral effects of COHb elevation does not prevent a theoretical and inferential discussion. There is sufficient theory available to begin to construct a model for the prediction of behavioral effects. Given the suggested model, there are also data available for extrapolation from other areas of research that can be used to construct at least a crude estimate of a behavioral dose-effects curve.

In the present chapter, a theoretical (mechanism of action) discussion will be given first, followed by a review of the literature. Extrapolations of a dose-effects curve will be reviewed as an effort to give numeric flesh to the theoretical bones. In each section a discussion will be made about what remaining work is needed before claims can be made of understanding the behavioral effects of short-term CO exposure.

Conspicuous by its absence will be any discussion of behavioral effects of long-term exposure to CO. There are few data and the theory is not as well considered as for short-term exposure.

2 PHYSIOLOGY — MECHANISMS

It has generally been assumed that the mechanism by which CO produces behavioral effects (whatever they may be) is by hypoxemia leading to tissue hypoxia (Coburn, 1979). It has also been suggested that CO dissolved in the blood and not bound to hemoglobin might diffuse into tissue and become cytotoxic (Piantadosi et al., 1987) but there has been no convincing evidence that this in fact occurs to an important extent (McMullen and Raub, 1991). It will therefore be assumed that the mechanism of CO effects on the central nervous system (CNS) is via hypoxemia, leading to tissue hypoxia.

2.1 Variables that Determine Tissue Hypoxia

2.1.1 Hypoxemia

The relationship between COHb and arterial hypoxemia is

$$CaO_2 = (1.39 \text{ Hb } \{SaO_2/100\})\,(1 - COHb/100) \qquad (10.1)$$

in which CaO_2 is the arterial oxygen (O_2) contents in ml/100ml (vol%), Hb is the total hemoglobin in g/ml of blood, SaO_2 is the percent O_2 saturation of arterial blood and COHb is in percent. With normal Hb and SaO_2, the expression reduces to approximately

$$CaO_2 = 19.6\,(1 - COHb/100) \qquad (10.2)$$

Equation 10.2 gives the amount of O_2 per 100 ml of arterial blood delivered to (in this case) the brain as a function of COHb.

2.1.2 Tissue Hypoxia

The effect of COHb on tissue hypoxia is, however, not as simple as the reduction in CaO_2. Not only is CaO_2 reduced during increased COHb, but extraction of O_2 from arterial blood by tissue becomes more difficult (Roughton and Darling, 1944).

The rate at which O_2 is metabolized by the brain ($BMRO_2$), in ml/min/100g of tissue, is given by

$$BMRO_2 = BBF\,(CaO_2 - CvO_2) \qquad (10.3)$$

in which BBF is the brain blood flow in ml/min/100g of tissue and CvO_2 is the venous blood O_2 content. The term (CaO_2–CvO_2) is the arterio-venous difference (AVD) or the amount of O_2 extracted from the arterial blood by the brain in ml/100ml of blood. During COHb elevation, BBF is observed to rise so that, on the average, the $BMRO_2$ is not altered until about 30% COHb, in dogs and sheep (Jones and Traystman, 1984; Chapter 4, this volume). The increased BBF, therefore, acts to compensate for the reduced CaO_2 as well as the increased difficulty in O_2 extraction, at least until COHb rises above 30%.

Apparently, the $BMRO_2$ is controlled closely by the physiological equivalent of a closed-loop regulatory system (Randall, 1962; Chapter 4, this volume). In such a system, if the system is stable and has well-matched time constants, there would be little change in the $BMRO_2$ as COHb increased. BBF would simply increase and perhaps extraction would increase to maintain a stable $BMRO_2$. Eventually, however, some sort of limit would be reached in the compensatory system and $BMRO_2$ would begin to drop. It is not clear what does, in fact, limit the regulation of $BMRO_2$, but apparently $BMRO_2$ does begin to drop as COHb approaches 30%. It is not clear whether the system begins to go out of regulation gradually, allowing $BMRO_2$ to slowly increase or whether the regulatory system breaks down precipitously in the individual, followed by a rapid drop in $BMRO_2$. The latter could be the case if each individual's regulatory limits were reached at a different COHb, because the mean curve would still appear to decline gradually. In any case, it would appear, upon cursory consideration, that until $BMRO_2$ began to decline, no behavioral decrements ought to be observed.

2.1.3 *Individual Differences*

The rise in BBF due to COHb increase has been observed indirectly in humans (Benignus et al., 1992). It was also observed that a considerable difference exists among individual subjects in the extent to which their BBF increased per unit increase in COHb. Some of the subjects appeared not to increase BBF at all. These differences were stable over 4 h. All differences were measured with respect to baseline, however, and the possibility of elevated baselines in the "noncompensators" cannot be ignored. It is noteworthy that there are large individual differences in many pulmonary and blood-gas behaviors in response to, for example, high altitude and carbon dioxide (CO_2) inhalation (Benignus, 1994) and it is therefore not unlikely that the BBF response to COHb could also be variable. If such differences can be reliably documented, it would appear that those persons whose BBF does not increase in response to COHb would be at greater risk than others because their $BMRO_2$ might decrease well before those whose BBF rises in a compensatory manner.

2.1.4 *Regional Differences*

The matter of characterizing brain oxygen consumption in the face of increased COHb is possibly further complicated by reports of regional differences in responsiveness of BBF (rBBF) to elevated COHb (Hanley et al., 1986; Koehler et al., 1982, 1984; Okeda et al., 1987; Chapter 4, this volume). In general, areas of the brain with high baseline blood flow developed larger increases than areas with low baseline flow. The functional significance of the above observations is not clear, however. It may be argued that areas with large baseline rBBF would require proportionally larger increases of rBBF to maintain regional $BMRO_2$ ($rBMRO_2$). Calculation of percent increase over baseline for the data of Koehler et al. (1984) reduced the variation across brain areas but there were still some differences. It would be difficult to determine whether the percent increases differed significantly from each other without the raw data. Without further information, the importance of the differences in compensatory increase in rBBF must be considered unresolved.

2.2 An Appropriate Independent Variable in CO Studies of Behavior

2.2.1 *Carboxyhemoglobin*

In the past, almost all investigators have related behavioral effects to COHb. If it may be assumed that the BBF and O_2-extraction responses of all subjects are the same, then COHb is uniquely related (although nonlinearly) to whole-brain $BMRO_2$ and thus would seem to be a reasonable independent variable. If, however, there are individual differences in the responsiveness of BBF or O_2 extraction to COHb elevation, then COHb is not uniquely related to $BMRO_2$ and may therefore be a poor independent variable for behavioral studies.

2.2.2 *Effect of Individual Differences*

If there is an important amount of individual difference in the responsiveness of BBF and O_2 extraction in the face of increased COHb, then the experimenter is faced with a much more complicated measurement problem in trying to quantify the independent variable. Some imaging method might be used concurrently with ongoing task performance, but the problems of (a) interference with task performance, (b) obtaining continuously varying numeric measures of $BMRO_2$, and (c) the expense of equipment are all formidable obstacles. Noninvasive measures of BBF, collected simultaneously with behavior, are free of most of the problems with imaging methods and might help quantification if O_2 extraction is relatively stable across individuals.

2.2.3 *Problem With COHb*

In the above, it has been tacitly assumed that COHb is known. Given the CO concentration in inhaled air and the value of various cardiopulmonary parameters in the individual subject, the COHb can be predicted with excellent accuracy (Chapter 3, this volume) using the Coburn-Forster-Kane equation (CFKE). The difficulty in practice is that many of the cardiopulmonary parameters cannot be known for the individual and are labile as a function of what the subject is doing, such as exercise level, emotional arousal, etc. This lability is especially troublesome for pulmonary ventilation (Benignus and Annau, 1994), and consequently, alveolar partial pressure of O_2. In an effort to solve the problem of unknown physiological parameters, a computer model has been devised (Benignus, 1994) to predict COHb as a continuous function of some 35 input and 60 output variables providing for continuous variation in the concentrations of CO, O_2, and CO_2.

Another problem with prediction of COHb occurs during very short exposures to high CO concentrations where the arterial COHB rises at a rapid rate to values considerably higher than observed in even peripheral venous blood samples (Benignus et al., 1994). A multicompartment model has been constructed (Smith et al., 1994) to make better arterial predictions. It has also been demonstrated that for very short, high-level exposures, aortic blood rises to even higher values than peripheral arterial blood in a pulsatile manner, synchronized with respiration (Abboud et al., 1974). These high concentrations of arterial COHb can become quite large, so that predictions must be used with considerable caution in such cases. High values of arterial COHb, even in short durations, can produce brain hypoxia with functional consequences.

The problem of COHb estimation does not occur if COHb is measured in the individual and the exposure levels of CO are low enough so that no appreciable differential arises between the arterial and venous COHb. If either of the above two conditions do not obtain, the problem of an adequate independent variable is even more complicated.

2.3 Required Research

The independent variable must be adequately measured or controlled. If there is consistent closed loop regulation of $BMRO_2$, then COHb would be a very nonlinear relative of $BMRO_2$ because it would have a high threshold followed by a more-or-less sudden effect onset. This would not prevent its being well characterized, however, merely inconvenient.

It would be more convenient and mechanistically informative if the relationship between $BMRO_2$ could be quantitatively described by a physiologically based model. Such a model could be used to predict $BMRO_2$ decreases in the many possible cases in which the closed-loop regulation would fail due to, for example, ill health or injury of the system. It would account for the high COHb threshold (if this were the case). It is within the grasp of physiological theory to devise such a model and it seems important to do so.

A well-constructed and correct model for relating COHb to $BMRO_2$ would work well for the average case or, perhaps, for the group average. If, however, there are large individual differences in the putative closed-loop regulatory system, then neither would COHb be closely related to $BMRO_2$, nor would $BMRO_2$ be predictable from the model. There is already some evidence that such large individual differences exist, although no well-demonstrated empirical answer to the question is available. Furthermore, if areas of the brain were to differ in the closeness to which $BMRO_2$ is regulated, the problem would become even more complicated. To reduce our uncertainty about such matters, it is important to learn to what extent individuals and/or brain areas differ in the responsiveness of BBF and O_2 extraction as COHb rises. Experiments on conscious laboratory animals have been done, but individual differences have not been quantified because of the requirements of repeated measures over time. The experiments can be performed on humans using noninvasive methods.

3 BEHAVIORAL EFFECTS

3.1 General Observations

In the previous section it was argued that the physiological mechanisms by which COHb produces behavioral effects was relatively well understood. Yet the relative importance of the various physiological factors was unknown. Furthermore, no method of making numerical predictions is available. Matters are even less well understood and reported findings are more confusing in the area of observed behavioral effects of COHb elevation.

3.1.1 Historical

Following early work with high-level CO exposure and more-or-less uncontrolled observations, two research reports on behavioral effects of CO in humans had historically important consequences. It may be argued that these two archetypical reports (a) raised the interest of both the scientific and regulatory communities in CO as a pollutant, and (b) apparently misled these communities to a surprising extent. The controversy generated by these reports is still in the process of being resolved.

Perhaps the most often-cited effect of behavioral (sensory) impairment of CO exposure is the report by McFarland et al. (1944) and reiterations of the same research (Halperin et al., 1959; McFarland, 1970). These investigators reported that dark-adapted visual thresholds were raised in a very systematic, dose-ordinal manner by COHb levels beginning as low as 4.5% and up to 20%. However, concurrent and subsequent studies found no such effect (Abramson and Heyman, 1944; Luria and McKay, 1979; Von Restorff and Hebisch, 1988) even though some of the COHb levels were up to 30%. McFarland himself, in a later study (McFarland, 1973), reported that no shift in thresholds was found even at 17% COHb. The entire visual dark adaptation curve was studied both psychophysically and electrophysiologically with 17% COHb and was found to be unaffected in any way (Hudnell and Benignus, 1989).

A probably more historically important report of CO effects on behavior was published by Beard and Wertheim (1967). These investigators reported that COHb of 2.7–12.5% impaired subjects' estimates of duration of signals in a statistically significant, dose-ordinal manner. Again, subsequent experiments, some of which were very close replications, could find no effect on time estimation whatever, even though some investigators employed COHb as high as 20% (O'Donnell et al., 1971; Otto et al., 1979; Stewart et al., 1972, 1973; Weir et al., 1973; Wright and Shephard, 1978).

It remains a puzzle as to why the above two groups found such systematic and plausible (from a purely empirical perspective) results when no one else has been able to replicate them. The research projects were not called archetypical in only the sense of their historical importance, however. The same story is repeated in many other research reports on CO's behavioral effects in humans. To be sure, there were methodological problems with both studies. McFarland et al. (1944) employed only 4 subjects and actually reported the data from only 1. There was no indication that the subjects or experimenters were blind to the exposure conditions or that appropriate fatigue controls had been tested. The work by Beard and Wertheim (1967) was much more carefully executed, testing 18 subjects who were blinded to their exposure (the experimenters were not) and reported reasonably analyzed data. Yet their results were simply not replicable. Many of the attempts at replication, however, had similar methodological problems that would have also biased their studies toward Type I errors and yet did not find significant effects.

Not only were the results reported by the above two archetypical experiments not replicable by other empirical work, but, from the preceding section on the mechanism of action of COHb effects on behavior, such low-level results should not be expected. If $BMRO_2$ is regulated until about 30% COHb is reached, it would require some rather extraordinary set of circumstances to produce effects as large and consistent as those reported by McFarland et al. (1944) and Beard and Wertheim (1967). Even if no BBF increase had occurred as COHb increased, the CaO_2 produced by COHb of 17% is about equivalent to the normocapnic hypoxic hypoxia level that produces first signs of sensory and behavioral effects (Benignus, 1994). Again, it would require an extraordinary set of circumstances to expect behavioral effects with as low as 2.5–4.5% COHb, even if the increased difficulty of O_2 extraction were considered.

It must be recalled that if reliable empirical evidence supports the notion of an effect, then no amount of theory that says it should not occur can overturn the observed effect. In fact, the above two archetypical reports were not supported either theoretically or empirically.

Coming at a time in which environmental concerns and activism were rising, the report of Beard and Wertheim (1967) exerted a great influence on the thinking of the times and the subsequent research. Both the scientific and the regulatory communities were, perhaps, a bit more credulous than warranted. Many of the nonreplications were at first dismissed as being methodologically flawed or even as simple statistical failures to find significant effects (Type II errors). Only after much and careful work has it become apparent that the problem lies with the original reports, not with the attempts at replication. The best interpretation is that the significant and systematic findings of the two archetypical researches were due to a combination of methodological flaws and statistical Type I errors (finding significant effects when none exist in the population).

3.1.2 Methodological Problems

Many of the research projects on the behavioral effects of CO exposure in humans were conducted in less than an ideal manner. Especially troubling were the frequent (a) lack of double-blind procedure, (b) questionable statistical practice, and (c) poor experimental control and measurement. Benignus (1992) attempted to explain the differences between studies of the effects of CO on behavior by considering the methodological problems of non-double-blinding and of Type-I nonconservative statistical practice (Muller and Benignus, 1992). There was no relationship between statistical problems and the results of the experiments, but there was a statistically significant and large effect of the blinding conditions. Only 6 out of 23 experiments that had employed a double-blind procedure found significant effects of CO on behavior, whereas 15 out of 20 experiments using single-blind methods found significant effects. There were no differences in the levels of COHb used. Apparently the experimenter can, by knowledge of exposure conditions, influence the outcome of his research. This finding explains some of the reports of effects of low-level COHb on behavior, but as will be subsequently seen, not all of them.

While the problems with statistical methods did not account for a significant difference in the number of reports of CO effects on behavior, there is no reason why adequate statistical analyses should not be done. One of the problems was testing the effect of CO exposure on many dependent variables measured on the same subjects without accounting for the thus-increased probability of finding a significant effect (Muller and Benignus, 1992). Other problems range from no significance tests at all to such minor difficulties as failure to adequately document what was done. These kinds of statistical inadequacies are not limited to the CO area or to older research, but persist in the area of modern health-effects research. Statistics, properly used, helps to decide whether an effect could have been due to mere chance, to document the nature of the effect, and to interpret the outcomes. Too often statistics are slavishly used without understanding or purpose because someone requires them to publish research results. This kind of analysis only harms the scientific enterprise.

3.2 Review of the CO Behavior Literature

The body of literature on CO effects on behavior in humans is large, including a total of approximately 73 research reports and numerous reviews and summaries. The body of research on laboratory animals and CO effects on behavior is comparatively small. To complicate matters further, most of the research reports measured many dependent variables. The number of citations needed to describe the research, when organized according to groups of dependent variables, is approximately 133. It may be fairly stated that none of the extant reviews, (including this one) regardless of the thoroughness, has included all work.

3.2.1 *General Observations*

Symptoms of low-level CO poisoning — headache and nausea — are cited (Klaassen, 1985) for COHb levels of 10–20%. These symptoms of CO exposure were not observed in a of COHb levels below 20% (Benignus et al.,

1987), but were reported in a study at COHb levels of 25–30% (Forbes et al., 1937). Sayers and Davenport (1930) described high-level CO poisoning. As COHb increases, the symptoms intensify and vomiting and unconsciousness sometimes occur at 35–45% COHb. Between 50 and 60% COHb, subjects become comatose and have increased and disrupted breathing patterns. Higher levels lead to death, depending on duration and treatment. In the extreme case, inhalation of 100% CO will result in unconsciousness within one or two breaths followed shortly by death.

3.2.2 Evaluative Tabulation of the Literature

In an attempt to succinctly summarize the CO behavior literature, valuative comparative tables will be used. Each table will be devoted to the findings regarding a particular family of dependent variables. More extensive discussion of each article may be found elsewhere (McMullen and Raub, 1991).

Table 1 is a list of the references that have already been reviewed above in connection with the historical archetypical visual sensitivity work of McFarland et al. (1944). It is included here for completeness and ease of comparison.

TABLE 1

Results of Studies on Visual Sensitivity

Study	# of Subs	Stat Notes[a]	Blind[b]	% COHb Range	Effects[c]
McFarland et al., 1944	4	1	N	4.5–19.7	Y, 4.5%
Abramson and Heyman, 1944	9	1	N	10–30	N
McFarland, 1973	27	1	S	6–17	N
Luria and McKay, 1979	18	2	S	9	N
Von Restorff and Hebisch, 1988	5	3	S	9–17	N
Hudnell and Benignus, 1989	21	3	D	17	N

[a] Statistical notes defined as follows: 1 = no statistical tests, 2 = multiple tests without corrections for increased α, 3 = appropriate testing strategy.

[b] Blinding in which N = no blinding, S = single blind, and D = double blind.

[c] Results in which Y = statistically significant or N = not statistically significant as declared by the author(s) of the report, followed by the % COHb at which the effect was declared significant. In the case of no significance tests being conducted, a Y or N with % COHb is given as the authors reported effects.

Critical flicker fusion, a measure of visual temporal resolution, was reported by three groups of researchers (see Table 2) to be affected by COHb. All of the studies were either single- or nonblind and all of them were tested using statistical methods tending toward Type I errors. Six other experiments were unable to find effects, even with higher COHb levels, even though most of the other studies also used statistical tests that would have

TABLE 2

Results of Studies on Critical Flicker Fusion

Study	# of Subs	Stat Notes[a]	Blind[b]	% COHb Range	Effects[c]
Von Post-Lingen, 1964	100	2	S	4–23	Y, 14%
Beard and Grandstaff, 1970	4	1	N	1.8–6.7	Y, ?%
Seppanen et al., 1977	22	2	S	5–12.7	Y, 5%
Lilienthal and Fugitt, 1946	5	1	N	12–17	N
Vollmer et al., 1946	17	2	S	7.5–17.5	N
Guest et al., 1970	8	2	D	8.9	N
O'Donnell et al., 1971	4	2	D	5.9–12.7	N
Fodor and Winneke, 1972	12	2	S	5.3	N
Winneke, 1974	18	1	S	10	N

[a] 1 = no statistical tests, 2 = multiple tests without corrections, 3 = appropriate strategy.
[b] N = no blinding, S = single blind and D = double blind.
[c] Y = significant or N = not significant as declared by the author(s) of the report, followed by % COHb at which the effect was declared significant.

tended to produce Type I errors. It seems reasonable to conclude that critical flicker fusion is not affected by COHb in the range tested.

Table 3 summarizes a number of studies of miscellaneous visual system functions. Visual acuity were reported to have been impaired in a study (Von Post-Lingen, 1964) that used only 4 subjects, no statistics, and was non-blind. Two other studies (Hudnell and Benignus, 1989; Stewart et al., 1975), both double-blind and using much higher COHb levels, found no acuity effects. Brightness threshold was reported to be impaired (Beard and Grandstaff, 1970; Ramsey, 1972) in non-double-blind studies, but was not affected when one of the two authors repeated the study in a double-blind manner (Ramsey, 1973). Three other reports of visual function decrements were found (Fodor and Winneke, 1972; Salvatore, 1974), all of which did not employ double-blind methods and most of which used statistics tending toward Type-I errors. Most of the other studies failed to find any effect on various visual functions even though higher levels of COHb were used. The most extensive and technologically advanced visual system evaluation (Hudnell and Benignus, 1989) using double-blind methods and appropriate statistical tests found no effects on any dependent variable even though tests were conducted with 17% COHb.

There is little literature on auditory effects of COHb elevation, but what there is, is all negative (see Table 4).

Only one single-blind study reported COHb effects on fine motor skills (Bender et al., 1972) below 20%. Another study (Winneke, 1974), also single blind, reported slight effects at 20%. The remainder of the 11 studies in this area found no effects on fine motor skills (see Table 5).

TABLE 3

Results of Miscellaneous Visual Effects

Study	# of Subs	Stat Notes[a]	Blind[b]	% COHb Range	Effects[c]
Beard and Grandstaff, 1970	4	1	N	3–7.5	Y, ?
Bender et al., 1972	42	2	S	7.3	Y, 7.3%
Fodor and Winneke, 1972	12	2	S	5.3	Y, 5.3%
Ramsey, 1972	20	2	S	5	Y, 5%
Salvatore, 1974	6	3	N	5.4	Y, 5.4%
Forbes et al., 1937	5	1	N	27–41	N
Vollmer et al., 1946	17	2	S	7.5–17.5	N
Stewart et al., 1970	11	2	D	1–20	N
Ramsey, 1973	60	2	D	7.6–11.2	N
McFarland, 1973	27	1	S	6–17	N
Wright et al., 1973	50	2	D	5.6	N
Stewart et al., 1975	27	3	D	1–20	N
Seppanen et al., 1977	22	2	S	4–12.7	N
Hudnell and Benignus, 1989	21	3	D	17	N

[a] 1 = no statistical tests, 2 = multiple tests without corrections, 3 = appropriate strategy.
[b] N = no blinding, S = single blind, and D = double blind.
[c] Y = significant or N = not significant as declared by the author(s) of the report, followed by % COHb at which the effect was declared significant.

TABLE 4

Results of Studies on Auditory Functions

Study	# of Subs	Stat Notes[a]	Blind[b]	% COHb Range	Effects[c]
Guest et al., 1970	8	2	D	8.9	N
Stewart et al., 1970	11	2	D	1–12	N
Haider et al., 1976	20	1	D	3–13	N

[a] 1 = no statistical tests, 2 = multiple tests without corrections, 3 = appropriate strategy.
[b] N = no blinding, S = single blind, and D = double blind.
[c] Y = significant or N = not significant as declared by the author(s) of the report, followed by % COHb at which the effect was declared significant.

One study (Weir et al., 1973) reported a slight decrement in reaction time at 20% COHb. The remaining studies were negative (see Table 6).

Until this point in the discussion, no convincing effect of COHb has been demonstrated at levels below 20%. Tracking performance (a task requiring the subject to use a manipulandum to either pursue a continuously

TABLE 5

Results of Studies on Fine Motor Skills

Study	# of Subs	Stat Notes[a]	Blind[b]	% COHb Range	Effects[c]
Bender et al., 1972	42	2	S	7.3	Y, 7.3%
Winneke et al., 1974	18	1	S	20	Y, 20%
Vollmer et al., 1946	17	2	S	7.5–17.5	N
Stewart et al., 1970	11	2	D	1–12	N
O'Donnell et al., 1971	9	2	D	3–12.4	N
Winneke et al., 1974	12	2	S	5.3	N
Weir et al., 1973	15	2	D	20	N
Wright et al., 1973	50	2	D	5.6	N
Stewart et al., 1975	27	3	D	1–20	N
Seppanen et al., 1977	22	2	S	4–12.7	N
Mihevic et al., 1983	16	3	S	5.5	N

[a] 1 = no statistical tests, 2 = multiple tests without corrections, 3 = appropriate strategy.
[b] N = no blinding, S = single blind, and D = double blind.
[c] Y = significant or N = not significant as declared by the author(s) of the report, followed by % COHb at which the effect was declared significant.

TABLE 6

Results of Studies on Reaction Time

Study	# of Subs	Stat Notes[a]	Blind[b]	% COHb Range	Effects[c]
Weir et al., 1973	25	2	D	7–20	Y, 20%
Forbes et al., 1937	5	1	N	27–41	N
Stewart et al., 1970	11	2	D	1–20	N
Fodor and Winneke, 1972	12	2	S	5.3	N
Ramsey, 1972	21	2	S	5	N
McFarland et al., 1973	27	1	S	6–17	N
Ramsey, 1973	60	2	D	4.6–11.2	N
Wright et al., 1973	50	2	D	5.6	N
Rummo and Sarlanis, 1983	7	2	S	7.6	N
Winneke, 1974	18	1	S	10	N
Luria and McKay, 1979	18	2	S	9	N
Harbin et al., 1988	55	3	D	5.3	N

[a] 1 = no statistical tests, 2 = multiple tests without corrections, 3 = appropriate strategy.
[b] N = no blinding, S = single blind, and D = double blind.
[c] Y = significant or N = not significant as declared by the author(s) of the report, followed by % COHb at which the effect was declared significant.

moving target or attempt to keep it from moving) provides a different story, however (see Table 7). Four well-conducted, carefully analyzed studies reported impaired tracking behavior at COHb levels in the 5–10% range. On the other hand, seven other studies found no effects. Especially interesting are two studies by Putz et al. (1976, 1979) that found decrements and were repeated as exactly as possible by Benignus et al. (1987, 1990). In one case (Benignus, 1987) decrements were found at 8.2% COHb, while in the second study (Benignus et al., 1990) the same investigator found no effects in a dose-effects study in which the maximum COHb was 17% and that included a COHb level at which previous effects were found. There is no answer to how this unreliability in findings could have come about, but what is most impressive is that the largest study (Benignus et al., 1990), which was also a dose-effects study including very high levels of COHb, was negative. It is, nonetheless, difficult to disregard the other evidence.

TABLE 7

Results of Studies on Tracking

Study	# of Subs	Stat Notes[a]	Blind[b]	% COHb Range	Effects[c]
Weir et al., 1973	25	2	D	7-20	Y, 20%
Putz et al., 1976	30	3	D	3–5.1	Y, 5.1%
Putz et al., 1979	30	3	D	3.5–4.6	Y, 4.6%
Benignus et al., 1987	22	3	D	8.2	Y, 8.2%
O'Donnell et al., 1971a	4	2	D	5.9–12.7	N
O'Donnell et al., 1971b	9	2	D	3–12.4	N
Winneke et al., 1972	12	2	S	5.3	N
Gliner et al., 1983	15	3	S	5.8	N
Schaad et al., 1986	10	2	S	20	N
Bunnell and Horvath, 1988	15	3	S	7–10	N
Benignus et al., 1990	74	3	D	5.6–17	N

[a] 1 = no statistical tests, 2 = multiple tests without corrections, 3 = appropriate strategy.
[b] N = no blinding, S = single blind, and D = double blind.
[c] Y = significant or N = not significant as declared by the author(s) of the report, followed by % COHb at which the effect was declared significant.

Studies of vigilance (see Table 8) provide a puzzle similar to the tracking studies. Here there is a 50–50 split in findings. What is more striking, however, is that in each case of a significant effect, the same group of investigators later failed to find an effect in attempts at replication. In one case (Horvath et al., 1971), an effect was found but not in two following studies (Christensen et al., 1977; Roche et al., 1981). Another case of direct replication failure by the same group of researchers (Fodor and Winneke, 1972; Winneke, 1974) was equally puzzling. Two studies found effects

(Groll-Knapp et al., 1972; Haider et al., 1976), although not always dose ordinal, and later the same researchers failed to replicate the results (Groll-Knapp et al., 1978). There was no relationship between the way the studies had been conducted or analyzed and the outcomes. Again, these results can neither be explained nor ignored.

TABLE 8

Results of Studies on Vigilance

Study	# of Subs	Stat Notes[a]	Blind[b]	% COHb Range	Effects[c]
Horvath et al., 1971	15	3	S	2.3–6.6	Y, 6.6%
Fodor and Winneke, 1972	12	2	S	5.3	Y, 5.6%
Groll-Knapp et al., 1972	21	1	N	3–7.6	Y, 3%
Haider et al., 1976	21	1	D	3–13	Y, 7.6%
Winneke, 1974	18	1	S	10	N
Christensen et al., 1977	10	3	D	4.8	N
Groll-Knapp et al., 1978	21	2	D	6–12	N
Roche et al., 1981	18	3	S	5	N

[a] 1 = no statistical tests, 2 = multiple tests without corrections, 3 = appropriate strategy.
[b] N = no blinding, S = single blind, and D = double blind.
[c] Y = significant or N = not significant as declared by the author(s) of the report, followed by % COHb at which the effect was declared significant.

Yet another dependent variable, continuous performance, presents a picture of unreplaceable findings of COHb effects (see Table 9). Light monitoring was affected in two studies by one group of researchers (Putz et al., 1976, 1979) in carefully executed and analyzed experiments but not in a direct replication by another group (Benignus et al., 1987) in an equally careful study nor in other related experiments (O'Donnell et al., 1971; Schaad et al., 1986). One group (Gliner et al., 1983) reported effects when the task was performed singly but not when performed simultaneously with tracking. No conclusions are immediately obvious in this area.

Time estimation effects (see the historical review above) are summarized in Table 10 and provide no reason to believe that any effects exist.

Cognitive tasks are frequently reported to be impaired by COHb (see Table 11) but the results are less equivocal than for tracking, vigilance, and continuous performance. One group of investigators (Groll-Knapp et al., 1978) reported memory deficits in a double-blind study and later could not replicate the effect (Haider et al., 1976). All other reports of cognitive effects were from single-blind experiments with other problems. Statistically questionable analyses were employed in two of them (Bender et al., 1972; Schulte, 1963. Results were non-dose ordinal (Mihevic et al., 1983) in another study. Similar variables as found significant by some experiments

TABLE 9

Results of Studies on Continuous Performance

Study	# of Subs	Stat Notes[a]	Blind[b]	% COHb Range	Effects[c]
Schulte, 1963	49	2	S	0–20	Y, 5%
Putz et al., 1976	30	3	D	3–5.1	Y, 5%
Putz et al., 1979	30	3	D	3.5–4.6	Y, 4.6%
Gliner et al., 1983	15	3	S	5.8	Y, 5.8
Insogna and Warren, 1984	9	3	S	2.1–4.2	Y, 4.2
O'Donnell et al., 1971	4	2	D	5.9–12.7	N
Benignus et al., 1977	52	3	D	4.6–12.6	N
Schaad et al., 1986	10	2	S	20	N
Benignus et al., 1987	22	3	D	8.2	N

[a] 1 = no statistical tests, 2 = multiple tests without corrections, 3 = appropriate strategy.
[b] N = no blinding, S = single blind, and D = double blind.
[c] Y = significant or N = not significant as declared by the author(s) of the report, followed by % COHb at which the effect was declared significant.

TABLE 10

Results of Studies on Time Estimation

Study	# of Subs	Stat Notes[a]	Blind[b]	% COHb Range	Effects[c]
Beard and Wertheim, 1967	18	2	S	2.7–12.5	Y, 2.7%
Stewart et al., 1972	27	2	D	1–20	N
Stewart et al., 1970	11	2	D	1–20	N
O'Donnell et al., 1971	9	2	D	3–12.4	N
Stewart et al., 1973	27	3	D	1–20	N
Weir et al., 1973	15	2	D	20	N
Stewart et al., 1975	27	2	D	1–20	N
Wright and Shephard, 1978	13	3	N	2–8	N
Otto et al., 1979	13	3	D	3.7–7.8	N

[a] 1 = no statistical tests, 2 = multiple tests without corrections, 3 = appropriate strategy.
[b] N = no blinding, S = single blind, and D = double blind.
[c] Y = significant or N = not significant as declared by the author(s) of the report, followed by % COHb at which the effect was declared significant.

were found not to be affected by others in related studies that were, however, not direct replications. The evidence for cognitive effects is much less impressive than for tracking, vigilance, and continuous performance.

If automobile driving is affected, it appears to be so only at about 20% COHb. The only study reporting effects at lower levels (Rummo and Sarlanis,

TABLE 11

Results of Studies on Miscellaneous Cognitive Tasks

Study	# of Subs	Stat Notes[a]	Blind[b]	% COHb Range	Effects[c]
Schulte, 1963	49	2	S	1–20	Y, 5%
Bender et al., 1972	42	2	S	7.3	Y, 7.3%
Groll-Knapp et al., 1978	10	2	D	11	Y, 11%
Mihevic et al., 1983	16	3	S	5.5	Y, 5.5%
Bunnell and Horvath, 1988	15	3	S	7–10	Y, 7%
Groll-Knapp et al., 1982	21	2	D	6–12	N
O'Donnell et al., 1971	4	2	D	5.9–12.7	N
Stewart et al., 1975	27	3	D	1–20	N
Haider et al., 1976	21	1	D	3–13	N
Groll-Knapp et al., 1982	21	2	D	10	N
Schaad et al., 1986	10	2	N	0–20	N

[a] 1 = no statistical tests, 2 = multiple tests without corrections, 3 = appropriate strategy.
[b] N = no blinding, S = single blind, and D = double blind.
[c] Y = significant or N = not significant as declared by the author(s) of the report, followed by % COHb at which the effect was declared significant.

TABLE 12

Results of Studies on Automotive Driving Tasks

Study	# of Subs	Stat Notes[a]	Blind[b]	% COHb Range	Effects[c]
Weir et al., 1973	12	2	D	7–20	Y, 20%
Rummo and Solanis, 1974	7	2	S	7.6	Y, 7.6%
Forbes et al., 1937	5	1	N	27–41	N
Wright et al., 1973	50	2	D	5.6	N
Wright and Shephard, 1978	10	2	D	7	N

[a] 1 = no statistical tests, 2 = multiple tests without corrections, 3 = appropriate strategy.
[b] N = no blinding, S = single blind, and D = double blind.
[c] Y = significant or N = not significant as declared by the author(s) of the report, followed by % COHb at which the effect was declared significant.

1974) was single blind and analyzed with statistics likely to produce Type I errors (see Table 12).

The laboratory animal literature is internally, remarkably consistent (see Table 13) in that all studies found marked effects of COHb, all of them used small numbers of subjects and few of them employed statistical analyses at all. What is also consistent is that none reported effects at lower levels than 12% COHb (even though many of the researchers tested at levels less

TABLE 13

Results of Studies on Schedule-Controlled Behavior

Study	# of Subs	% COHb Range	Effects[a]
Annau, 1975	5	9–58	Y, 20%
Ator, 1982	15	3.5–55	Y, 32%
Ator et al., 1976	4	8–54	Y, 37%
Beard and Wertheim, 1967	?	15–55	Y, 14%
Fountain et al., 1986	22	9–58	Y, 45%
Mullin and Krivanek, 1982	6	12–54.9	Y, 12.2%
Purser and Berrill, 1983	3	1–32	Y, 16%
Schrot and Thomas, 1986	3	34–53	Y, 45%
Schrot et al., 1984	3	44–66	Y, 45%
Smith et al., 1976	4	15–40	Y, 20%

[a] Y = significant or N = not significant as declared by the author(s) of the report, followed by % COHb at which the effect was declared significant.

than this) and the large majority were at or above 20%. This body of findings has been used to make the point that rats are less sensitive to COHb impairment of the brain than are humans. Subsequent discussion in this chapter will dispute this claim.

3.2.3 Summary of Human Behavioral Effects of CO

Table 14 is a summary of the statistically significant findings discussed above. If consideration is limited to double-blind studies only, examination of Table 14 and the preceding review reveals that there are only three general areas in which the effects of CO exposure seem to produce effects in some cases: tracking, vigilance, and continuous performance. All of the other dependent variables were not affected in the large majority of experiments and those few studies that did report effects had serious methodological flaws. Even in the above three areas of study there were serious problems with replication of each of the reports of significant effects.

Among double-blind studies, the most reliably reported effect of COHb elevation is tracking ($p = 0.57$), followed by continuous performance ($p = 0.40$) and lastly vigilance ($p = 0.33$). If tracking and vigilance are considered as particular kinds of continuous performance, then the probability of a reported effect is $p = 0.47$. Thus, even if one restricts consideration to double-blind studies, there are still about 50% of the studies reporting significant effects and 50% finding none. This state of affairs is not helped by consideration of other aspects of the studies, such as statistics, number of subjects tested, or level of COHb employed (McMullen and Raub, 1991). There are too many reported statistically significant findings to ignore and too many nonreplications to believe them.

TABLE 14

Summary of Research on Behavioral Effects[a] of COHb Elevation

Dependent Variable	Non-Double Blind	Double Blind
Absolute visual threshold	1/4 = 0.20	0/1 = 0.00
Critical flicker fusion	3/7 = 0.43	0/2 = 0.00
Misc. visual functions	5/9 = 0.56	0/5 = 0.00
Misc. auditory functions	0/0	0/3 = 0.00
Fine motor skills	2/6 = 0.33	0/5 = 0.00
Reaction time	0/7 = 0.00	1/5 = 0.20[b]
Tracking	0/4 = 0.00	4/7 = 0.57
Vigilance	3/5 = 0.60	1/3 = 0.33
Continuous performance	3/4 = 0.75	2/5 = 0.40
Time estimation	1/2 = 0.50	0/6 = 0.00
Misc. cognitive functions	4/5 = 0.80	1/6 = 0.17
Automobile driving	1/2 = 0.50	1/3 = 0.33[b]

[a] Effects are summarized as the ratio of studies reporting significant effects to the total number of studies and as proportions.
[b] Effects found only at 20% COHb.

3.2.4 Extrapolations from Other Findings

An effort was made to provide estimates of dose-effects curves for COHb and behavior by Benignus (1994) using meta analyses and quantitative extrapolations. A number of assumptions were made about the similarity of tasks, species, and physiology. The steps taken were as follows.

For extrapolation across species, the literature of behavioral effects in rats was summarized quantitatively by transforming the dependent variable in each study to a new metameter in which the scale was 1.0 for baseline and descending for impairment toward 0.0 for no remaining performance (a proportion-of-baseline score). Following this, all studies that could be so transformed were pooled and a curve was fitted to the transformed, pooled data, yielding an estimate of the dose-effects curve for rats. It is known that rats become hypothermic in a short time when COHb is elevated, but humans do not. Therefore, the effect of COHb on behavior could be due, in part, to hypothermic effects on the brain in rats, but not in humans. To estimate the hypothermic effects separately and account for them, the literature was quantitatively analyzed as above, for hypothermic-behavior effects and for COHb-hypothermia effects, yielding two effects curves. Considering that the effects of hypothermia and COHb might be via different mechanisms, the hypothermia for each COHb level was estimated and the estimated effects of hypothermia subtracted from the rat COHb-behavior effects curve. The result of the process was considered an extrapolation from rat to human because the known differences between the species were accounted for.

Another set of meta analyses and extrapolations was done in the same project to estimate the human COHb-behavior dose-effects curve from the literature on hypocapnic hypoxic hypoxia. In this case, it was necessary to estimate the effects curves for behavioral effects of hypocapnic hypoxic hypoxia and for hypocapnia separately, and then correct the hypoxia curve for hypocapnia. Because hypocapnia is not produced by CO hypoxia, the thusly corrected curve was considered an extrapolation from hypocapnic hypoxic hypoxia to CO hypoxia, both estimated in humans.

Figure 1 is a plot of the two extrapolations (dashed line) from CO hypoxia in rats to humans and (solid line) from hypocapnic hypoxic hypoxia in humans to CO hypoxia in humans. The fact that the two curves overlay each other so closely, and yet originated from such different sources of data, is impressive. It would seem that if assumptions or methods were erroneous, the probability of the two curves being so nearly the same would be low.

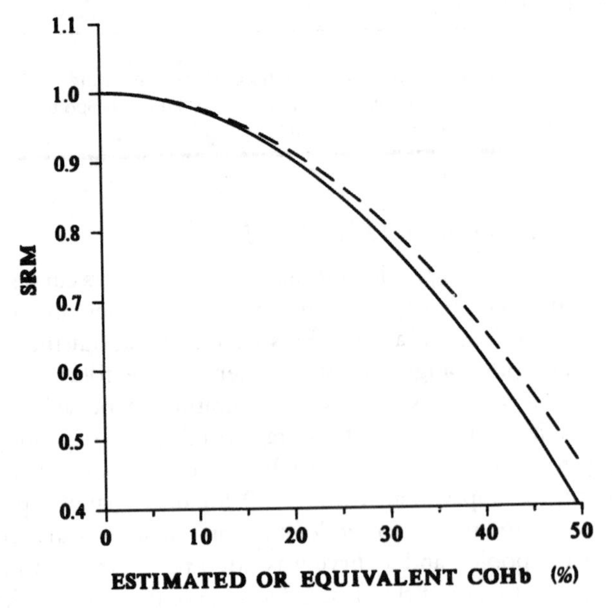

FIGURE 1
Estimates of human dose-effects curve for COHb and behavior. Dashed line extrapolated from rats, solid line extrapolated from hypocapnic hypoxic hypoxia in humans. SRM = standard response metameter.

The estimated curve (solid line) in Figure 1 was plotted against all of the double-blind COHb-behavior data (see above) that could be converted to the proportion-of-baseline metameter (see Figure 2). The filled (solid) points in Figure 2 were from studies in which effects were declared significant. The extrapolation seems to fit the data from the literature fairly well. It was noted (Benignus, 1994) that the scatter in Figure 2 was not dissimilar to the scatter observed around the curves for hypocapnic hypoxic hypoxia and for CO hypoxia in rats. If, however, lines were drawn through the points that were reportedly significantly different from controls, the resulting dose effects curve would be both steeper and have a lower threshold than the curves of Figure 1. It is also noteworthy that the reportedly statistically significant points in Figure 2 are mostly grouped around low COHb levels, while higher COHb levels were not statistically significant.

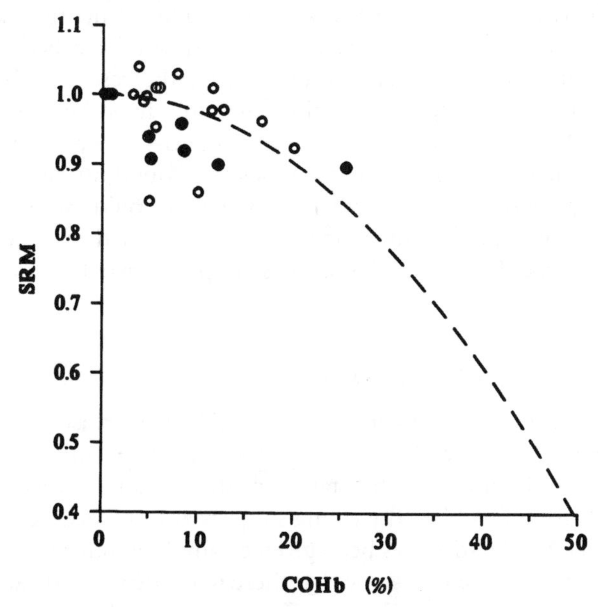

FIGURE 2
Estimated dose-effects curve (from hypocapnic hypoxic hypoxia) and data from human studies of COHb effects on behavior. Solid points were reported to be statistically significant by the original authors. SRM = standard response metameter.

3.2.5 Possible Explanations

If the extrapolations (Benignus, 1994) are given weight of consideration, then the statistically significant findings in the above review are made more questionable because the extrapolations appear to agree with the studies finding no behavioral effects. In view, however, of the potentially grave consequences of deciding that behavioral decrements of approximately 10% of baseline ought not to occur until COHb approaches 18–25% (Benignus, 1994), it would seem prudent to consider the matter further.

If some subjects do not compensate for the reduced CaO_2 due to COHb increase by increased BBF, those subjects would have earlier reductions in $BMRO_2$ and thus larger effects at low levels of COHb. By sampling fluctuations, some studies could have included more “noncompensators” than others and thereby obtained significant effects. The probability of such hypothetical sampling fluctuations is difficult to calculate until an estimate is available of the proportion of such "noncompensators" in the population. While this is a highly speculative discussion of a possible explanation for the differences among reports of effects, it is a testable line of reasoning and could account for the literature.

Another possible explanation of the variance in the reports of effects is the "file-drawer effect" (Rosenthal, 1979). There is reason to believe that, especially in the climate of environmental activism during which much of the research was done, an experiment that produced no significant effect when others had already reported them would not have been published. Such nonpublication might have resulted from the researcher's perplexity and statistical uncertainty — accepting the null hypothesis (Muller and Benignus, 1992) — or from the reluctance of editors to accept such findings. If there were many well-executed researches that found no significant effect, but were not published, the review of the literature could take on a different appearance and it could be much more plausibly assumed that the significant effects were Type I errors. This also is a speculative line of reasoning, however.

3.2.6 Conclusions About the Findings

At the present stage of information, it would appear reasonable to assume that the mean dose effects curve for COHb and behavior is very near to those depicted in Figure 1. Such a statement should not, however, be applied to any population except healthy young subjects at rest, because no empirical or theoretical consideration has been given to anything but the above group. It also seems appropriate to assume that there is a wide range of susceptibility, leading to some findings of COHb effects that would not agree with the curves of Figure 1. It is not reasonable to speculate about how great or how probable such hypothetical susceptibility might be.

3.3 Required Research

From a review of the behavioral literature, it would appear that further attempts to confirm or disconfirm significant findings would be nonproductive. To disconfirm a particular reported finding in a convincing manner would require more than a few reports of negative results. To be sure, if a result could be demonstrated to be repeatable by more than one experimenter and in a number of situations, then it would be worth finding. The probability of such demonstrating such an effect seems too low to pursue, however.

Devising and estimating some independent variable closely related to $BMRO_2$ (see discussions in Section 2.3) would allow the relation of behavioral effects to a variable that might be more highly correlated than COHb. This is true only if there are substantial individual differences in the way that subjects compensate for decreased CaO_2, however. The knowledge exists and the methods are available to make models of the physiology and test them in a behavioral context. Until such models are devised and tested, it appears that resolution to the COHb-behavior conflict will not easily be arranged.

Areas of research that have never been tapped in the COHb behavior or physiology area are (a) development and aging, (b) brain or O_2-delivery impairment, (c) simultaneous physical work and behavioral tasks to drive up competition for O_2, and (d) simultaneous exposure to other pollutants to approximate real-world environments. Any one or any combination of the above could lead to much lower effects thresholds and/or steeper dose effects functions. The importance of these data to many environments need not be emphasized. Thus considered, the area of COHb-behavioral research is only beginning to be explored.

4 SUMMARY

CO inhalation produces COHb, which results in arterial hypoxemia (reduced CaO_2). As CaO_2 declines, BBF rises in a compensatory manner, keeping $BMRO_2$ from changing until about 30% COHb, thus apparently avoiding brain tissue hypoxia. There may be important individual differences in the compensatory response mechanisms. If this is so, then COHb is not a good independent variable. Some independent variable closely related to $BMRO_2$ may be devised that could be estimated from CaO_2 and BBF.

Experiments designed to document behavioral effects of COHb elevation have been troubled with unreplicability. While there are many methodological problems in the extant literature, they cannot account for the difficulties. There appears to be a core of reports of effects on continuous performance that were methodologically sound but simply cannot be replicated.

Extrapolations from rats to humans and from hypocapnic hypoxic hypoxia to CO hypoxia have been used to estimate a dose-effects function for behavior and COHb (see Figure 1). These estimates agree with each other and roughly with data concerning reduction in $BMRO_2$. According to these considerations, a behavioral decrement of 10% should not occur in the average subject until COHb exceeds 18–25%. This is in sharp disagreement with some experimental results (all sometimes unreplicable) that report effects as low as 5% COHb (see Figure 2).

No resolution of the above conflicting information is readily available. Perhaps there are large individual differences that are sufficiently probable so that sometimes sampling variation results in a sample of higher than usual susceptibility to COHb. Perhaps nonsignificant findings are much more frequent, but are simply not reported. It is probable that in the average, healthy, young, sedentary adult, no detectable COHb effects on behavior will occur until COHb exceeds 18–25%.

Research needs do not include simply repeating behavioral experiments of the kind that are extant in the literature. Physiological mechanistic modeling and research should be done with a view to determining the magnitude and effect of individual differences and devising an appropriate independent variable. The issues of age of subjects, health of subjects and nonsedentary subjects should also be explored.

Barring the advent of a very well documented and replicable behavioral demonstration of COHb behavioral effects, physiological understanding of mechanism appears to be the only hope of resolution of the uncertainty. Until the time that either of the above occurs, CO exposure must be treated with considerable caution. Too little is known about the average, normal, young, healthy, at rest subject's effects, much less about cases that do not fit into these categories and are yet exposed to CO in their normal environment.

REFERENCES

Abboud, R., Andersson, G., and Coburn, R.F., Evaluation of stagnant pulmonary capillary blood during breath holding in dogs, *J. Appl. Physiol.*, 37, 397, 1974.

Abramson, E. and Heyman, T., Dark adaptation and inhalation of carbon monoxide, *Acta Physiol. Scand.*, 7, 303, 1944.

Annau, Z., The comparative effect of hypoxic and carbon monoxide hypoxia on behavior, in *Behavioral Toxicology,* Weiss, B. and Laties, V.G., Eds., Plenum Press, New York, 1975, 105.

Ator, N.A., Modulation of behavioral effects of carbon monoxide by reinforcement contingencies, *Neurobehav. Toxicol. Teratol.*, 4, 51, 1982.

Ator, N.A., Merigan, W.H., and McIntire, R.W., The effects of brief exposures to carbon monoxide on temporally differential responding, *Environ. Res.*, 12, 81, 1976.

Beard, R.R. and Grandstaff, N., Carbon monoxide exposure and cerebral function, *Ann. N.Y. Acad. Sci.*, 174, 385, 1970.

Beard, R.R. and Wertheim, G.A., Behavioral impairment associated with small doses of carbon monoxide, *Am. J. Public Health,* 57, 2012, 1967.

Bender, W., Goethert, M., and Malorny, G., Effect of low carbon monoxide concentrations on psychological functions, *Staub-Reinhalt Luft,* 32, 54, 1972.

Benignus, V.A., A model to predict carboxyhemoglobin and pulmonary parameters after exposure to O_2, CO_2, and CO, *Aviat. Space Environ. Med.*, 66, 369, 1995.

Benignus, V.A. and Annau, Z., Carboxyhemoglobin formation due to carbon monoxide exposure in rats, *Toxicol. Appl. Pharmacol.*, 128, 151, 1994.

Benignus, V.A., Behavioral effects of carbon monoxide: Meta analyses and extrapolations, *J. Appl. Physiol.*, 76, 1310, 1994.

Benignus, V.A., Hazucha, M.J., Smith, M.V., and Bromberg, P.A., Prediction of carboxyhemoglobin formation due to transient exposure to carbon monoxide, *J. Appl. Physiol.*, 76, 1739, 1994.

Benignus, V.A., Importance of experimentor-blind procedure in neurotoxicology, *Neurotoxicol. Teratol.*, 15, 45, 1992.

Benignus, V.A., Neurotoxicity of environmental gases, in *Handbook of Neurotoxicology, Vol. II, Effects and Mechanisms,* Chang, L.W. and Dyer, R.S., Eds. 1994, 1005.

Benignus, V.A., Kafer, E.R., Muller, K.E., and Case, M.W., Absence of symptoms with carboxyhemoglobin levels of 16–23%, *Neurotoxicol. Teratol.,* 9, 345, 1987.

Benignus, V.A., Muller, K.E., Barton, C.N., and Prah, J.D., Effect of low level carbon monoxide on compensatory tracking and event monitoring, *Neurotoxicol. Teratol.,* 9, 227, 1987.

Benignus, V.A., Muller, K.E., Pieper, K.S., and Prah, J.D., Compensatory tracking in humans with elevated carboxyhemoglobin, *Neurotoxicol. Teratol.,* 12, 105, 1990.

Benignus, V.A., Otto, D.A., Prah, J.D., and Benignus, G., Lack of effects of carbon monoxide on human vigilance, *Percept. Motor Skills,* 45, 1007, 1977.

Benignus, V.A., Petrovick, M.K., Newlin-Clapp, L., and Prah, J.D., Carboxyhemoglobin and brain blood flow in humans, *Neurotoxicol. Teratol.,* 14, 285, 1992.

Bunnell, D.E. and Horvath, S.M., Interactive effects of physical work and carbon monoxide on cognitive task performance, *Aviat. Space Environ. Med.,* 59, 1133, 1988.

Christensen, C.L., Gliner, J.A., Horvath, S.M., and Wagner, J.A., Effects of three kinds of hypoxias on vigilance and performance, *Space Environ. Med.,* 48, 491, 1977.

Coburn, R.F., Mechanisms of carbon monoxide toxicity, *Prev. Med.,* 8, 310, 1979.

Fodor, G.G. and Winneke, G., Effect of low CO concentrations on resistance to monotony and on psychomotor capacity, *Staub-Reinhalt Luft,* 32, 46, 1972.

Forbes, W.H., Dill, D.B., DeSilva, H., and Van Deventer, F.M., The influence of moderate carbon monoxide poisoning upon the ability to drive automobiles, *J. Ind. Hyg. Toxicol.,* 19, 598, 1937.

Fountain, S.B., Raffaele, K.C., and Annau, Z., Behavioral consequences of intraperitoneal carbon monoxide administration in rats, *Toxicol. Appl. Pharmacol.,* 83, 546, 1986.

Gliner, J.A., Horvath, S.M., and Mihevic, P.M., Carbon monoxide and human performance in a single and dual task methodology, *Aviat. Space Environ. Med.,* 54, 714, 1983.

Groll-Knapp, E., Haider, M., Hoeller, H., Jenkner, H., and Stidl, H.G., Neuro- and psychophysiological effects of moderate carbon monoxide exposure, in *Multidisciplinary Perspectives in Event-Related Brain Potential Research,* Otto, D., Ed., EPA Report No. EPA-600/9-77-043, U.S. Environmental Protection Agency, Washington, D.C., 1978, 424. (Available from NTIS, Springfield, VA, PB-297137.)

Groll-Knapp, E., Haider, M., Jenkner, H., Liebich, H., Neuberger, M., and Trimmel, M., Moderate carbon monoxide exposure during sleep: Neuro- and psychophysiological effects in young and elderly people, *Neurobehav. Toxicol. Teratol.,* 4, 709, 1982.

Groll-Knapp, E., Wagner, H., Hauck, H., and Haider, M., Effects of low carbon monoxide concentrations on vigilance and computer-analyzed brain potentials, *Staub-Reinhalt Luft,* 32, 64, 1972.

Guest, A.D.L., Duncan, C., and Lawther, P.J., Carbon monoxide and phenobarbitone: A comparison of effects on auditory flutter fusion threshold and critical flicker fusion threshold, *Ergonomics,* 13, 587, 1970.

Haider, M., Groll-Knapp, E., Hoeller, H., Neuberger, M., and Stidl, H., Effects of moderate carbon monoxide dose on the central nervous system-electrophysiological and behaviour data and clinical relevance, in *Clinical Implications of Air Pollution Research,* Finkel, A.J. and Duel, W. C., Eds., Publishing Sciences Group, San Francisco, 1976, 217.

Halperin, M.H., McFarland, R.A., Niven, J.I., and Roughton, F.J.W., The time course of the effects of carbon monoxide on visual thresholds, *J. Physiol. (London),* 146, 583, 1959.

Hanley, D.F., Wilson, D.A., and Traystman, R.J., Effect of hypoxia and hypercapnia on neurohypophyseal blood flow, *Am. J. Physiol.,* 250, H7, 1986.

Harbin, T.J., Benignus, V.A., Muller, K.E., and Barton, C.N., The effects of low-level carbon monoxide exposure upon evoked cortical potentials in young and elderly men, *Neurotoxicol. Teratol.,* 10, 93, 1988.

Horvath, S.M., Dahms, T.E., and O'Hanlon, J.F., Carbon monoxide and human vigilance: A deleterious effect of present urban concentrations, *Arch. Environ. Health,* 23, 343, 1971.

Hudnell, H.K. and Benignus, V.A., Carbon monoxide exposure and human visual detection thresholds, *Neurotoxicol. Teratol.,* 11, 363, 1989.

Insogna, S. and Warren, C.A., The effect of carbon monoxide on psychomotor function, in *Trends in Ergonomics/Human Factors,* Vol. I, Mital, A., Ed., Elsevier/North Holland, Amsterdam, 1984, 331.

Jones, M.D. and Traystman, R.J., Cerebral oxygenation of the fetus, newborn and adult, *Sem. Perinatol.,* 8, 205, 1984.

Klaassen, C.D., Nonmetallic environmental toxicants: Air pollutants, solvents and vapors, in *The Pharmacologic Basis of Therapeutics,* Goodman, A.G. and Gillman, A., Eds., MacMillan, New York, 1985, 1628.

Koehler, R.C., Jones, M.D., and Traystman, R.J., Cerebral circulatory response to carbon monoxide and hypoxic hypoxia in the lamb, *Am. J. Physiol.,* 243, H27, 1982.

Koehler, R.C., Traystman, R.J., Zeger, S., Rogers, M.C., and Jones, M.D., Jr., Comparison of cerebrovascular response to hypoxic and carbon monoxide hypoxia in newborn and adult sheep, *J. Cereb. Blood Flow Metab.,* 4, 115, 1984.

Lilienthal, J.L., Jr. and Fugitt, C.H., The effect of low concentrations of carboxyhemoglobin on the "altitude tolerance" of man, *Am. J. Physiol.,* 145, 359, 1946.

Luria, S.M. and McKay, C.L., Effects of low levels of carbon monoxide on visions of smokers and nonsmokers, *Arch. Environ. Health,* 34, 38, 1979.

McFarland, R.A., Low level exposure to carbon monoxide and driving performance, *Arch. Environ. Health,* 25, 355, 1973.

McFarland, R.A., The effects of exposure to small quantities of carbon monoxide on vision, *Ann. N.Y. Acad. Sci.,* 174, 301, 1970.

McFarland, R.A., Roughton, F.J.W., Halperin, M.H., and Niven, J.I., The effects of carbon monoxide and altitude on visual thresholds, *J. Aviat. Med.,* 15, 381, 1944.

McMullen, T.B. and Raub, J.A., Eds., *Air Quality Criteria for Carbon Monoxide,* EPA-600/8-90/045F, U.S. Environmental Protection Agency, Washington, D.C., 1991.

Mihevic, P.M., Gliner, J.A., and Horvath, S.M., Carbon monoxide exposure and information processing during perceptual-motor performance, *Int. Arch. Occup. Environ. Res.,* 51, 355, 1983.

Muller, K.E. and Benignus, V.A., Increasing scientific power with statistical power, *Neurotoxicol. Teratol.,* 14, 211, 1992.

Mullin, L.S. and Krivanek, N.D., Comparison of unconditioned reflex and conditioned avoidance tests in rats exposed by inhalation to carbon monoxide, 1,1,1-trichlorethane, toluene or ethanol, *Neurotoxicology,* 3, 126, 1982.

O'Donnell, R.D., Chikos, P., and Theodore, J., Effect of carbon monoxide exposure on human sleep and psychomotor performance, *J. Appl. Physiol.,* 31, 513, 1971a.

O'Donnell, R.D., Mikulka, P., Heinig, P., and Theodore, J., Low level carbon monoxide exposure and human psychomotor performance, *Toxicol. Appl. Pharmacol.,* 18, 593, 1971b.

Okeda, R., Matsuo, T., Kuroiwa, T., Nakai, M., Tajima, T., and Takahashi, H., Regional cerebral blood flow of acute carbon monoxide poisoning in cats, *Acta Neuropathol.,* 72, 389, 1987.

Otto, D.A., Benignus, V.A., and Prah, J.D., Carbon monoxide and human time discrimination: Failure to replicate Beard-Wertheim experiments, *Aviat. Space Environ. Med.,* 50, 40, 1979.

Piantadosi, C.A., Sylvia, A.L., and Jobsis-Vandervliet, F.F., Differences in brain cytochrome responses to carbon monoxide and cyanide *in vivo, J. Appl. Physiol.,* 62, 1277, 1987.

Purser, D.A. and Berrill, K.R., Effects of carbon monoxide on behavior in monkeys in relation to human fire hazard, *Arch. Environ. Health,* 38, 308, 1983.

Putz, V.R., Johnson, B.L., and Setzer, J.V., A comparative study of the effects of carbon monoxide and methylene chloride on human performance, *J. Environ. Pathol. Toxicol.,* 2, 97, 1979.

Putz, V.R., Johnson, B.L., and Setzer, J.V., Effects of CO on Vigilance Performance: Effects of Low Level Carbon Monoxide on Divided Attention, Pitch Discrimination and the Auditory Evoked Potential, U.S. Department of Health, Education, and Welfare, National Institute for Occupational Safety and Health, Cincinnati, 1976. (Available from NTIS, Springfield, VA, PB-274219.)

Ramsey, J.M., Carbon monoxide, tissue hypoxia and sensory psychomotor response in hypoxaemic subjects, *Clin. Sci.,* 42, 619, 1972.

Ramsey, J.M., Effects of single exposures of carbon monoxide on sensory and psychomotor response, *Am. Ind. Hyg. Assoc.,* 34, 212, 1973.

Randall, J.E., *Elements of Biophysics,* Year Book Medical Publishers, Chicago, 1962, 91.

Roche, S., Horvath, S., Gliner, J., Wagner, J., and Borgia, J., Sustained visual attention and carbon monoxide: Elimination of adaptation effects, *Hum. Factors,* 23, 175, 1981.

Rosenthal, R., The "file drawer problem" and tolerance for null results, *Psychol. Bull.,* 86, 638, 1979.

Roughton, F.J.W. and Darling, R.C., The effect of carbon monoxide on the oxyhemoglobin dissociation curve, *Am. J. Physiol.,* 141, 17, 1944.

Rummo, N. and Sarlanis, K., The effect of carbon monoxide on several measures of vigilance in a simulated driving task, *J. Saf. Res.,* 6, 126, 1974.

Salvatore, S., Performance decrement caused by mild carbon monoxide levels on two visual functions, *J. Saf. Res.,* 6, 131, 1974.

Sayers, P.R. and Davenport, S.J., Review of Carbon Monoxide Poisoning, Public Health Bulletin No. 195, U.S. Government Printing Office, Washington, D.C., 1930.

Schaad, G., Kleinhans, G., Piekarski, C., Seebas, M., and Gorges, W., Ergonomische aspekte zur optimierung der versorgung von schutzraumen mit atemluft in notsituationen, *Wehrmed. Monatschr.,* 1, 13, 1986.

Schrot, J. and Thomas, J.R., Multiple schedule performance changes during carbon monoxide exposure, *Neurobehav. Toxicol. Teratol.,* 8, 225, 1986.

Schrot, J., Thomas, J. R., and Robertson, R. F., Temporal changes in repeated acquisition behavior after carbon monoxide exposure, *Neurobehav. Toxicol. Teratol.,* 6, 23, 1984.

Schulte, J.H., Effects of mild carbon monoxide intoxication, *Arch. Environ. Health,* 7, 30, 1963.

Seppanen, A., Hakkinen, V., and Tenkku, M., Effect of gradually increasing carboxyhemoglobin saturation on visual perception and psychomotor performance of smoking and nonsmoking subjects, *Ann. Clin. Res.,* 9, 314, 1977.

Smith, M.V., Hazucha, M.J., Benignus, V.A., and Bromberg, P.A., Effect of regional circulation patterns on observed COHb levels, *J. Appl. Physiol.,* 77, 1659, 1994.

Smith, M.D., Merigan, W.H., and McIntire, R.W., Effects of carbon monoxide on fixed-consecutive-number performance in rats, *Pharmacol. Biochem. Behav.,* 5, 257, 1976.

Stewart, R.D., Hosko, P.E., Peterson, J.E., and Mellender, J.W., The effect of carbon monoxide on time perception, manual coordination, inspection and arithmetic, in *Behavioral Toxicology,* Weiss, B. and Laties, V.G., Eds., Plenum Press, New York, 1975, 29.

Stewart, R.D., Newton, P.E., Hosko, M., and Peterson, J.E., Effect of carbon monoxide on time perception, *Arch. Environ. Health,* 27, 155, 1973.

Stewart, R.D., Newton, P.E., Hosko, M.J., Peterson, J.E., and Mellender, J.W., The effect of carbon monoxide on time perception, manual coordination, inspection and arithmetic, in *Behavioral Toxicology,* Weiss, B. and Laties, V.G., Eds., New York, Plenum, 1972, 29.

Stewart, R.D., Peterson, J.E., Baretta, E.D., Bachand, R.T., Hosko, M., and Herrmann, A.A., Experimental human exposure to carbon monoxide, *Arch. Environ. Health,* 21, 154, 1970.

Vollmer, E.P., King, G.B., Birren, J.E., and Fisher, M.B., The effects of carbon monoxide on three types of performance at simulated altitudes of 10,000 and 15,000 feet, *J. Exp. Psychol.,* 36, 244, 1946.

Von Post-Lingen, M.-L., The significance of exposure to small concentrations of carbon monoxide, *Proc. R. Soc. Med.,* 57, 1021, 1964.

Von Restorff, W. and Hebisch, S., Dark adaptation of the eye during carbon monoxide exposure in smokers and nonsmokers, *Aviat. Space Environ. Med.,* 59, 928, 1988.

Weir, F.W., Rockwell, T.H., Mehta, M.M., Attwood, D.A., Johnson, D.F., Herrin, G.D., Anglin, D.M., and Safford, R.R., An Investigation of the Effects of Carbon Monoxide on Humans in the Driving Task, Ohio State University Research Foundation, Columbus, OH, 1973. (Available from NTIS, Springfield, VA, PB-224646.)

Winneke, G., Effects of methylene chloride and carbon monoxide as assessed by sensory and psychomotor performance, in *Behavioral Toxicology: Early Detection of Occupational Hazards,* Xintaras, C., Johnson, B.L., and deGroot, I., Eds., DHEW Publication No. (NIOSH) 74-126, Department of Health, Education and Welfare, National Institute for Occupational Safety and Health, Cincinnati, 1974. (Available from NTIS, Springfield, VA, PB-259322.)

Wright, G.R., Randell, P., and Shephard, R.J., Carbon monoxide and driving skills, *Arch. Environ. Health,* 27, 349, 1973.

Wright, G.R. and Shephard, R.J., Carbon monoxide exposure and auditory duration discrimination, *Arch. Environ. Health,* 33, 226, 1978.

CHAPTER 11

Delayed Sequelae in Carbon Monoxide Poisoning and the Possible Mechanisms

Eric P. Kindwall

CONTENTS

1 INTRODUCTION

The clinical course following acute carbon monoxide (CO) poisoning varies, depending on the age of the patient, the length of exposure, the concentration of CO inhaled, the amount of physical activity of the victim,

0-8493-4796-3/96/$0.00+$.50

the form and timing of treatment afforded the patient after rescue, and probably other factors related to biochemical mechanisms in the patient's internal milieu at the time of poisoning. For example, the presence of alcohol may increase the severity (Winston et al., 1974).

Mild poisoning produces no measurable long-term aftereffects, but as the severity of the exposure increases, the likelihood of developing mild to severe neurologic sequelae increases, assuming the patient survives. Disability following poisoning falls into two general categories. The first is a simple continuum of incapacitation, where the patient may regain consciousness but has immediate obvious neurologic disabilities, some of which may improve slowly with time. The other category includes patients who regain consciousness and seem to recover completely for a few days or weeks but then suffer delayed deterioration and long-term disability. This period of apparent complete recovery, where the patient often goes back to work and resumes full-time normal activities, is referred to as the "lucid interval." It appears in 10 to 36% of severe poisonings (Min, 1986). Older patients are more likely to suffer delayed sequelae. Delayed deterioration appearing 10 to 40 d later may progress dramatically, even to the point of death.

To be completely candid, the exact mechanisms for delayed deterioration following CO poisoning are unknown. There most probably are more than one. Until about 5 years ago, most physicians felt that the symptoms of CO poisoning were entirely explained by hypoxic anoxia secondary to an inability of carboxyhemoglobin to carry oxygen (O_2) to the tissues. Now, however, because of a great deal of careful research, the problem of CO poisoning resembles a vast jigsaw puzzle, with many pieces seemingly in place, but with the remainder not fully defined. The pieces of the puzzle that may help explain the appearance of the lucid interval followed by subsequent deterioration are the subject of this chapter. Any theory as to why this lucid interval can occur has to explain how the central nervous system can seem to function normally or nearly normally for a period of time, even though damage has occurred that may later prove fatal.

We shall explore what is known about the pathophysiology of CO poisoning and the development of both biochemical and anatomic lesions in light of our present knowledge.

2 SYMPTOMATOLOGY

The symptoms of CO poisoning are protean and, as noted, may present immediately or following a severe poisoning, present later. They are typically characterized by headache that can become severe, disorientation, memory defects, muscular weakness, coma, and death. The patient can also have fever and leukocytosis. Cortical blindness, which may occur more often in children, tends to be transitory and is sometimes accompanied by a *belle indifference* (Kindwall, unpublished data). Pupillary reactions remain intact despite visual loss. Immediate and then delayed symptoms may also include aphasia and apraxia with hallucinations, muscular rigidity with cogwheeling, and incontinence of urine and feces. If the patient has attempted suicide, the organic brain syndrome following poisoning may be misinterpreted as depression (Werner et al., 1985). The carboxyhemoglobin measured on presentation at the emergency room typically bears little relationship to the seriousness of symptoms (Sokal, 1975). Olsky and Woods, in reviewing 100 consecutive CO poisonings admitted to Lutheran General Hospital in Park Ridge, IL, also found that admission carboxyhemoglobin levels bore no statistical relationship to outcome (Olsky and Woods, 1983).

More reliable, but not infallible, in determining severity is the mental status exam and/or history of unconsciousness, and in particular, the length of the exposure. In serious cases, the arterial pH is usually quite low, at 7.2 or less, and the blood sugar is often significantly elevated (Leikin et al., 1988; Sokal and Kralkowska, 1985). The CPK and LDH may be elevated as well (Olsky and Woods, 1983). We have often found potassium to be low in serious poisonings, for reasons that have not been well explained.

3 PATHOPHYSIOLOGY

For many years it was thought that simple hypoxia caused by the inability of blood saturated with CO to carry O_2 was responsible for all of the symptoms and signs found in CO poisoning. That this is not so was dramatically driven home by the study of Goldbaum et al. (1975). The researchers poisoned dogs by enclosing them in a box with 13% CO. The dogs all died. At death, however, the carboxyhemoglobin levels ranged between 54 and 90%. They then partially exsanguinated a second group of 5 dogs, producing a 68% reduction in circulating hemoglobin, but then infused Dextran and Ringer's solution to render them normovolemic. The dogs suffered no noticeable impairment. A third group of 5 dogs was bled to a 68% loss of circulating blood but then infused with packed cells from donor dogs that had 80% carboxyhemoglobin. This produced a circulating level of 57 to 64% carboxyhemoglobin in the dogs. They suffered no apparent ill effect. Thus, it seems clear that the serious or potentially fatal symptoms and signs produced following severe CO poisoning are not mediated solely by hypoxia, but are due to some form of direct cellular toxicity.

The evidence for cellular toxicity is not new. Warburg (1926) discovered the action of the cytochrome system by blocking its function with CO.

J.B.S. Haldane (1927) noted that invertebrates, microbes, and plants are immune to CO intoxication. This would suggest that differences in their metabolic pathways might point to critical metabolic processes unique to vertebrates. He also found that high levels of O_2 could be protective in mammals. He placed rats in a tank and pressurized them to 3 atmospheres absolute (ATA) with pure oxygen. He then added another atmosphere of pure CO, which resulted in the rats inhaling a mixture containing 250,000 ppm CO. Under the protection of superabundant amounts of oxygen, CO caused the rats no symptoms and seemed to behave as an inert gas. Only when a second full atmosphere of CO was added to produce a total CO pressure of 2 ATA did the animals succumb.

Ball et al. (1951) studied the behavior of cytochrome a_3 oxidase in beef heart and found that it combined more readily with O_2 as compared with CO in a 9:1 ratio. Therefore, it would appear from that study that the cytochrome system would be unaffected unless there was nearly lethal hypoxia present.

More recently, Brown and Piantadosi (1990), using reflectant spectrophotometry, have shown that CO binds to cytochrome C oxidase (cytochrome a, a_3) *in vivo*. Of interest is that although the rats were ventilated with 90% O_2 at 1 ATA, this was not enough to raise the pO_2 in the mitochondria sufficiently to block attachment of CO to cytochrome a_3. The animals did develop carboxyhemoglobinemia, which lowered the available O_2.

Geyer et al. (1976) exposed rats to 90% O_2 and 10% CO at 1 ATA, which proved lethal. He then placed bloodless rats perfused with a fluorocarbon-Tris buffer blood substitute and found the rats suffered no ill effect after 17 h exposure. The concentration of fluorocarbon, however, provided an intraarterial O_2 content of 7.04 vol % of fully available O_2 in physical solution (Geyer, 1979). This provides more O_2 than is used by the brain. Thus, these animals were fully oxygenated throughout their exposure and presumably the amount of O_2 available was sufficient to protect the cytochrome system or other metabolic functions within the mitochondria from the deleterious effects of CO. The 17-h exposure would have been sufficient to allow a saturation equilibrium of all the tissues to occur. The water content of the perfusate was 70%. If we assume the solubility of CO in water and plasma to be somewhat similar, it would appear probable that CO was carried into the tissues.

4 PATHOLOGY

Previously, only those patients who died could provide us a clue as to the location of the lesions within the central nervous system (CNS) as seen at autopsy. In recent years, computerized axial tomography and magnetic resonance imaging have allowed us more insight into the development of sublethal changes in the human brain (Chang et al., 1992; Choi et al., 1992, 1993; De Reuck et al., 1992; Hopkins et al., 1993; Jaeckle and Nasrallah, 1985, Kee et al., 1992;). For many years, it was simplistically felt that CO produced lesions unique to that poison in the CNS, chiefly involving the basal ganglia. Specifically, destruction of the globus pallidus of the striatum was felt to be pathognomonic of CO poisoning. Careful review of the neurologic literature, however, reveals that hypoxic anoxia stemming from many causes, but coupled also with a necessary period of hypotension or ischemia, can produce the same general pathologic lesions. The tendency to equate this effect with CO probably stems from the fact that CO is such a ubiquitous cause of anoxia, especially in those cases where death is delayed long enough to allow gross cellular necrosis, that it dominates the literature. Other forms of hypoxic asphyxia often produce death before frank necrosis becomes evident. Song et al. (1983) studied the pathogenesis of the selective lesion of the globus pallidus in cats poisoned with CO. They looked at systemic blood pressure, venous pressure, blood flow in the left common carotid, and local blood flow in the globus pallidus, putamen, and claustrum. Of 28 cats, half (14) developed pallidal lesions. The only parameter that correlated with globus pallidus pathology was local blood flow. In those cats with pallidal lesions, local blood flow decreased to 67.3% of the initial value, while in the others it increased to 188% of normal. This conforms with the notion that hypotension is required for CO damage to the globus pallidus as well as other brain structures.

The areas affected by a combination of hypoxic anoxia and hypotension tend to be selective, depending on the severity. Neuronal death may occur, but there seems to be a definite order in which the various areas of the brain are affected.

With regard to where the most damage caused by CO poisoning is found, it seems to be chiefly demyelinization of the white matter. However, in severe poisoning, neuronal death also occurs (Nabeshima et al., 1991; Sawa et al., 1981; Song et al., 1983). Nabeshima et al. (1991) found that by 3 d post-poisoning, the number of pyramidal cells in the hippocampal CA1 subfield in mice was lower than that of controls. However, neurodegeneration in the parietal cortex, area 1, was not observed until 7 d after exposure. They concluded that delayed amnesia induced by CO, as measured in mice using a passive avoidance task, was the result of delayed neuronal death in the hippocampal CA1 subfield and dysfunction in the acetylcholinergic neurons in the frontal cortex, the striatum, and/or the hippocampus.

5 THE HYPOTHESES

The question arises as to what the ultimate mechanism is for damage to brain tissues following CO poisoning. It is clear that hypoxia plays a central role, as demonstrated by most investigators (Jain, 1990a,b). However, simple anoxia in the absence of ischemia or hypotension does not produce the same results (Thom, 1990). Also, damage appears to occur *following* poisoning or after rescue as the area is reperfused with O_2 (Thom, 1988).

Thom was the first to point out that it was only after exposure to CO ceased that lipid peroxidation could be detected in the rat brain (Thom, 1988). He found that both CO hypoxia and a period of hypotension, even though very brief, were required to produce lipid peroxidation. As the carboxyhemoglobin levels fell, the products of lipid peroxidation rose, as estimated by measuring conjugated dienes. This supports the view that the damage occasioned by CO is indeed a reperfusion injury and appears only when O_2 is reintroduced. Thom also discovered that O_2 at 3 ATA terminated lipid peroxidation, whereas 100% O_2 at 1 ATA had no effect, and the effects at 2 ATA were only moderate (Thom, 1990).

Brown and Piantadosi utilized *in vivo* reflectant spectrophotometry through the intact rat skull to study the *in vivo* binding of CO to cytochrome C oxidase (Brown and Piantadosi, 1990). While the animals were being ventilated with CO, cytochrome a, a_3 absorption at 605 nm increased in the parietal cortex of anesthetized rats during carboxyhemoglobin formation. Arterial blood pressure varied inversely with carboxyhemoglobinemia. During CO exposure, decreases were found in blood pressure, O_2 content, and cytochrome a, a_3 oxidation level. These changes could be partly reversed, even though CO was present, by compressing animals to 3 ATA in oxygen. After removing CO from the inspired gas at 3 ATA, optical and physiological parameters recovered completely. The researchers concluded that hyperbaric oxygen at 3 ATA caused dissociation of the cytochrome a_3–CO complex.

In later work, Brown and Piantadosi (1992) confirmed that CO exposure significantly decreased blood pressure and cytochrome oxidation level. They also measured a drop in phosphocreatine and a rise in lactase/pyruvate, glucose, and succinate. During exposure to CO, they found that the intracellular pH dropped. In recovery, normobaric oxygen (NBO) improved the cytochrome oxidation state to 80% of normal after 90 min, but there was complete recovery after hyperbaric oxygen (HBO) at 2.5 ATA. Intracellular pH continued to fall for 45 min while the animals received HBO and did not completely recover by 90 min. The same was true for phosphocreatine. After HBO, phosphocreatine and intracellular pH improved within 45 min but still remained slightly below control at 90 min. They concluded that intracellular uptake of CO impairs cerebral energy metabolism despite the elimination of CO from the blood. This was further underscored when they found that simple hypoxemia produced an oxidation level of cytochrome oxidase of about 50%, while CO exposure produced an oxidation level of about 40% of control. Thus, the specific toxic effects of the gas increase the effects of simple hypoxia.

An apparent consequence of impairment of mitochondrial electron transport is the production of partially reduced oxygen species. Zhang and Piantadosi (1992) demonstrated that hydrogen peroxide (H_2O_2) production in the forebrain was increased on reoxygenation following CO poisoning. Since catalase is involved in H_2O_2 production, evidence pointed to the mitochondria being the source of the H_2O_2. Using a salicylate probe, the salicylate hydroxylation products 2,3- and 2,5-dihydroxybenzoic acid (DHBA) were increased 3.4-fold immediately after CO exposure and 3-fold after 120 min reoxygenation. Because these by-products were not found in postmitochondrial fractions, the authors proposed that the partially reduced oxygen species (PROS) generated in the brain after CO hypoxia originate primarily in the mitochondria. They suggest that PROS contribute to CO-mediated neuronal damage during reoxygenation after severe CO poisoning.

Hippocampal damage in humans was confirmed by Hopkins et al. (1993) when they showed that MRI studies found the mean area of hippocampus for CO-poisoned subjects was different from that of control subjects ($p < 0.05$). No differences were found for the mean areas of the parahippocampal gyrus and the temporal lobe. The hippocampal changes correlated with significant memory deficits, as measured by the Denman verbal, nonverbal, and full-scale memory quotients. These memory changes were significant at the 0.002, 0.01, and 0.001 levels, respectively.

As previously noted, Thom (1988) had shown that lipid peroxidation occurs in the brain after CO exposure has been concluded. In 1992 he reported that xanthine dehydrogenase is converted to xanthine oxidase (XO) in CO poisoning (Thom, 1992). XO has been implicated in postischemic brain injury because treatment with XO inhibitors improves tissue salvage. XO is an important source of free radicals in reperfused tissues (Kontos, 1989).

In Thom's experiments with rats, sulfhydryl-irreversible XO increased from a control level of 15% to a peak of 36% 90 min following CO poisoning, while the products of lipid peroxidation doubled. Reversible XO was 3–6% of the total enzyme activity over this span of time but increased to 31% between 90 min and 2 h following removal of the CO. Thom found that 2 h after poisoning, 66% of the total enzyme activity was due to reversible and irreversible XO. When rats had been depleted of XO by a tungsten diet, no lipid peroxidation was found (Thom, 1992).

Fisher et al. (1993) also reported reperfusion injury following long ischemia. They observed that pO_2-dependent hydroxyl radical production lay behind postischemic pulmonary edema and that superoxide dismutase reduced edema formation 60–80% after air- and N_2-ventilated ischemia, whereas treatment with catalase protected only N_2-ventilated ischemia.

Although the mechanism of injury appears to be related to reperfusion injury and lipid peroxidation, the areas of the brain attacked by CO intoxication are quite selective. Neuronal cell death can occur within hours to 7 d, but most authors agree that the main damage occurs in the cerebral white matter (Jain, 1990a). In that sense, the areas affected are not fundamentally different from other anoxic encephalopathies (Brucher, 1967).

Lesions of the basal ganglia are commonly described and, historically, gross destruction of the globus pallidus has been considered pathognomonic of CO poisoning. However, more modern studies show that the globus pallidus can be more severely damaged in non-CO anoxia. Lapresle and Fardeau (1967) found that in 22 fatal cases of CO poisoning, basal ganglia lesions were present in 16. Basal ganglia changes in hypoxia victims are not recognized unless the patient survives for more than 24 h (Lapresle and Fardeau, 1967). Hippocampal lesions, typically in Ammon's horn, are found in about half the cases of CO poisoning (Jain, 1990a). As noted earlier, the white matter is the most severely affected, with destruction of myelin in the centra semiovale, the corpus collosum, internal capsule, and the optic tracts. Axis cylinders seem to be preserved to a great extent, but the breakdown of myelin is profound. This is followed by proliferation of lipid phagocytes and fibrous astrocytes (Jain, 1990a).

It has been noted for many years that neuropsychiatric symptoms may appear following a period of apparent recovery in severe CO poisoning. Jacob (1939) first suggested that white matter degeneration and demyelination was the neuropathological basis for such delayed deterioration.

Courville (1957) noted that smaller foci of myelin loss gradually fuse into larger foci, which in turn tend to fuse into irregular conglomerate areas. Myelin loss seems to be more diffuse in the deeper portions of the white matter but, although irregular and sometimes elongated "arms" tend to reach out toward the cortex, the subcortical arcuate fibers are preserved. The same author also observed that oligodendroglia cannot be readily demonstrated. He could find only faint outlines of acutely swollen oligodendroglia, even in well-preserved portions of the white matter. In areas of focal necrosis, the oligodendroglia completely disappear, with the absence of their characteristic round nuclei.

Courville (1957) concluded that the oligodendroglia were easily crippled by anoxia and are very vulnerable to ischemia. Astrocytes persisted in these demyelinated areas, however. Courville found them in crippled form, but Jain reports them to be robust (Jain, 1990a). Courville found the cerebral cortex gray matter to be essentially unaltered with regard to its architecture. He felt that all of these changes occurred in CO poisoning as well as in other forms of cerebral anoxia, but that the disappearance of capillaries in the necrotic foci indicated that vasospastic processes produced by anoxemia of the vasomotor center caused the long penetrating cortical arteries to be predominantly affected. For this reason the subcortical white matter was damaged most (Courville, 1957).

In tissue culture that had been exposed to CO bubbled through it, Walum et al. (1985) studied cell respiration. They found that rat astrocytes, chick skeletal muscle, and chick heart cells were all inhibited (13–29%) by CO as compared to nitrogen controls. Mouse neuroblastoma cells were found not to be inhibited. However, astrocytes were more sensitive than neuroblastoma cells.

Using single photon emission computed tomography (SPECT) with technetium-99m hexamethylpropylene amine oxime, Choi and co-workers (1992) studied brain perfusion following CO poisoning in 6 patients. Based on SPECT findings showing diffuse patchy patterns of hypoperfusion, which improved on follow-up images, there was good correlation between clinical outcome and the SPECT images. They concluded that, in patients with CO poisoning, cerebrovascular changes causing hypoperfusion may produce delayed neurologic sequelae.

Many have observed that the carboxyhemoglobin level is not a good indicator of the systemic toxicity of CO and cannot be used to predict whether or not delayed sequelae will occur. The levels of CO in physical solution in the blood are quite low, as it is easily bound to circulating hemoglobin. However, when a pregnant woman is exposed to CO, the fetus forms carboxyhemoglobin (Friberg et al., 1959).

In animal studies using an experimental pneumoperitoneum, the air in the peritoneal cavity equilibrates with CO when the animal is exposed to CO. When rabbits were exposed to 1000 ppm of CO, COHb equilibrium was reached in 90 min, whereas the concentration of CO in the pneumoperitoneum took 15 h to equilibrate and 10 h to disappear (Gothert and Malorny, 1969). The fact that it takes so long for the tissues to equilibrate accords with clinical experience that long exposures, even at low levels, produce the most serious results. A very brief exposure, even with a high carboxyhemoglobin, may be treated easily without sequelae (Sokal, 1975). This has also been seen in our experience.

The CNS seems to tolerate a great deal of neuronal damage or loss because of its extensive redundancy. This appears to extend even to well-defined nuclei. Sawa et al. (1981) describe the case of a 52-year-old female with a 12-h exposure to CO who initially was unable to talk but responded to pain. She had bilateral Babinski signs. After 7 d in the hospital, she was discharged as normal mentally except that her friends felt that "her sparkle had gone." At 14 d post-poisoning she became listless, apathetic, and withdrawn, and 10 d later she was disoriented with severe memory impairment. She continued to deteriorate, and by 35 d was incontinent, with cogwheel rigidity of her musculature. By 40 d she was unable to carry out routine neurophysiological tests. The first CT scan, at 34 d post-poisoning, revealed punched-out lesions in her globus pallidus. Four weeks later, a repeat CT showed no change from the initial exam. The patient recovered gradually, and at 11 months her neurological examination was normal. A repeat CT scan 10 months after the poisoning showed mild enlargement of the anterior aspect of the third ventricle, but the previously demonstrated lesions in the globus pallidus on both sides could no longer be identified with certainty. Thus, it would appear that an apparent complete recovery is possible with severe damage to the globus pallidus even as seen on CT. Plum also described an autopsy-proven globus pallidus lesion in a fatal CO poisoning that produced no distinctive symptomatology. Others coming to autopsy showing pallidal lesions but who had shown no symptoms during life had experienced a remote anoxic exposure (Plum et al., 1962).

The reasons for the involvement of the globus pallidus following CO poisoning remain unknown, but it is possible that its blood supply is more tenuous and therefore more susceptible to transient hypotension with ischemia, as described earlier (Song et al., 1983). Of interest is that Sawa's patient did not demonstrate Parkinsonism, despite evidence of globus pallidus lesions. In a case described by Jaeckle and Nasrallah (1985), globus pallidus lesions were found on CT in a patient with severe Parkinsonism following CO poisoning, although the scan revealed no evidence of substantia nigra or other basal ganglia involvement. Smaller symmetric hypolucencies were found in the posterior limbs of both internal capsules. This patient, however, lost all his symptoms, including symptoms of a severe depression as measured by the Hamilton rating scale, when given 800 mg of L-Dopa plus 170 mg of Carbidopa per day. A CT scan performed 6 months later showed that the internal capsule lesions were less well visualized.

A study of 42 patients with computerized axial tomography by Choi and co-workers (1993) revealed that the commonest findings were low-density lesions in the cerebral white matter. In 33 patients, low-density globus pallidus lesions were seen bilaterally. Of note was that the prognosis depended only on the low-density white matter lesions rather than on those of the globus pallidus. There seemed to be significant correlation between the cerebral white matter changes in the initial CT scan and the development of delayed neurologic sequelae. This was particularly true in middle-aged or older patients. The CT findings did not correlate, however, with recovery from these delayed neurologic sequelae.

Magnetic resonance imaging in 15 patients obtained by Chang et al. (1992) confirmed the CT findings showing that 4–9 weeks following exposure to CO, cerebral white matter is the main target, with a reversible demyelinating process. These authors also felt that delayed neurologic sequelae are more characteristic of acute CO poisoning than other types of anoxia. Follow-up MR imaging in 4 patients showed that 3 of the 4 exhibited a decrease in extent and intensity of white matter lesions which accompanied lessening of clinical symptoms.

As noted, carboxyhemoglobin levels, neurological examination immediately following recovery, and even pretreatment psychological testing are not good predictors of delayed encephalopathy (Thom et al., 1992). The only mention of a seemingly reliable indicator has been suggested by He et al. (1993). This is the use of somatosensory evoked potentials (SEP), pattern reversal visual evoked potentials (VEP), and brainstem auditory evoked potentials (BAEP). They studied 109 healthy adults and 88 patients with acute CO poisoning. They found that a consistent abnormality involving N20 and subsequent peaks in the SEP, an unusual prolongation of P100 latency in the VEP, or a prolongation of III-V interpeak latency in the BAEP, as well as recurrences of evoked potential abnormalities after initial recovery, all were indicators of unfavorable outcome in acute CO poisoning. They were sensitive indicators of delayed encephalopathy.

A most unusual paper published by Foncin and LeBeau (1978) described the clinical case of a 24-year-old female who died of CO poisoning after a 10-d downhill course. This case was unusual in that a cerebral biopsy was obtained on day 3 of hospitalization. The specimen consisted of cortex and white matter taken from the right frontal region. Autopsy specimens are usually taken after some delay, and detailed ultrastructural study by electron-microscopy is often limited by post-mortem changes. In this case, the tissue could be fixed immediately. Electron microscopy revealed that the white matter was selectively attacked with destruction of the myelin coverings of the axons. In the deep cortex, the myelin lesions were less intense, and in the remainder of the cortex, the myelin covering appeared normal. Most striking, however, was the damage to the oligodendrocytes, which showed pyknotic or absent nuclei and homogenization of the cytoplasm with empty mitochondria. The astrocytes were remarkably normal, with integrity of their nuclear cytoplasm. The blood vessels showed occasional areas of endothelial tumefaction with normal nuclei but in all other respects were normal. The neuronal axons appeared unchanged, with normal synapses. In summary, the biopsy taken on the 3rd day showed a very severe lesion that selectively destroyed oligodendrocytes in the white matter as a recent event without signs of reaction in the adjacent myelin. This agrees with the findings of Lapresle and Fardeau (1967), the pathology corresponding to their group III. In their experience, six cases out of six presented with a diphasic clinical course, with leukoencephalopathy and necrosis being the only finding.

Foncin and LeBeau (1978) point out that a selective attack on the oligodendroglia has rarely been advanced as an explanation for the delayed effects of CO intoxication. Meyer (1928) was the first to point out oligodendroglial lesions where the nuclei were pyknotic or had entirely disappeared. Plum et al. (1962) ruled out vascular and edematous causes as well as similarities to allergic encephalomyelitis and also agreed with the role of selective attack on the oligodendrocytes. With regard to cyanide poisoning, which also attacks the cytochrome system, Ferraro (1933) and Lumsden (1950) also felt that the primary attack was of the oligodendrocyte. Bass (1968) confirmed this using other methods. On the other hand, Hirano et al. (1967), in their ultrastructural studies following acute inhalation of hydrogen cyanide, could only show axonal lesions without a specific attack on the myelin or the oligodendrocytes.

Additional evidence to support selectivity of oligodendroglial pathology was advanced by Foncin and LeBeau (1978), who pointed out that they found little damage to capillaries and that the absence of early necrosis made an ischemic mechanism improbable.

It is a tantalizing hypothesis to think that the selective destruction of oligodendrocytes, with the primary pathology occurring in the nucleus, might explain the lucid interval seen in 10–40% of the cases of severe CO poisoning. The oligodendrocyte lays down or produces myelin in the brain (Kandel et al., 1991). There appears to be a destruction of a metabolic center in the oligodendrocyte mitochondria that makes its effect felt in the periphery of the cell only after a certain delay. This would require, however, that the myelin in the CNS would have to be supported by oligodendroglial cells, as well as being elaborated by them, because later necrosis of the myelin with phagocytosis of the debris is usually noted. Thus, it would seem that demyelinization is not simply a lack of replacement of normally metabolized myelin (Plum et al., 1962). It does not seem to be that selective attack of the oligodendrocyte is based simply on metabolic rate because, metabolically, the neuron is more active and the oligodendrocyte shows intermediate activity. Other glial cells that are less active are unaffected (Pope, 1958).

If the observations of those workers who have studied oligodendrocyte pathology are correct, destruction of the oligodendroglial nuclei would seem to be a proximate cause. At this time the peculiar biochemical or enzymatic Achilles heel of the oligodendrocyte remains unknown.

In summary, both neurons and myelin can be destroyed by CO. The enormous redundancy and alternate pathways available in the CNS may make possible partial or complete functional compensation for neuronal loss. If enough cells are destroyed, recovery does not occur. Most evidence seems to point to demyelinization of the white matter as the underlying cause of delayed neurologic sequelae following CO poisoning, at least those symptoms that present later than a week after poisoning.

Aside from the possible use of cortical evoked potentials to predict delayed deterioration, choice of treatment to prevent late sequelae must be based on clinical judgment. Matthieu et al. (1985) suggest that referral to the hyperbaric chamber should be considered when (1) the patient is comatose, (2) there are abnormal clinical findings referring either to the sensorium or the EKG, or (3) patients have been unconscious during exposure, irrespective of whether they are conscious on admission and have a normal clinical appearance.

REFERENCES

Ball, E.G., Strittmatter, C.F., and Cooper, O., The reaction of cytochrome oxidase with CO, *J. Biochem.,* 193, 635, 1951.

Bass, N.H., Pathogenesis of myelin lesions in experimental cyanide encephalopathy: Microchemical study, *Neurology,* 18, 167, 1968.

Brown, S.D. and Piantadosi, C.A., *In vivo* binding of carbon monoxide to cytochrome-c oxidase in rat brain, *J. Appl. Physiol.*, 68, 604, 1990.

Brown, S.D. and Piantadosi, C.A., Recovery of energy metabolism in rat brain after carbon monoxide hypoxia, *J. Clin. Invest.*, 89, 666, 1992.

Brucher, J.M., Neuropathological problems posed by carbon monoxide poisoning and anoxia, in Buhr, H. and Ledingham, I.M., Eds., *Carbon Monoxide Poisoning,* Elsevier, Amsterdam, 1967, 75.

Chang, K.H., Han, M.H., Kim, H.S., Wie, B.A., and Han, M.C., Delayed encephalopathy after acute carbon monoxide intoxication: MR imaging features and distribution of cerebral white matter lesions, *Radiology,* 184, 117, 1992.

Choi, I.S., Kim, J.H., Choi, Y.C., and Lee, S.S., and Lee, M.S., Evaluation of outcome after acute carbon monoxide poisoning by brain CT, *J. Korean Med. Sci.*, 8, 78, 1993.

Choi, I.S., Lee, M.S., Lee, Y.J., and Kim, J.H., Lee, S.S., Kim, W.T., Technetium-99m HM-PAO SPECT in patients with delayed neurologic sequelae after carbon monoxide poisoning, *J. Korean Med. Sci.*, 7, 11, 1992.

Courville, C.B., The process of demyelination in the central nervous system: (IV) Demyelination as a delayed residual of carbon monoxide asphyxia, *J. Nerv. Ment. Dis.*, 125, 534, 1957.

De Reuck, J., Decoo, D., Vienne, J., Striejckmans, K., and Lemahieu, I., Significance of white matter lucencies in post-hypoxic-ischemic encephalopathy: Comparison of clinical status and of computed and positron emission tomographic findings, *Eur. Neurol.*, 32, 344, 1992.

Ferraro, A., Experimental toxic encephalomyelopathy: Diffuse sclerosis following subcutaneous injection of potassium cyanide, *Psychiatr. Q.*, 7, 167, 1933.

Fisher, P.W., Huang, Y.T., Kennedy, T.P., and Piantadosi, C.A., PO_2-dependent hydroxyl radical production during ischemia-reperfusion lung injury, *Am. J. Physiol.*, 265, 279, 1993.

Foncin, J.F. and LeBeau, J., [French] Myélinopathie par intoxication oxycarbonée: Neuropathologie ultrastructurale, *Acta Neuropathol.* (Berlin), 43, 153, 1978.

Friberg, L., Nyström, Å., and Swanberg, H., Transplacental diffusion of carbon monoxide in human subjects, *Acta Physiol. Scand.*, 45, 363, 1959.

Geyer, R.P., Personal communication, 1979.

Geyer, R.P., Taylor, K., Eccles, R., and Duffett, E., Survival of bloodless rats in 10% carbon monoxide, *Fed. Proc.*, 35, 828, 1976 [Abstract only].

Goldbaum, L.R., Ramirez, R.G., and Absalom, K.B., What is the mechanism of carbon monoxide toxicity? *Aviat. Space Environ. Med.*, 46, 1289, 1975.

Göthert, M. and Malorny, G., Zur Verteilung von Kohlenoxid zwischen Blut und Gewebe, *Arch. Toxikol.*, 24, 260, 1969.

Haldane, J.B.S., Carbon monoxide as a tissue poison, *Biochem. J.*, 21, 1068, 1927.

He, F., Liu, X., Yang, S., Zhang, S., Xu, G., Fang, G., and Pan, X., Evaluation of brain function in acute carbon monoxide poisoning with multimodality evoked potentials, *Environ. Res.*, 60, 213, 1993.

Hirano, A., Levine, S., and Zimmerman, H.M., Experimental cyanide encephalopathy: Electron microscopic observations of early lesions in white matter, *J. Neuropathol. Exp. Neurol.*, 26, 200, 1967.

Hopkins, R.O., Weaver, L.K., and Kessner, R.P., Long-term memory impairments in hippocampal magnetic resonance imaging in carbon monoxide poisoned subjects, *Undersea Hyperbaric Med.*, 20 (Suppl.), 15, 1993.

Jacob, H., Über die diffuse Hemisphärenmarkerkrankung nach Kohlenoxydvergiftung bei Fällen mit klinisch intervallärer Verlaufsform, *Z. Neurol. Psychiatr.*, 139, 161, 1939.

Jaeckle, R.S. and Nasrallah, H.A., Single case study: Major depression in carbon monoxide induced Parkinsonism: Diagnosis, computerized axial tomography and response to L-Dopa, *J. Nerv. Ment. Dis.*, 173, 503, 1985.

Jain, K.K., Pathology of carbon monoxide poisoning, in *Carbon Monoxide Poisoning*, Warren H. Green, St. Louis, 1990a, Chap. 6.

Jain, K.K., Pathophysiology of carbon monoxide poisoning, in *Carbon Monoxide Poisoning*, Warren H. Green, St. Louis, 1990b, Chap. 5.

Kandel, E.R., Schwartz, J.H., and Jessell, T.M., *Principles of Neural Science*, 3rd ed., Elsevier, New York, 1991.

Kee, H.C., Moon, H.H., Hak, S.P., Bong, A.W., and Man, C.H., Delayed encephalopathy after acute carbon monoxide intoxication: MR imaging features and distribution of cerebral white matter lesions, *Radiology*, 184, 117, 1992.

Kontos, H.A., Oxygen radicals in CNS damage, *Chem. Biol. Interact.*, 72, 229, 1989.

Lapresle, J. and Fardeau, M., The central nervous system and carbon monoxide poisoning, in *Progress in Brain Research*, Vol. 24, Buhr, H. and Ledingham, I. M., Eds., Elsevier, Amsterdam, 1967, 31.

Leikin, J.B., Goldenberg, R.M., Edwards, D., and Zell-Kantor, M., Metabolic predictors of carbon monoxide poisoning, *Vet. Hum. Toxicol.*, 30, 40, 1988.

Lumsden, D., Cyanide leukoencephalopathy in rats: An observation on the vascular and ferment hypothesis of demyelinating diseases, *J. Neurol., Neurosurg. Psychiatr.*, 113, 1, 1950.

Mathieu, D., Nolf, M., Durocher, A., Saulnier, F., Frimat, P., Furon, D., and Wattel, F., Acute carbon monoxide poisoning: Risk of late sequelae and treatment by hyperbaric oxygen, *Clin. Toxicol.*, 23, 315, 1985.

Meyer, A., Experimentelle Erfahrungen über die Kohlenoxydvergiftung, *Z. Ges. Neurol. Psych.*, 112, 187, 1928.

Min, S.K., A brain syndrome associated with delayed neuropsychiatric sequelae following acute carbon monoxide intoxication, *Acta Psychiatr. Scand.*, 73, 80, 1986.

Nabeshima, T., Katoh, A., Ishimaru, H., Yoneda, Y., Ogita, K., Murase, K., Ohtsuka, H., Inari, K., Fukuta, T., and Kameyama, T., Carbon monoxide-induced delayed amnesia, delayed neuronal death and change in acetylcholine concentration in mice, *J. Pharmacol. Exp. Ther.*, 256, 378, 1991.

Olsky, M. and Woods, J.R., Lutheran General Hospital, Park Ridge, IL, unpublished communication, 1983.

Plum, F., Posner, J.B., and Hain, R.F., Delayed neurological deterioration after anoxia, *Arch. Intern. Med.,* 110, 18, 1962.

Pope, A., Implication of histochemical studies for metabolism of the neuroglia, in *Biology of Neuroglia,* Windle, W. F., Ed., Charles C. Thomas, Springfield, IL, 1958, 211.

Sawa, G. M., Watson, C.P.N., Terbrugge, K., and Chiu, M., Delayed encephalopathy following carbon monoxide intoxication, *Can. J. Neurol. Sci.,* 8, 77, 1981.

Sokal, J.A., Lack of correlation in biochemical effects on rats and blood carboxyhemoglobin concentrations in various conditions of single acute exposure to carbon monoxide, *Arch. Toxicol.,* 34, 331, 1975.

Sokal, J.A. and Kralkowska, E., Relationship between exposure duration, carboxyhemoglobin, blood glucose, pyruvate, and lactate, and the severity of intoxication in 39 cases of acute carbon monoxide poisoning in man, *Arch. Toxicol.,* 57, 196, 1985.

Song, S.V., Okeda, R., Funata, N., and Higashino, F., An experimental study of the pathogenesis of the selective lesion of the globus pallidus in acute carbon monoxide poisoning in cats, *Acta Neuropathol.* (Berlin), 61, 232, 1983.

Thom, S.R., Antagonism of carbon monoxide-mediated brain lipid peroxidation by hyperbaric oxygen, *Toxicol. Appl. Pharmacol.,* 105, 340, 1990a.

Thom, S.R., Carbon monoxide-mediated brain lipid peroxidation in the rat, *J. Appl. Physiol.,* 68, 997, 1990b.

Thom, S.R., Dehydrogenase conversion to oxidase and lipid peroxidation in brain after carbon monoxide poisoning, *J. Appl. Physiol.,* 73, 1584, 1992.

Thom, S.R., Experimental carbon monoxide-mediated brain lipid peroxidation and the effect of oxygen therapy, *Ann. Emerg. Med.,* 17, 403, 1988.

Thom, S.R., Taber, R. L., Mendiguren, I., Clark, J.M., and Fisher, A.B., Delayed neuropsychiatric sequelae following CO poisoning and the role of treatment with 100% O_2 or hyperbaric oxygen: A prospective randomized clinical study, *Undersea Biomed. Res.,* 19 (Suppl.), 47, 1992 [Abstract only].

Walum, E., Varnbo, I., and Peterson, A., Effects of dissolved carbon monoxide on the respiratory activity of perfused neuronal and muscle cell cultures, *Clin. Toxicol.,* 23, 299, 1985.

Warburg, O., Über die wirkung des Kohlenoxyds auf den Stoffwechsel der Hefe, *Biochem. Z.,* 177, 471, 1926.

Werner, B., Bäck, W., Åkerblom, H., and Barr, P.O., Two cases of acute carbon monoxide poisoning with delayed neurological sequelae after a "free" interval, *Clin. Toxicol.,* 23, 249, 1985.

Winston, J.M., Creighton, J.M., and Roberts, R.J., Alteration of carbon monoxide and hypoxic hypoxia-induced lethality following phenobarbitol, chlorpromazine or alcohol pre-treatment, *Toxicol. Appl. Pharmacol.,* 30, 458, 1974.

Zhang, J. and Piantadosi, C.A., Mitochondrial oxidative stress after carbon monoxide hypoxia in the rat brain, *J. Clin. Invest.,* 90, 1193, 1992.

CHAPTER 12

Treatment of Carbon Monoxide Poisoning

Suzanne R. White

CONTENTS

0-8493-4796-3/96/$0.00+$.50

1 INTRODUCTION

Carbon monoxide (CO) poisoning is the leading cause of toxicologic death in the United States, with 3800 fatalities reported annually. Worldwide it remains the most lethal toxin in every community in which it has been studied (Runciman and Gorman, 1993). In addition to the high mortality rates associated with acute exposure to CO, significant long-term morbidity exists as well. Most notably, numerous studies have documented delayed neuropsychiatric sequelae in a significant percentage of CO survivors (Hart et al., 1988; Klawans et al., 1982; Lugaresi et al., 1990; Sauk et al., 1981; Smith and Brandon, 1983; Werner et al., 1985).

This chapter will focus primarily on the treatment of both acute and chronic CO poisoning. It should be kept in mind that our present knowledge regarding therapy for CO poisoning is limited for several reasons. First, effective medical treatment is ideally guided by predictors for either positive or negative outcomes following exposure to toxic substances. For example, blood or urine levels of toxins, combined with characteristic signs or symptoms of toxicity, often aid in the institution of appropriate therapy. Unfortunately, symptoms relating to CO exposure are notoriously vague, and some studies estimate that the diagnosis is missed in 30% of cases (Barret et al., 1985). Furthermore, carboxyhemoglobin levels neither correlate with toxicity nor predict the risk for development of long-term effects (Lasater, 1986; Sokal et al., 1984). Although other predictors of long-term neuropsychiatric sequelae are proposed — loss of consciousness (Olson, 1984), cerebral edema on brain computed tomography (Ikeda et al., 1978), elevated blood glucose (Penney, 1988), or a history of a "soaking" type exposure (Bogusz et al., 1975) — their sensitivity and specificity are unproven. As a result, how to best treat patients with such warning signs remains controversial. Second, appropriate therapy for poisoned patients is ideally guided by an understanding of the toxic mechanisms of that poison. Unfortunately, even though CO has most likely been present since the beginning of time, and has been studied clinically for over 100 years, an adequate understanding of its toxic mechanisms eludes us. Finally, treatment guidelines should ideally be based on prospective, well-controlled, peer-reviewed studies. Unfortunately, there is a dearth of such studies as they relate to CO poisoning treatment in the literature.

Despite these limitations, a general approach to treatment will be described. As an overview, treatment is based on the cessation of tissue hypoxia, the removal of CO from the body, the consideration of potential neuroprotective interventions, and the management of the long-term sequelae of CO poisoning. First a review of historical, often failed treatments for CO poisoning will be presented, followed by a discussion of promising neuroprotective agents. Finally, a clinical approach to the CO-poisoned patient will be outlined.

2 HISTORICAL PERSPECTIVE

Oxygen (O_2) therapy has been the mainstay of treatment for CO poisoning since it was first used therapeutically by Linas and Limousin (1868). Haldane subsequently was able to experimentally demonstrate that mice exposed to "carbonic oxide" were unaffected if O_2 was provided during the exposure. In this seminal work, Haldane concluded that "the higher the oxygen tension the less dependent an animal is on its red corpuscles as oxygen carriers, since the oxygen simply dissolved in the blood becomes considerable when the oxygen tension is high" (Haldane, 1895). Indeed, surface O_2 decreases the half-life of CO from 320 min to 80 min. Unfortunately, surface O_2 alone has not been entirely effective in the treatment of CO poisoning, particularly with regard to the prevention of delayed neuropsychiatric sequelae. This realization has prompted researchers and physicians to search for other treatment modalities.

Numerous fascinating therapies for CO poisoning have not proved to be effective, and are mentioned here only for historical interest. Methylene blue, succinic acid, persantine, iron and cobalt preparations, and ascorbic acid have all been tried, without benefit (Jain, 1990). In animals, cytochrome c, theorized to activate cytochrome oxidase upon supplementation, has not been associated with clinical improvement (Gros and Leandri, 1956). Hydrogen peroxide (H_2O_2) infusions do reduce carboxyhemoglobin content in experimental animals, but the absence of human experience with this chemical and the danger of air embolism preclude its clinical use (Bentolila et al., 1973). While ultraviolet radiation was proposed to facilitate the dissociation of COHb from erythrocytes during transit through skin capillaries and to decrease mortality in animals (Koza, 1930), these effects were not able to be corroborated in a subsequent animal trial (Estler, 1935). Intravenous procaine hydrochloride does not improve the anoxia of CO poisoning in humans (Amyes et al., 1950) and intravenous lidocaine, advocated based on its facilitation of neuronal recovery after cerebral ischemia in experimental animals, has not yet been employed in humans (Evans et al., 1989).

Hyperbaric oxygen (HBO) was first suggested as therapy for CO poisoning in 1901 by Mosso (1901). However, the first clinical use of HBO in the treatment of human CO poisoning did not occur until 1960. Currently, HBO is the mainstay of therapy in severe CO poisoning and is discussed in more detail in Chapter 13.

3 EFFECTIVE NEUROPROTECTIVE AGENTS

3.1 Hyperbaric Oxygen (HBO)

In addition to the obvious effects of increasing both the amount of dissolved O_2 in the blood and the rate of displacement of CO from hemoglobin,

HBO may have other beneficial effects. A brief summary of newer developments in the pathophysiology of CO-mediated brain injury follows, as it provides the basis for understanding many of the other non-HBO treatment modalities discussed later in this chapter. After removal from the CO environment, animals demonstrate marked changes in neutrophil structure and function (Thom, 1993b). Abnormal adherence to brain endothelial cell receptors quickly occurs, possibly as a result of endothelial damage. Upregulation of endothelial intercellular adhesion molecules (ICAM) is demonstrated on endothelial cells as a result of activation by inflammatory mediators. ICAMs bind beta 2 integrins located on polymorphonuclear cells (PMNs), resulting in aggregation of PMNs onto endothelial cell surfaces. Subsequent degranulation of PMNs results in release of destructive proteases, which cause further oxidative injury. HBO at 2.5–3.0 ATA reversibly inhibits PMN CD18 beta 2 integrin activation and therefore decreases adherence of PMNs to endothelial cells (Thom, 1993a). *In vivo*, HBO at 3 ATA also prevents functional neurological impairment in rats (Tomaszewski et al., 1992). Despite the above promising HBO discoveries by Thom and others in animal models and their significant contribution to our understanding of CO-induced microvascular injury, there are to date no prospective, blinded, well-controlled studies based on neuropsychiatric testing in humans that demonstrate a decrease in the incidence of neuropsychiatric sequelae when HBO vs. normobaric oxygen (NBO) is utilized. Hence the role of HBO vs. NBO in the treatment of the patient with CO poisoning is at this point unproved. In addition, the lack of ready availability of hyperbaric chambers in some rural areas and the risk of transporting critically ill patients over long distances underscore the need for continued research in the areas of nonhyperbaric treatment of CO poisoning.

3.2 Allopurinol and N-Acetylcysteine

There is considerable evidence that reactive oxygen metabolites mediate neurologic injury in models of CO poisoning. Lipid peroxidation is documented in rats as a result of exposure to CO at a concentration sufficient to cause unconsciousness. Products of lipid peroxidation are increased by 75% over the baseline values 90 min after CO exposure. Unconsciousness is associated with a brief period of hypotension, so brief that in itself it causes no apparent insult. Lipid peroxidation occurs only after the animals are returned to CO-free air, and there is no direct correlation with the carboxyhemoglobin level (Thom, 1990). Xanthine oxidase (XO) has a central role in this process. During the above described CO-induced PMN degranulation, the released proteases convert xanthine dehydrogenase to XO. The XO then generates superoxide free radicals and lipid peroxidation occurs (Thom, 1992). The XO enzyme is a NAD-dependent dehydrogenase that under ischemic conditions converts to an oxidase, utilizes molecular O_2 rather than NAD as an energy source, and generates the superoxide radical and H_2O_2

(Figure 1). These products in turn cause tissue injury, the brain being particularly susceptible with its low content of catalase and glutathione peroxidase (Marklund et al., 1982). The restoration of xanthine dehydrogenase functional activity is accomplished through the use of XO inhibitors (allopurinol) (Toledo-Pereyra et al., 1974) and sulfhydryl donors (NAC) (Stewart et al., 1982) in non-CO-mediated neuronal injury.

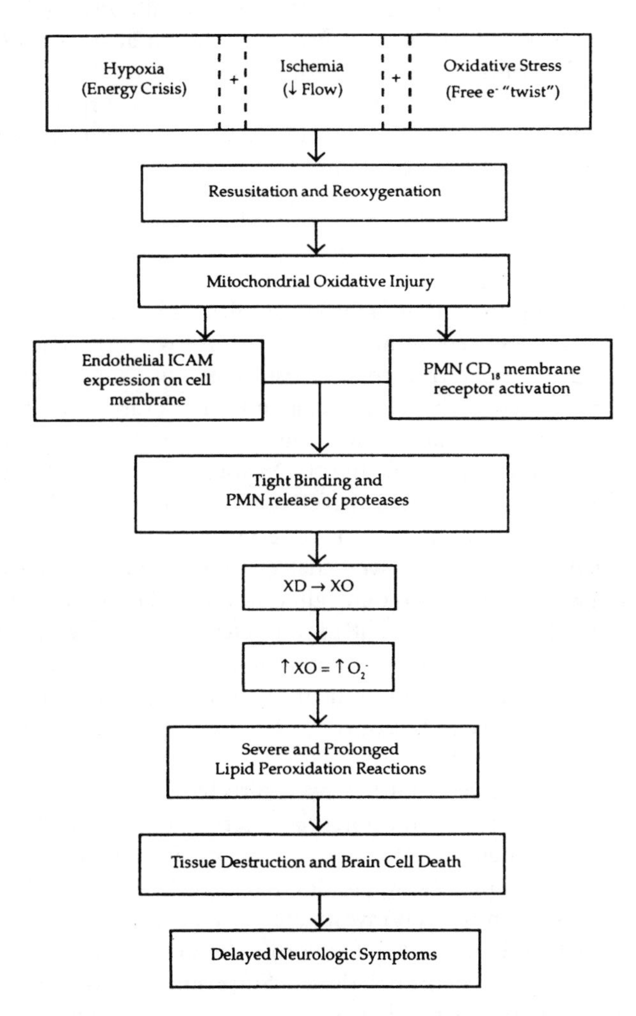

FIGURE 1
Mechanisms of carbon monoxide toxicity. (Adapted from *Oxygen Currents,* HBO Medicine Program newsletter, Emerson, S., Ed., 1 (2), 1993. With permission.)

Similarly, the above two agents have been used in the treatment of CO-induced neuronal injury. Thom demonstrated decreased conversion of xanthine dehydrogenase to XO with decreased lipid peroxidation in rats pretreated with allopurinol (Thom, 1992). Only one human case report demonstrating effectiveness of such combined therapy for the treatment of CO poisoning is reported. A 26-year-old male with a COHb level of 25% 40 h post-exposure who was comatose for 4 d with cerebral edema on CT was treated with both an XO inhibitor (allopurinol) and a sulfhydryl donor, N-acetylcysteine (NAC). NAC was given intravenously over a 20-h period and allupurinol was given orally for 2 weeks. Eight hours after the completion of this regimen, the patient became responsive and gradually improved over the next 3 weeks. "Neurological and mental examination at 6 weeks follow-up were normal" (Howard et al., 1987). Although no formal neuropsychiatric testing was reported, this type of therapy may perhaps provide a basis for further study. Such agents may someday serve as useful adjuncts with HBO in limiting free-radical mediated injury.

3.3 Insulin

In humans and animals, numerous studies have shown that elevated blood glucose is associated with worsened neurologic outcome after brain ischemia caused by stroke or cardiac arrest. Similarly, acute severe CO poisoning is characterized by hyperglycemia and this elevation has been linked to increased severity of brain dysfunction in the rat (Penney et al., 1990). Indeed, animal studies also show that CO exposure raises blood glucose in a dose-dependent manner, and hyperglycemia is an independent predictor of neurologic outcome. A few similar observations have been made in CO-poisoned patients. Penney observed that elevated admission blood glucose was associated with worse neurologic outcome after CO poisoning in patients (Penney, 1988). Leikin et al. (1988) found elevated blood glucose in most patients presenting with carboxyhemoglobin saturation above 25%. Furthermore, anecdotal evidence in the human literature suggests that the neurologic outcome in diabetics poisoned with CO is generally worse than in nondiabetics (Pulsinelli, 1983).

Considerable basic research has been directed at identifying molecular mechanisms of tissue injury and potential interventions to allow the preservation or rescue of neurons after stroke and cardiac arrest. For example, because of the association between elevated blood glucose and poor outcome, insulin has recently been investigated as a potential therapeutic agent in various models of brain and spinal cord ischemia, and indeed appears to substantially ameliorate neuronal death. Surprisingly, this neuron-sparing effect is known to be independent of insulin-induced reductions in blood glucose, and is hypothesized to be mediated through cell signal transduction mechanisms, in common with other growth factors.

To date, very few studies have approached the problem of neurologic damage induced by CO by utilizing molecular-based concepts similar to those now being applied to the problems of brain ischemic-anoxia. Hence, the question of whether insulin ameliorates neuronal injury secondary to CO toxicity was investigated in rats who were exposed to a CO LD50 of

2400 ppm for 90 min. Survivors were treated for 4 h with: (a) normal saline infusion, (b) continuous infusion of glucose to clamp blood levels at 250–300 mg/dl, and (c) continuous infusion of glucose to maintain blood levels at 250–300 mg/dL with intraperitoneal injection of 4 units per kg regular insulin. Neurologic scoring was performed at times 0, 5.5, 24, 48, 72, and 96 h after using a standardized system modified from Lundy et al. (1985). It was noted that significant neurologic deficit occurred in all groups after the CO exposure and treatment period. Induced hyperglycemia after CO exposure was associated with significantly worsened neurologic scores as compared to saline-treated controls. Insulin therapy simultaneous with induced hyperglycemia significantly improved neurologic scores at all times despite maintenance of comparable hyperglycemia with respect to the group treated only with glucose (Figure 2). No significant difference in mortality was found between treatment groups (White and Penney, 1994).

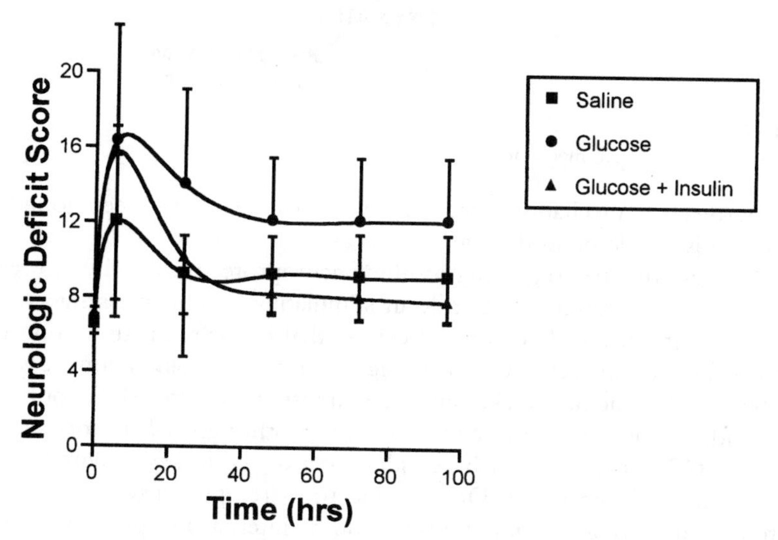

FIGURE 2
Effects of insulin and glucose treatment on neurologic outcome after carbon monoxide poisoning.

Several theories have evolved regarding the post-receptor binding protective effect of insulin on the neuron. Insulin has been shown to provide neuromodulatory inhibition of synaptic transmission *in vivo* and *in vitro*. As an inhibitor of glial uptake of gamma amino butyric acid (GABA), insulin may increase the availability of this inhibitory neurotransmitter and may decrease neuronal firing, beneficially reducing cell metabolism (Bouhaddi et al., 1988). The additional effect of sodium extrusion from the cell that affords subsequent protection against water accumulation may prevent neuronal swelling (Stahl, 1988). Furthermore, Voll (1988) has suggested that an insulin-induced elevation of brain catecholamines through both inhibition of catecholamine uptake and stimulation of release might be a contributory

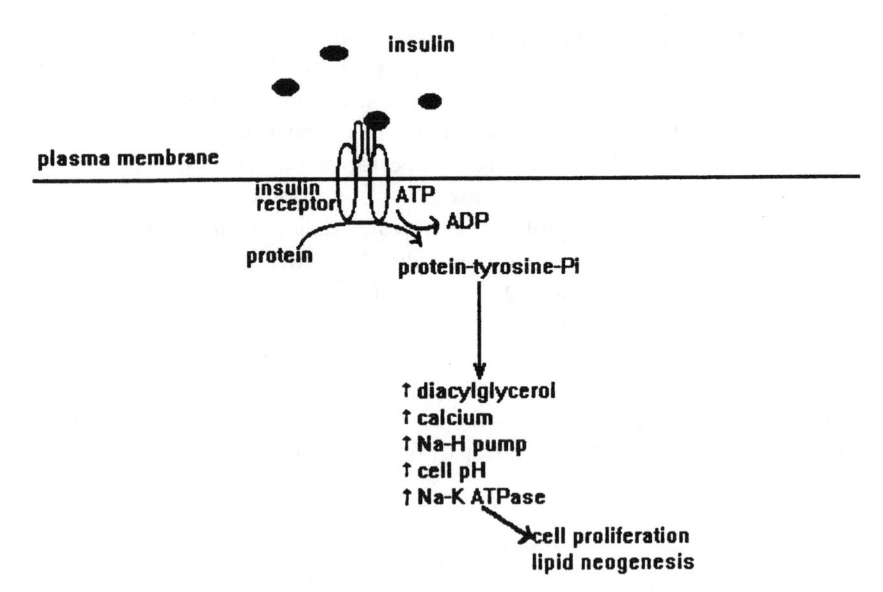

FIGURE 3
Insulin's neuroprotective mechanisms.

neuroprotective mechanism because catecholamines have been found to attenuate ischemic brain damage.

Of these theories regarding insulin's neuroprotective activity, however, the most recent highlights its role in stimulating second messengers, and emphasizes its potential genomic effects — that is, the regulation of protein synthesis, enzymatic activity, and the signaling of cell proliferation. It is well established that the neonatal brain is rich in insulin-like growth factor receptors. Indeed, insulin is similar in structure to other growth factors such as PDGF, EGF, and IGF-1. Such peptides are involved in basic neuron development and differentiation. Once bound to its receptor, like other growth factors, insulin triggers signal transduction by internal autophosphorylation of tyrosine on the insulin receptor, which subsequently enhances further phosphorylation reactions of other tyrosine-containing substrates by tyrosine kinase, also located on the insulin receptor (Figure 3).

This type of tyrosine phosphorylation is important in signaling pathways for such growth factors and products of protooncogenes. Insulin is a progression growth factor in the replication (GOGI) phase and works synergistically with other growth factors to generate both competence and progression of cells. Through tyrosine phosphorylation of phosphokinase C, other second messengers such as diacyglycerol are formed that cause increased intracellular calcium, activation of the sodium–hydrogen pump, and increased intracellular pH. This pH change in turn activates the sodium potassium ATPase pump that signals cell proliferation (MaCara, 1985). Insulin also regulates specific mRNA levels through diacylglycerol (Standaert and Pollet, 1988) and may increase mRNA efflux from the nucleus via nuclear

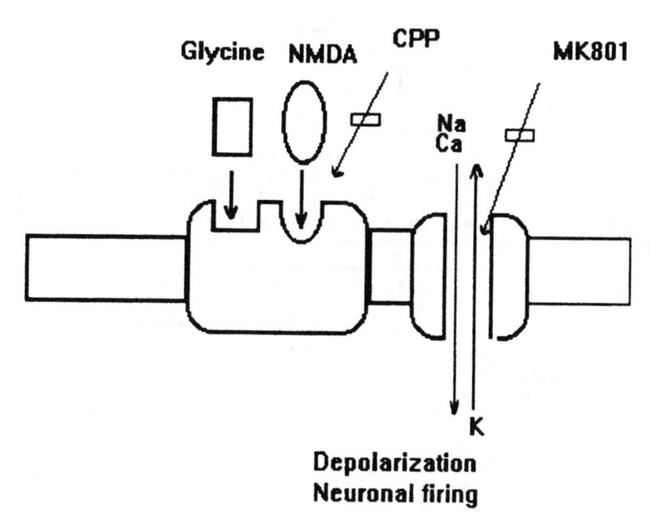

FIGURE 4
The NMDA receptor complex.

triphosphatase activation (Goldfine, 1987; Goldfine et al., 1985). Of even greater importance is the fact that insulin also stimulates lipid neogenesis. It appears, therefore, that the effects of insulin are fundamental with regard to cell signaling, proliferation, replication, and repair following injury. These processes are crucial to cells such as neurons, which normally are terminally differentiated and contain little if any capacity to replicate or to synthesize repair lipids.

Certainly, anatomic correlates to the above proposed mechanisms are in place. For example, it has been demonstrated that the location of insulin and IGF-I receptors correlates with phosphotyrosine products in the brain (Moss et al., 1990). Moreover, the basal ganglia, areas typically found to be damaged by CO, possess low levels of insulin receptors. Although the initial results of animal studies such as those above may provide building blocks for clinical work, any recommendations regarding the use of insulin in humans as treatment for CO poisoning will, of course, await further studies.

3.4 NMDA Receptor Antagonists

Recent evidence implicates the endogenous excitatory amino acids such as N-methyl D-aspartate (NMDA) in ischemic neurodegeneration (Beneviste et al., 1984; Jorgensen and Diemer, 1982; Simon et al., 1984) (Figure 4). Moreover, the NMDA receptor antagonist, MK801, prevents non-CO-induced ischemic neurodegeneration in Mongolian gerbils (Gill et al., 1987). Successive CO exposures induce a consistent pattern of degeneration of hippocampal CA1 pyramidal neurons, a selective neuronal death that resembles that seen with other models of cerebral ischemia.

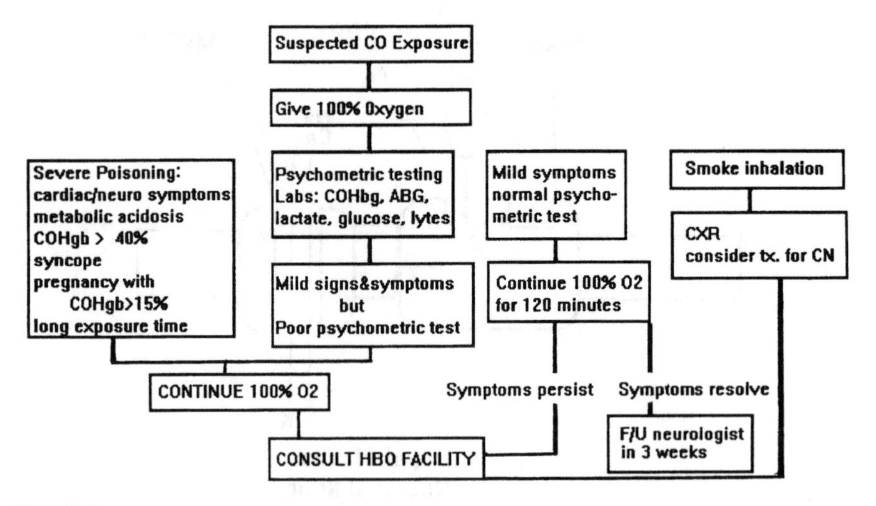

FIGURE 5
Treatment of CO poisoning. (Adapted from *Oxygen Currents,* HBO Program Newsletter, Emerson, S., Ed., 1 (2), 1993. With permission.)

This observation has prompted the study of NMDA receptor antagonists in CO poisoning in mice. Ishimaru et al. (1991) pretreated animals with a competitive NMDA antagonist, CPP; a noncompetitive NMDA antagonist, MK-801; a glycine binding site antagonist, 7-CK; a polyamine binding site antagonist, ifenprodil; glycine, and saline. At 7 d post-exposure the number of hippocampal CA1 pyramidal cells was quantified using an image analyzer. A decrement of 20% in the number of hippocampal CAl pyramidal cells was noted relative to the control group. Those animals receiving high doses of MK-801, 7-CK, and CPP had significant reduction in neuronal damage. No clear protective effect was obtained with ifenprodil. Interestingly, glycine, a facilitory neurotransmitter at the NMDA receptor complex, did not exaggerate the CO-induced neuronal damage as might be expected. Although no neurologic outcome correlates or survival data are reported, this work may provide valuable mechanistic and possibly future therapeutic insights.

3.5 Hypothermia

Hypothermia was found to be beneficial in the management of CO poisoning by Sluitjer (1963) — an effect that was thought to be secondary to the increased amounts of dissolved O_2 found in the bloodstream at lower temperatures. However, Pierce et al. (1972) were unable to demonstrate any synergistic effect when hypothermia was used in conjunction with HBO in a dog model. An interesting report of the use of mechanical ventilation and hypothermia in patients with abnormal motor activity or coma to treat CO toxicity noted complete reversal of these manifestations in 3 patients when therapy was initiated within the first 24 h. No beneficial effects were noted in a 4th patient who did not receive hypothermic treatment until 5 d after exposure. HBO was not available to these patients (Boutros and Hoyt, 1976).

4 APPROACH TO THE PATIENT WITH CO POISONING

4.1 GENERAL

The most critical step in managing the patient poisoned with CO is the cessation of tissue hypoxia. This involves supplementation with 100% O_2, delivered by either a tight-fitting continuous positive airway pressure (CPAP) mask or by endotracheal intubation. Intubation may be necessary in the patient with chronic obstructive pulmonary disease, to avoid CO_2 retention secondary to high concentrations of O_2. The use of 5% CO_2 mixed with 95% O_2 (carbogen) has been proposed by some to facilitate the release of CO from hemoglobin by increasing ventilatory response. Given the fact that carbogen does not normalize blood pH or lactate, nor does it improve survival in animals relative to those breathing 100% O_2 alone (Schwerma et al., 1948), this therapy is of questionable value. Furthermore, because the possibility of CO_2 retention with its subsequent worsening of existing acidosis during this therapy exists, such treatment would be contraindicated in the patient with chronic obstructive pulmonary disease (COPD), concurrent poisoning with respiratory depressants, or altered mental status. The duration of O_2 therapy is guided by a knowledge of CO half-life and allows for a margin of safety. Generally this would involve at least 120 min of therapy on 100% O_2, longer if the patient is gravid (see "pregnancy" below). An early chest X-ray is mandatory to assess for evidence of pulmonary edema resulting from CO or other inhaled toxins.

As airway control and oxygenation proceed, attention should be directed toward the cardiovascular system. Continuous cardiac monitoring is advisable and a 12-lead ECG should be obtained to assess for subclinical cardiac ischemia. Should arrhythmias, ischemia, or hemodynamic instability occur despite therapy with 100% O_2, the patient should be considered a candidate for HBO therapy. Myocardial depression and arrhythmias may occur at extremely low arterial pH, which is generally secondary to severe lactic acidosis. Despite this, full correction of the acidosis with sodium bicarbonate is not advisable as this could result in a further shift of the oxyhemoglobin dissociation curve to the left, and thus impair the unloading of O_2 to hypoxic tissues.

Removal of CO from the body is best accomplished through displacement by O_2, with either NBO or HBO. Other less conventional therapies have been used anecdotally with favorable outcomes, and include both exchange transfusion (Yee and Brandon, 1983) and extracorporeal oxygenation (Radushevich and Koroteeva, 1976). However, these invasive approaches would be recommended only in unusual circumstances, for example, if HBO therapy were not readily available.

Assessment for associated trauma, burns, or inhalation injury is imperative early in the hospital course. Should the patient have been CO poisoned during the inhalation of smoke from a fire, numerous other products of combustion may be contributing to the metabolic and pulmonary derangements seen. Of particular interest is cyanide, a lethal combustion product that is commonly elaborated when plastics or synthetic materials burn. The patient with CO poisoning who remains significantly acidotic despite treatment with O_2 after smoke inhalation should be suspected to have cyanide toxicity. Some authors advocate empiric treatment of fire victims with the sodium thiosulfate component of the Lilly cyanide antidote kit. The methemoglobin-forming agents such as amyl nitrite and sodium nitrite have been traditionally withheld if CO poisoning is suspected in order to avoid further hemoglobinopathy and worsened hypoxia.

Moore et al. (1987) demonstrated a 25% increase in mortality in sodium nitrite-treated animals with CO poisoning compared to the untreated controls. Despite this, a recent human study demonstrated that 5 of 7 patients with CO poisoning were safely treated with the antidote kit in its entirety (Kirk et al., 1993). However, these patients had only moderate COHb levels with a mean value of 26%. Coupled with the fact that only a small number of patients were studied, this form of empiric treatment for cyanide poisoning in victims of smoke inhalation cannot yet be widely recommended. Other proposed cyanide antidotes such as hydroxycobalamin require that large volumes of fluid be given (2.4 l per average dose), are costly, and are not readily available at this time in the United States. Therefore, given the safety of the sodium thiosulfate component of the Lilly kit when given alone, this is the only antidote advisable at this time for the treatment of the patient doubly poisoned with CO and cyanide.

Once the patient has been stabilized, consideration of the use of possible neuroprotective agents including HBO must ensue. If the patient is awake, neuropsychiatric testing must be performed. Abbreviated neuropsychologic tests have been developed specifically for the CO poisoned patient. Examples of tasks performed by the patient include placing pegs in a board, rapid finger tapping, memorization, construction, number processing, and subjective stress response (Seger and Welch, 1994). Should the patient perform abnormally on the CO neuropsychiatric screening battery (CONSB), have a history of a soaking-type exposure or loss of consciousness, have any abnormal neurologic findings, exhibit cerebral edema on CAT scan, or have evidence of cardiovascular involvement, HBO is indicated. Other patients who would be considered candidates for HBO would include the pregnant patient with a carboxyhemoglobin level greater than 15%, the patient with a history of coronary artery disease and a carboxyhemoglobin level greater than 20%, the asymptomatic patient with a level greater than 40% COHb, or the patient with recurrent or persistent symptoms despite 6 h of therapy with NBO (Figure 5).

4.2 Pregnancy

The fetus is particularly vulnerable to the effects of CO, which readily crosses the placenta and is even more tightly bound to fetal hemoglobin. The fetus also reaches higher peak COHb levels than does the mother. Fetuses that survive a significant CO poisoning may be left with limb malformation, hypotonia, areflexia, persistent seizures, mental and motor disability, and microcephaly (Ginsberg and Myers, 1974, 1976). The only prospective, multicenter study of acute CO poisoning in pregnancy recently reported adverse outcomes in 60% of children whose mothers suffered severe CO toxicity. Of those babies born to mothers with mild to moderate CO exposure, normal physical exams and neurobehavioral development were reported (Koren et al., 1991). Since CO elimination from the fetus is prolonged (7-10 h), it is generally accepted that HBO therapy is useful at lower maternal COHb levels than would be acted upon in the nongravid patient. In addition, surface O_2 therapy should be extended to four to five times the normal duration. Although controversial, HBO has been reported to be safe in pregnancy (Brown et al., 1992), despite theoretical dangers of fetal hyperoxia in animal models (Ferm, 1964; Fujikura, 1964; Miller et al., 1971). A recent report of 44 women undergoing HBO during pregnancy for CO exposure suggests that HBO is safe and should be considered, although miscarriages did occur, and 6 patients were lost to follow-up (Elkaharrat et al., 1991). It should be noted that HBO was implicated in the induction of labor in 1 pregnant patient; however, the pregnancy was near term when the CO exposure occurred (Farrow et al., 1991). Indications for HBO therapy in the pregnant patient are as follows:

1. Maternal COHb level >15–20% at any time during the exposure.
2. Any neurological signs or symptoms.
3. Evidence of fetal distress (fetal tachycardia, decreased beat-to-beat variability, late decelerations).
4. If maternal neurologic symptoms or fetal distress persist 12 h after initial therapy, additional HBO treatments may be necessary.

4.3 Children

Crocker and Walker reviewed 28 pediatric patients with CO poisoning, noting that all with carboxyhemoglobin levels of greater than 15% were symptomatic. Furthermore, all with carboxyhemoglobin levels greater than 24% had syncope (Crocker and Walker, 1985). A second study looking at infants less than 2 years of age found some patients to be asymptomatic at carboxyhemoglobin levels greater than 15% (Rudge, 1993). In addition, a recent study demonstrated decreased CO half-life in children treated with NBO relative to the half-life in adults (Decker and Wiley, 1994). These disparate findings are difficult to interpret, but younger children have traditionally been thought to be more susceptible to CO poisoning based on more rapid metabolic rates and higher O_2 demands. Delayed sequelae in the children in the above study by Crocker was reported to have occurred in 3 of 28 patients; however, formal neuropsychiatric testing was not performed. Sequelae reported include chronic headaches, memory impairment, and difficulty in school. Indeed, it is difficult to assess the true incidence of this phenomenon in children since neuropsychiatric testing is often impossible to perform in this population. The two aforementioned clinical studies in children showed favorable outcomes in those patients treated with HBO. Until further knowledge and experience is gained in this area, children should be treated at least as aggressively as adults who are CO poisoned.

5 MANAGEMENT OF THE SEQUELAE OF CO POISONING

Delayed sequelae from CO poisoning is devastating and occurs in 10–30% of persons recovering from the acute exposure. Parkinsonism, the most frequent long-term neurologic complication, has a grim prognosis. Unfortunately, conventional therapy with agents such as dopa has been disappointing. Another centrally acting dopaminergic agonist, bromocriptine, may be more promising. Nine patients (mean age 61 years; suffering from CO-induced Parkinsonism) who were given bromocriptine (5–30 mg/day) displayed improvement in Webster's scores while under treatment (De Pooter et al., 1991). Clearly no definitive conclusions regarding bromocriptine therapy can be made, but perhaps this study will provide a basis for future investigations.

Repetitive HBO has been advocated by some to improve the long-term neurologic deficits from CO toxicity, even if instituted weeks after the initial CO exposure (Gibson et al., 1991; Myers et al., 1981, 1983, 1985; Raphael et al., 1989; Smith and Brandon, 1973). Such practice, which is advocated by several treatment centers in the United States, lacks validation by well-controlled, blinded clinical studies that utilize neuropsychiatric testing data. Interestingly, behavioral treatment has been successful when guided by formal neuropsychiatric testing. In certain patients, indirect measures of learning better predict treatment utility (Heinrichs, 1990). Patients who present to health care facilities late or have suffered recurrent or chronic low-level exposures provide a particular challenge to the clinician. Any proposed therapeutic approach to such patients should be considered carefully, given the fact that no definitive clinical or animal studies in this area exist.

6 CONCLUSIONS

There is much yet to be learned about the treatment of CO poisoning. Potential areas for further research would include the delineation of clear-cut indications for the use those neuroprotective agents already shown to have some beneficial effect (i.e., HBO, insulin, sulfhydryl donors, allopurinol, NMDA receptor antagonists, and hypothermia). Further efficacy studies in humans for such agents are also needed. In addition, how to best use these agents synergistically remains to be seen. Other questions that remain unanswered are the safety profile and indications for HBO in children and pregnancy, the true incidence of adverse fetal outcome from CO exposure during pregnancy, and both the true incidence of and how best to test for delayed neuropsychiatric sequelae in children.

REFERENCES

Amyes, E.W., Ray, J.W., and Brockman, N.W., Carbon monoxide anoxia; intravenous administration of procaine hydrochloride in the treatment of acute and chronic effects, *JAMA,* 142, 1054, 1950.

Barret, L., Danel, V., and Faure, J., Carbon monoxide poisoning, a diagnosis frequently overlooked, *Clin. Toxicol.,* 23, 309, 1985.

Beneviste, H., Drejer, J., Schousboe, A., and Diemer, N. H., Elevation of extracellular concentrations of glutamate and aspartate in rat hippocampus during transient cerebral ischaemia monitored by intracerebral microdialysis, *J. Neurochem.,* 43, 1369, 1984.

Bentolila, P., Tran, G., and Olive G., Essai de traitment de l'intoxication oxycarbonee par perfusion de solutions diluees de paroxyde d'hysrogene, resultats obtnus chez le lapin, *Therapie,* 28, 1043, 1973.

Bogusz, M., Cholewa, L., Pach, J., and Mlodkowska, K., A comparison of two types of acute carbon monoxide poisoning, *Arch. Toxicol.,* 33, 141, 1975.

Bouhaddi, K., Thomopoulos, P., Fages, C., Khelil, M., and Tardy, M., Insulin effect on GABA uptake in astroglial primary cultures, Neurochem. Res., 13, 1119, 1988.

Boutros, A. R. and Hoyt, J. L., Management of carbon monoxide poisoning in the absence of hyperbaric oxygenation chamber, *Crit. Care Med.,* 4, 144, 1976.

Brown, D.B., Mueller, G.L., and Golich, F.C., Hyperbaric oxygen treatment for carbon monoxide poisoning in pregnancy: A case report, *Aviat. Space Environ. Med.,* 63, 1011, 1992.

Crocker, P.J. and Walker, J.S., Pediatric carbon monoxide toxicity, *J. Emerg. Med.,* 3, 443, 1985.

Decker, J. and Wiley, J.F., Carboxyhemoglobin elimination half-life in children, paper presented at the North American Congress of Clinical Toxicology, Salt Lake City, Utah, September, 1994.

De Pooter, M.C., Leys, D., Godefroy, O., DeReuck, J., and Peter, H., Parkinsonian syndrome caused by carbon monoxide poisoning. Preliminary results of the treatment with bromocriptine, *Rev. Neurol.,* 147, 399,1991.

Elkaharrat, D., Raphael, J.D., Korach, J.M., Jars-Guincestre, M.C., Chastang, C., Harborn, C., and Gajdos, P., Acute carbon monoxide intoxication and hyperbaric oxygen in pregnancy, *Int. Care Med.,* 17, 289, 1991.

Estler, W., Experimentelle untersuchungen uber die anwendung der bestrahlung mit ultravoilettem licht zur behandlung der kohlenoxidvergiftung, *Arch. llyg. Bakt.,* 115, 152, 1935.

Evans, D.E., Catron, P.W., McDermott, J.J., et al., Effect of lidocaine after experimental cerebral ischemia induced by air embolism, *J. Neurosurg.*, 70, 97, 1989.
Farrow, J.R., Davis, G.J., Roy, T.M., et al., Acute carbon monoxide intoxication and hyperbaric oxygen in pregnancy, *Int. Care Med.*, 17, 289, 1991.
Ferm, V.H., Teratogenic effects of hyperbaric oxygen, *Proc. Soc. Exp. Biol. Med.*, 116, 975, 1964.
Fujikura, T., Retrolental fibroplasia and prematurity in newborn rabbits induced by maternal hyperoxia, *Am. J. Obstet. Gynecol.*, 90, 854, 1964.
Gibson, A.J., Davis, F.M., and Ewer, T., In practice: Delayed hyperbaric oxygen therapy for carbon monoxide intoxication — Two case reports, *N. Z. Med. J.*, 104, 64, 1991.
Gill, R., Foster, A. C., and Woodruff, G. N., Systemic administration of MK-801 protects against ischemia-induced hippocampal neurodegeneration in the gerbil, *J. Neurosci.*, 3343, 1987.
Ginsberg, M.D. and Myers, R.E., Fetal brain damage following maternal carbon monoxide intoxication, *Acta Obstet. Gynecol.*, 53, 309,1974.
Ginsberg, M.D. and Myers, R.E., Fetal brain injury after maternal carbon monoxide intoxication, *Neurology*, 26, 15,1976.
Goldfine, I.D., The insulin receptor: Molecular biology and trans membrane signaling, *Endocrinology*, 8, 235, 1987.
Goldfine, I.D., Purrello, F., Vigneri, R., and Clawson, G.A., Direct regulation of nuclear functions by insulin: Relationship to mRNA metabolism, in *Molecular Basis of Insulin Action*, Czech, M.P., Ed., Plenum, New York, 1985, 329.
Gros, J.F. and Leandri, P., Traitment de l'intoxication oxycarbonee par le cytochrome, *Presse Med.*, 64, 1356, 1956.
Haldane, J., The relation of the action of carbonic oxide to oxygen tension, *J. Physiol.*, 18, 201, 1895.
Hart, I.K., Kennedy, P.G., Adams, J.H., et al., Neurological manifestations of carbon monoxide poisoning, *Postgrad. Med. J.*, 64, 213, 1988.
Heinrichs, R.W., Relationship between neuropsychological data and response to behavioral treatment in a case of carbon monoxide toxicity and dementia, *Brain Cognit.*, 14, 213, 1990.
Howard, R.J., Blake, D.R., Pall, H., Williams, A., and Green, I.D., Allopurinol/n-acetylcysteine for carbon monoxide poisoning, *Lancet*, 2, 628,1987.
Ikeda, T., Kondo, T., Mogami, H., Miura, T., Mitomo, M., Shimazaki, S., and Sugimoto, T., Computerized tomography in cases of acute carbon monoxide poisoning, *Med. J. Osaka Univ.*, 29, 253, 1978.
Ishimaru, H., Katoh, A., Suzuki, H., Fukuta, T., Kameyama, T., and Nabeshima, T., Effects of N-methyl-D aspartate receptor antagonists on carbon monoxide induced brain damage in mice, *J. Pharm. Exp. Ther.*, 261, 349, 1991.
Jain, K.K., *Carbon Monoxide Poisoning*, Warren H. Green, St Louis, 1990, 140.
Jorgensen, M.B. and Diemer, N.H., Selective neuron loss after cerebral ischaemia in the rat: Possible role of transmitter glutamate, *Acta Neurol. Scand.*, 66, 536, 1982.
Kirk, M.A., Gerace, R., and Kulig, K.W., Cyanide and methemoglobin kinetics in smoke inhalation victims treated with the cyanide antidote kit, *Ann. Emerg. Med.*, 22, 1413, 1993.
Klawans, H.L., Stein, B.W., Tanner, C.M., et al., A pure Parkinsonian syndrome following acute carbon monoxide intoxication, *Arch. Neurol.*, 39, 302, 1982.
Koren, G., Sharav, T., Pastuszak A., Garrettson, L.K., Hill, K., Samson, I., Rorem, M., King, A., and Dolgin, J.E., A multicenter, prospective study of fetal outcome following accidental carbon monoxide poisoning in pregnancy, *Reprod. Toxicol.*, 5, 397, 1991.
Koza, F., Die kohlenmonoxidevergiftungund deren neurartige therapie mit bestrahlung, *Med. Klinik.*, 26, 422,1930.
Lasater, S.R., Carbon monoxide poisoning, *Can. Med. Assoc. J.*, 134, 991, 1986.
Leikin, J.B., Goldenberg, R.R., Edwards, D., and Zell-Kantor, M., Metabolic predictors of carbon monoxide poisoning, *Vet. Hum. Toxicol.*, 30, 40, 1988.
Linas, A.J. and Limousin, S., *Bull. Mem. Soc. Ther.*, 2, 32, 1868.
Lugaresi, A., Montagna, P., Morreale, A., et al., "Psychic akinesia" following carbon monoxide poisoning, *Eur. Neurol.*, 30, 167, 1990.

Lundy, E.F, Dykstra, J., Luyckx, B., Zelenock, G.B., and D'Alecy, L.G., Reduction of neurologic deficit by 1,3-butanediol induced ketosis in Levine rats, *Stroke,* 16, 855, 1985.

MaCara, I.G., Oncogenes, ions, and phospholipids, *Am. J. Physiol.,* 248, C3, 1985.

Marklund, S.L., Westman, G., Lundgren, E., and Roos, G., CuZn superoxide dismutase, Mn superoxide dismutase, catalase and glutathione peroxidase in normal and neoplastic human cell-lines and normal human tissues, *Cancer Res.,* 42, 1955, 1982.

Miller, P.D., Telfored, I.D., and Haas, G.R., Effects of hyperbaric oxygen on cardiogenesis in the rat, *Biol. Neonat.,* 17, 44, 1971.

Moore, S.J., Norris, J.C., Walsh, D.A., et al., Antidotal use of methemoglobin-forming cyanide antagonists in concurrent carbon monoxide/cyanide intoxication, *J. Pharm. Exp. Ther.,* 242, 70, 1987.

Moss, A.M., Unger, J.W., Moxley, R.T., and Livingston, J.N., Location of phosphotyrosine-containing proteins by immunocytochemistry in the rat forebrain corresponds to the distribution of the insulin receptor, *Proc. Natl. Acad. Sci. U.S.A.,* 87, 4453, 1990.

Mosso, A., La mort apparente du coer secours l'empoisonneement par l'oxyd de carbonee, *Arch. Ital. Biol.,* 35, 75, 1901.

Myers, R.A., Mitchell, J.T., and Cowley, R.A., Psychometric testing and carbon monoxide poisoning, *Disaster Med.,* 1, 279, 1983.

Myers, R.A., Snyder, S.K., and Emhoff, T.A., Subacute sequelae of carbon monoxide poisoning, *Ann. Emerg. Med.,* 14, 1163, 1985.

Myers, R.A., Snyder, S.K., Linberg, S., et al., Value of hyperbaric oxygen in suspected carbon monoxide poisoning, *JAMA,* 246, 248, 1981.

Olson, K.R., Carbon monoxide poisoning: Mechanisms, presentation, and controversies in management, *J. Emerg. Med.,* 1, 233, 1984.

Peirce, E.C., Zacharias, A., Alday, J.M., Hoffman, B.A., and Jacobson, J.H., Carbon monoxide poisoning: Experimental hypothermic and hyperbaric studies, *Surgery,* 72, 229, 1972.

Penney, D.G., Hyperglycemia exacerbates brain damage in acute-severe carbon monoxide poisoning, *Med. Hypoth.,* 27, 241, 1988.

Penney, D.G., Helfman, C.C., Hull, J.C., Dunbar J.C., and Verman, K., Elevated blood glucose is associated with poor outcome in the carbon monoxide poisoned rat, *Toxicol. Lett.,* 54, 287, 1990.

Pulsinelli, W.A., Levy, D.E., Sigsbee, B., Scherer, P., and Plum, F., Increased damage after ischemic stroke in patients with hyperglycemia with or without established diabetes mellitus, *Am. J. Med.,* 74, 540, 1983.

Radushevich, V.P. and Koroteeva, E.L., Parallel blood circulation with oxygenation of blood in severe poisoning with carbon monoxide fumes, *Vestn. Khir.,* 116, 131, 1976.

Raphael, J., Elkharrat, D., Jars-Guincestre, M., et al., Trial of normobaric and hyperbaric oxygen for acute carbon monoxide intoxication, *Lancet,* 19, 414,1989.

Rudge, F.W., Carbon monoxide poisoning in infants: Treatment with hyperbaric oxygen, *South. Med. J.,* 86, 334, 1993.

Runciman, W.W. and Gorman, D.F., Carbon monoxide poisoning: From old dogma to new uncertainties, *Med. J. Aust.,* 158, 439, 1993.

Sauk, G.M., Watson C.P., Tebragge, K., et al., Delayed encephalopathy following carbon intoxication, *Can. J. Neurol. Sci.,* 8, 77, 1981.

Schwerma, H., Ivy, A.C., Friedman, H., et al., Study of resuscitation from justalethal effects of exposure to carbon monoxide, *Occup. Med.,* 5, 24, 1948.

Seger, D. and Welch, L., Carbon monoxide controversies: Neuropsychologic testing, mechanism of toxicity, and hyperbaric oxygen, *Ann. Emerg. Med.,* 24, 242, 1994.

Simon, R.P., Swan, J.H., Griffith, T., and Meldrum, B.S., Blockade of N-methyl-D-aspartate receptors may protect against ischaemic damage in the brain, *Science,* 226, 850, 1984.

SIuitjer, M.E., The treatment of carbon monoxide poisoning by administration of oxygen at high pressure, *Proc. R. Soc. Med.,* 56, 1002, 1963.

Smith, J.S. and Brandon, S., Morbidity from acute carbon monoxide poisoning at three-year follow-up, *Br. Med. J.,* 1, 279, 1983.

Sokal, J.A., Majka, J., and Palus, J., The content of carbon monoxide in the tissues of rats intoxicated with carbon monoxide in various conditions of acute exposure, *Arch. Toxicol.*, 56, 106, 1984.

Stahl, W., The Na-K-ATPase of nervous tissue, *Neuro. Int.*, 8, 449, 1988.

Standaert, M.L. and Pollet, R.J., Insulin-glycerolipid mediators and gene expression, *FASEB J.*, 2, 2453, 1988.

Stewart, J.R., Blackwell, W.H., Crute, S.L., Loughlin, V., Hess, M.L., and Greenfield, L.J., Prevention of myocardial ischemia/reperfusion injury with oxygen-free radical scavengers, *Surg. Forum*, 33, 317, 1982.

Thom, S.R., Carbon monoxide-mediated brain lipid peroxidation in the rat, *J. Appl. Physiol.*, 68, 997, 1990.

Thom, S.R., Dehydrogenase conversion to oxidase and lipid peroxidation in the brain after carbon monoxide poisoning, *J. Appl. Physiol.*, 73, 1584, 1992.

Thom, S.R., Functional inhibition of neutrophil beta 2 integrins by HBO in CO-mediated brain injury, *Toxicol. Appl. Pharmacol.*, 123, 248, 1993a.

Thom, S.R., Leukocytes in carbon monoxide-mediated brain oxidative injury, *Toxicol. Appl. Pharmacol.*, 123, 234, 1993b.

Toledo-Pereyra, L.H., Simmons, R.L., and Najarian, J.S., Effect of allopurinol on the preservation of ischemic kidneys perfused with plasma or plasma substitutes, *Ann. Surg.*, 180, 780, 1974.

Tomaszewski, C., Rudy, J., Wathen, J., Brent, J., Rosenberg, N., and Kulig, K., Prevention of neurologic sequelae from carbon monoxide by hyperbaric oxygen in rats, *Ann. Emerg. Med.*, 21, 631, 1992.

Voll, C.L. and Auer, R.N., The effect of postischemic blood glucose levels on ischemic brain damage in the rat, *Ann. Neurol.*, 24, 638, 1988.

Werner, B., Back, W., Akerblom, H., et al., Two cases of acute carbon monoxide poisoning with delayed neurologic sequelae after a "free" interval, *Clin. Toxicol.*, 23, 249, 1985.

White, S.R. and Penney, D.G., Effects of insulin and glucose treatment on neurologic outcome after carbon monoxide poisoning, *Ann. Emerg. Med.*, 23, 606, 1994.

Yee, L.M. and Brandon, G.K., Successful reversal of presumed carbon monoxide-induced semicoma, *Aviat. Space Environ. Med.*, 54, 641, 1983.

CHAPTER 13

OPTIONS FOR TREATMENT OF CARBON MONOXIDE POISONING, INCLUDING HYPERBARIC OXYGEN THERAPY

Stephen R. Thom

CONTENTS

0-8493-4796-3/96/$0.00+$.50

1 INTRODUCTION

The principal goal of this chapter is to review the various methods of treatment that have been attempted for carbon monoxide (CO) poisoning. As hypoxia has been a recognized characteristic of CO poisoning since the work of Claude Bernard (1857), treatment with oxygen (O_2) has been a central theme. Haldane (1895) underscored the hypoxic character of CO poisoning by demonstrating that unconsciousness in mice exposed to CO at a partial pressure of approximately 1 atm could be prevented by concurrently exposing the animals to hyperbaric oxygen (HBO) at 2 atmospheres absolute (ATA). Since that time a great deal of research has been performed to investigate various treatments and, in some cases, to ascertain their mechanisms of action.

2 EMERGENCY CARE

The treatment of a patient suffering from CO poisoning must initially focus on maintaining the patient's airway, ventilation, oxygenation, and adequate perfusion. These actions lay a proper foundation for the emergency management of CO poisoning, but do not represent specific treatment measures. Indeed, it might be argued that truly specific treatment can only come when all of the mechanisms responsible for CO-mediated pathology are elucidated.

A cornerstone to the treatment of patients with CO poisoning is the use of supplemental O_2. Bernard (1857) showed that CO will bind with hemoglobin and precipitate an hypoxic stress. Linas and Limousin (1868) were the first to record their use of O_2 in treating CO poisoning. Historically, emphasis has been placed on O_2 treatment because the rate of carboxyhemoglobin (COHb) dissociation is proportional to the arterial oxygen partial pressure [PaO_2] (Pace et al., 1950). Hence, initial treatment of patients with CO poisoning should aim to achieve an FiO_2 of 1.0.

3 ATTEMPTS AT TREATMENT

A variety of therapeutic modalities have been used to treat CO poisoning. Frequently, combinations of agents have been administered as a treatment "cocktail." Warburg (1926) demonstrated that CO would bind to cytochromes and depress their function in the respiratory chain. In the 1950s, data from several European investigators led to a belief that infusion of cytochromes would stimulate electron transport function (Gros and Leandri, 1956; Stelter, 1953). For a short period of time, infusion of multivitamins and cytochromes was used in clinical treatment (Roche et al., 1968). Other redox-active agents and intermediary metabolites were also purported to be of benefit in CO poisoning, although their use was never broadly recognized. These agents included succinic acid, ascorbic acid, and methylene blue (Bussaberger, 1934; Gershon et al., 1961; Prisco, 1941).

There have been numerous schemes for eliminating COHb from the body. Bentolila (1973) showed that infusion of hydrogen peroxide would hasten dissociation of CO from hemoglobin, although the risk of vascular sclerosis and gas embolism prevented clinical application. Extracorporeal oxygenation and exchange transfusions have been utilized in some clinical cases (Radushevich and Koroteeva, 1976; Yee and Brandon, 1983). However, this invasive methodology probably holds little advantage over providing a patient with supplemental oxygenation except in settings where a patient's vital signs cannot be supported. Others have proposed use of perfluorochemical emulsions, which would circumvent the functional anemia caused by COHb (Geyer and Haggard, 1976; Levine and Tremper, 1985). Use of O_2 mixtures containing 3 to 5% CO_2 (carbogen) was recommended in the 1920s, based on a number of animal and clinical observations, as a method for increasing the ventilatory rate and thus hastening CO dissociation from hemoglobin (Henderson and Haggard, 1920, 1922). This treatment presented the danger of respiratory acidosis and subsequent studies failed to find a difference in recovery rates between dogs treated with 1 ATA O_2 and carbogen (Schwerma et al., 1948; Walton et al., 1926).

The pathophysiological consequences of CO poisoning are numerous. Hypotheses regarding possible mechanisms for the insults to the central nervous system have led investigators to suggest use of adjunctive agents as either prophylactic or therapeutic treatment, such as persantine (Frimmer and Hegner, 1963) and bromocriptine (Starkstein et al., 1989). Cerebral edema is sometimes encountered in CO poisoning victims, and, more generally, may occur following a variety of hypoxic-ischemic encephalopathies. Measures that were formerly thought to have neuroprotective effects in this setting, and were used in CO poisoning, include moderate dehydration, barbiturate sedation, and neuromuscular blockade (Krantz et al., 1988). For a time, hypothermia was also used as an adjunctive measure (Craig et al., 1959; Sluitjer, 1963). However, in a dog model, hypothermia was found to be of no benefit in reducing mortality (Pearce et al., 1972). Impaired temperature regulation, either hyper- or hypothermia, appears to be associated with higher mortality in clinical cases of CO poisoning (Kaltwasser et al., 1977).

4 HYPERBARIC OXYGEN (HBO) TREATMENT

4.1 Carboxyhemoglobin Dissociation

End and Long (1942) showed in animals that exposure to HBO offered a physiological benefit in acute CO poisoning by hastening the dissociation of CO from hemoglobin. They also showed that the decrease in COHb level was correlated with restoration of consciousness. A number of groups have since demonstrated that increased partial pressures of O_2 will hastened COHb dissociation in humans (Britten and Myers, 1985; Pace et al., 1950; Peterson and Stewart, 1970; Weaver et al., 1994). A rather large variation in COHb half-life has been observed both within, and among, research groups for the effect of O_2 at 1 ATA (Table 1). However, all of the measurements of COHb half-life under HBO have been in rather close agreement with the theoretical estimation based on the equation of Coburn et al. (1965).

TABLE 1

Human Carboxyhemoglobin Half-Life Variation with Oxygen

Oxygen Partial Pressures (ATA)	Calculated Values	Peterson[a]	Pace[b]	Britten[c]	Weaver[d]
0.21	252	320	249	—	—
1	35	80	47	131	74
2.5	31	—	22	—	—
3	29	23	—	43	—

Note: Values are half-life in minutes; Column 2 calculated using equation of Coburn et al. (1965) based on following assumptions: Equilibrium constant for reaction between CO and HbO_2 = 218, average partial pressure of O_2 in lung capillaries = 100 mmHg, endogenous CO production = 0.007 ml/min, diffusivity of CO in lung = 30 ml/min. mmHg, alveolar ventilation rate = 4000 ml/min, blood volume = 5500 ml.

[a] Peterson and Stewart, 1970.
[b] Pace et al., 1950.
[c] Britten and Myers, 1985.
[d] Weaver et al., 1994.

4.2 Restoration of Mitochondrial Function

The first controlled animal experiment to investigate the effect of HBO on mortality was done by Pearce et al. (1972), who showed a significant benefit in a dog model. The benefit was thought to be due to a more rapid elimination of COHb, as well as an effect on "some type of tissue poisoning." Approximately 10 to 15% of the total amount of absorbed CO is bound to extracellular heme-containing proteins (Coburn, 1970). Included among this group of proteins are myoglobin, reduced cytochromes, guanylate cyclase and nitric oxide synthase (Brune et al., 1990; Chance et al., 1970; Coburn and Mayers, 1971; White and Marletta, 1992).

With regard to the possible benefits of HBO therapy, some attention has been focused on the question of whether CO may impair the function of cytochrome oxidase *in vivo*, and whether HBO can hasten the dissociation of CO and restore electron transport. Chance et al. (1970) demonstrated that CO could inhibit electron transport in intact mitochondria. Although the affinity of CO for cytochrome oxidase was low relative to the affinity of O_2, they showed that binding by CO could occur when mitochondria became anoxic — a situation that might occur *in vivo* when tissue perfusion was temporarily interrupted. Once CO became bound, a relatively prolonged impairment of oxidative phosphorylation would be expected because the rate of CO dissociation was found to be quite slow (Brown and Piantadosi, 1989; Chance et al., 1970). HBO has been shown to markedly accelerate the dissociation of CO from cytochrome oxidase in an animal model, and therefore improve oxidative metabolism (Brown and Piantadosi, 1989).

4.3 Inhibition of Neutrophil Adherence

A recently identified benefit of HBO pertains to the involvement of polymorphonuclear leukocytes (PMN) in CO-mediated neurological injury (Thom, 1993a). In animal studies, CO poisoning has been shown to trigger PMN to adhere to the brain microvasculature. Proteases from activated PMN convert endothelial xanthine dehydrogenase to xanthine oxidase, which generates oxygen-based free radicals that cause brain lipid peroxidation. HBO prevented PMN adherence to the microvasculature and therefore, dehydrogenase conversion to oxidase and lipid peroxidation (Thom, 1993b). In separate studies with rats poisoned with CO according to the same pattern, Tomaszewki et al. (1992) found that CO poisoning caused an impairment in learning, and that HBO treatment prevented this functional deficit.

The mechanism of action of HBO has been found to be based on its ability to inhibit the function of membrane receptors on PMN called B2 integrins (Thom, 1993b). The function of B2 integrins is to cause an essentially irreversible adherence of activated PMN to the microvasculature. There are a number of receptor–counter-receptor pairs of ligands that coordinate the immunological surveillance function of PMN (von Andrian et al., 1991). Receptors called selectins are constitutively present with PMN and interact with carbohydrate moieties on endothelial cells to slow the passage of PMN in the microcirculation. This interaction does not appear to be inhibited by HBO, nor does HBO inhibit other essential functions such as degranulation, production of oxygen-based free radicals, chemotaxis, or bacterial killing (Gadd et al., 1990; Thom, 1993b; Thom et al., 1986).

The way in which HBO inhibits B2 integrin function, and presumably the basis for its selectivity in not impairing other PMN functions, appears to be the inhibition of a membrane-bound guanylate cyclase in PMN (Chen et al.). Receptor guanylate cyclase enzymes span the cell membrane (Wong and Garbers, 1992) These enzymes bind one or more peptides on the extracellular surface and transmit this information to the intracellular compartment by synthesizing cyclic GMP on the inside surface of the cell membrane. The tertiary structure of the extracellular region of these proteins are maintained by six or more disulfide bridges (Wong and Garbers, 1992). Moreover, the catalytic (intracellular) function also depends on disulfide bridging among guanylate cyclase monomers (Lowe and Fendly, 1992; Wong and Garbers, 1992). HBO inhibits enzyme function by breaking one or more disulfide bonds (Thom et al., 1995; Chen et al., 1996).

4.4 Clinical Use of HBO

The place of HBO therapy in treating acute CO poisoning is not firmly established. It remains a point of contention whether knowledge and experience gleaned from past clinical observations are an adequate justification for treatment, or whether a modern, prospective study on the efficacy of HBO in CO poisoning is required. Clearly, there are prominent clinicians who differ strongly on this point (Myers, 1984; Olson, 1984).

Prior to 1989, all clinical studies of treatment efficacy were retrospective in nature. Most described the experiences of the authors, listed the outcome of patients, and often recommended a particular approach to treatment based on a comparison of clinical results with those recorded in the literature. We recently compiled a list of studies reported in peer-reviewed journals where 50 or more patients were treated (Hardy and Thom, 1994). This list has been reproduced in Table 2.

One of the major shortfalls in clinical evaluations of patients has been the lack of a standard definition for determining the severity of CO poisoning. This problem limits our ability to compare the results of different reports in the literature. It has become clear that the magnitude of the COHb level does not predict clinical outcome (Choi, 1983; Garland and Pearce, 1967; Min, 1986; Myers et al., 1985; Norkool and Kirkpatrick, 1985; Smith and Brandon, 1973). Often, severe exposures have been identified by the occurrence of a transient or persistent alteration in consciousness, or abnormal vital signs (Choi, 1983; Crocker and Walker, 1985; Garland and Pearce, 1967; Min, 1986; Smith and Brandon, 1973). In this regard, therefore, authors who have used a consistent method for estimating poisoning severity, and who have compared outcomes between patients treated with different PaO_2s, or different protocols, have provided the most useful information for estimating the efficacy of HBO (Table 2) (Gorman et al., 1992; Goulon et al., 1969; Myers et al., 1985; Raphael et al., 1989). Factors that may influence the apparent efficacy of HBO therapy include the delay in time from poisoning to treatment, possibly the number of HBO treatments, and also the partial pressure of O_2 that was used. From the perspective of mechanism, and in so far as impairment of PMN B2 integrins may be involved, we have shown that a pressure of 2.0 ATA O_2 only partially inhibits B2 integrin-dependent binding by human PMN, but 2.8 or 3.0 ATA causes virtually complete inhibition that lasts for approximately 8 to 12 h (Thom et al., 1994).

TABLE 2

Outcome of Carbon Monoxide Poisoning from the Literature

Ref.	No. Patients	Mortality (%)	Morbidity (%)	DNS (%)	Treatment
Henderson and Haggard, 1922	85	19	9	n.r.	CO_2/O_2
Shillito et al., 1936	21,143	31	0.3	0.2	n.r.
Simpson, 1963	103	31	33	n.r.	CO_2/O_2 or O_2
Bour et al., 1966	290	26	15	7	O_2
Goulon et al., 1969	147	14	19	0.7	HBO within 6 h
	53	30	21	35.8	O_2 or HBO after 6 h
Lamy and Hauguet, 1969	50	0	8	2	HBO
Smith and Brandon, 1970	206	34	9	1	~50% none, or O_2
Smith and Brandon, 1973	74	14	11	43	44% none, 56% O_2
Choi, 1983	2360	2	8	3	n.r.
Mathieu et al., 1985	230	2	8	4	88% HBO, 12% O_2
Myers et al., 1985	131	2	0	0	HBO
	82	0	0	12	O_2
Norkool and Kirkpatrick, 1985	115	10	4	2.6	HBO
Min, 1986	2967	n.r.	n.r.	3	1% with HBO, 99% n.r.
Krantz et al., 1988	79	30	14	n.r.	O_2
Raphael et al., 1989	127	5	8	47.3	HBO (one tx)
	125	5	11	35.1	HBO (two tx)
Gorman et al., 1992	8	0	63	33	O_2
	32	0	34	10	HBO (one tx)
	68	0	13	0	HBO (> 2 tx)

Note: Reports are listed that describe 50 or more severely poisoned patients. Morbidity includes neurological and cardiac sequelae present during hospitalization. Delayed neuropsychological sequelae (DNS) are listed separately. In addition to supportive care, treatments included ambient pressure 100% O_2 (O_2), carbogen — 5% CO_2, 95% O_2 (CO_2/O_2), and hyperbaric oxygen at 2 to 3 ATA (HBO). n.r. = not reported.

From Hardy, K.R. and Thom, S.R., *Clinical Toxicology,* by courtesy of Marcel Dekker, Inc.

The first clinical use of HBO for CO poisoning was in 1960 (Smith and Sharp, 1960). Goulon et al. (1969) reported a retrospective study that demonstrated a reduction in mortality if HBO was administered within 6 h after discovery of the victim. Mortality was 13.5% if treatment was begun within 6 h and 30.1% if HBO treatment was delayed for more than 6 h.

The efficacy of HBO in diminishing neurological morbidity also appears to be adversely affected by a delay in time from when a patient is removed from the CO environment to the time when treatment is administered. The importance of a delay to treatment was first described by Goulon et al. (1969) (Table 2). A high incidence of delayed neurological sequelae, 47%, was reported in recent prospective, randomized study (Raphael et al., 1989). However, in this study neurological sequelae were based on patient's answers on a self-assessment questionnaire, and not direct examinations, which limits the strength of the study. Assuming the observations were valid, this study is instructive because changes occurred despite use of HBO treatment. The mean time to randomization (and presumably to treatment) in the study was 6 h; thus, the poor patient outcome was consistent with Goulon's (1969) earlier findings and the treatment pressure was 2.0 ATA. We recently reported a prospective, randomized trial in which the incidence of delayed neurological sequelae was compared between patients treated with ambient pressure or HBO (2.8 ATA) (Thom et al.). Sequelae were defined as new symptoms plus a deterioration in scores on an objective psychometric examination. This was a study of patients who might be deemed to have suffered only mild poisoning. Patients had measurable levels of COHb and symptoms of poisoning, but they had not sustained a period of unconsciousness and their vital signs were normal. All patients were treated within 6 h of poisoning. Twenty-three percent (7 of 30 patients) treated with ambient pressure O_2 developed delayed neurological sequelae, whereas none of 30 (0%) ($p<0.05$) treated with HBO developed neurological sequelae.

5 SUMMARY

Beyond emergency stabilization, including administration of 100% O_2 at 1 ATA, the only treatment measure that has been shown to be beneficial in both animal and clinical studies is HBO. There may be several mechanisms of action to explain its benefit. If one considers use of HBO within the limits of 6 h from the time of poisoning, and with use of a partial pressure of 2.8 ATA O_2 or more, it appears to be effective in improving clinical outcome following CO poisoning.

Currently, we make our decisions regarding utilization of HBO based on recommendations published by the Hyperbaric Oxygen Therapy Committee (1992) of the Undersea and Hyperbaric Medical Society. Thus, for the subset of patients at greatest risk, aggressive treatment including HBO is carried out. This group includes patients who are comatose, or who have suffered a brief syncopal episode. Our recent study regarding the benefit of HBO for diminishing the incidence of delayed neurological sequelae was carried out in patients with symptoms of CO poisoning (headache, nausea/vomiting, lethargy, confusion, obtundation, or near-syncope) but without frank loss of consciousness. We believe that it is premature to broaden indications for HBO to these patients until further studies are performed.

Risks from HBO are minimal. Barotrauma can usually be prevented by proper pretreatment screening, cautious compression, and slow decompression. The incidence of central nervous system O_2 toxicity is low and it is estimate to occur in approximately 1 out of 10,000 treatments (Davis et al., 1988). Concerns associated with transferring a critically ill patient to a facility capable of delivering HBO are legitimate, but complications are rare (Sloan et al., 1989). Hence, HBO should not be withheld from an appropriate patient based only on this concern.

Recommendations for treatment of CO-exposed pregnant women have triggered considerable debate, but there is no strong scientific evidence to consider a departure from general treatment recommendations. The obvious concerns with this group of patients are whether HBO may diminish the fetal risk from CO poisoning and whether HBO itself presents a teratogenic risk. Animal studies have demonstrated that hyperoxia can increase the incidence of fetal anomalies or resorption. However, these studies employed O_2 at higher pressures or for longer period of time than treatment protocols currently in use (Ferm, 1964; Telford et al., 1969). Other studies utilizing standard treatment protocols have failed to show an increased incidence of teratogenicity (Cho and Yun, 1982; Gilman et al., 1983). Moreover, CO itself presents a teratogenetic risks (Ginsberg and Myers, 1976; Hemminki and Vineis, 1985; Norman and Halton, 1990). Clinical reports of pregnant women treated with HBO because of CO poisoning have described generally good maternal and fetal outcomes (Brown et al., 1992; Elkharrat et al., 1991; Hollander et al., 1987; Koren et al., 1991; VanHoesen et al., 1989). Perhaps the best counsel for management of the CO-intoxicated pregnant patient, based on current information, is that if clearly indicated HBO should not be withheld from the mother on the basis of concern for hyperoxic effects on the fetus. The mother should be informed that there is increased risk for fetal abnormalities on the basis of the CO exposure, but it is unclear whether HBO treatment will necessarily remove this risk.

REFERENCES

Bentolila, P., Tran, G., and Olive, G., Essai de traitment de l'intoxication oxycarbonee par perfusion de solutions diluees de paroxyde d'hysrogene. Resultats obtenus chez le lapin, *Therapie,* 28, 1943, 1973.

Bernard, C., *Le Cons Sur les Effects des Substances Toxiques et Medicamenteuses,* J. B. Bailliere, Paris, 1857.

Bour, H., Pasquier, P., and Bertrand-Handy, J., Le coma oxycarbone, *Sem. Hop. Paris,* 40, 1839, 1966.

Britten, J.S. and Myers, R.A.M., Effects of hyperbaric treatment on carbon monoxide elimination in humans, *Undersea Biomed. Res.,* 12, 431, 1985.

Brown, D.B., Mueller, G.L., and Golich, F.C., Hyperbaric oxygen treatment for carbon monoxide poisoning in pregnancy: A case report, *Aviat. Space Environ. Med.,* 63, 1011, 1992.

Brown, S.D. and Piantadosi, C.A., Reversal of carbon monoxide-cytochrome C oxidase binding by hyperbaric oxygen *in vivo, Adv. Exp. Biol. Med.,* 248, 747, 1989.

Brune, B., Schmidt, K.U., and Ullrich, V., Activation of soluble guanylate cyclase by carbon monoxide and inhibition by superoxide anion, *Eur. J. Biochem.,* 192, 683, 1990.

Bussaberger, R.A., Glutathione and methylene blue in carbon monoxide poisoning, *Proc. Soc. Exp. Biol. Med.,* 31, 598, 1934.

Chance, B., Erecinska, M., and Wagner, M., Mitochondrial responses to carbon monoxide toxicity, *Ann. N.Y. Acad. Sci.,* 174, 193, 1970.

Chen, W., Banick, P., and Thom, S.R., Membrane-associated guanylate cyclase and B2 integrin function in rat polymorphonuclear leukocytes: Inhibition of enzyme activity and cell adherence by hyperbaric oxygen, *J. Pharm. Exp. Therap.*, in press.

Cho, S.H. and Yun, D.R., The experimental study on the effect of the hyperbaric oxygenation on the pregnancy wastage of the rats in acute carbon monoxide poisoning, *Seoul J. Med.*, 23, 66, 1982.

Choi, I.S., Delayed neurologic sequelae in carbon monoxide intoxication, *Arch. Neurol.*, 40, 433, 1983.

Coburn, R.F., The carbon monoxide body stores, *Ann. N.Y. Acad. Sci.*, 174, 11, 1970.

Coburn, R.F., Forster, R.E., and Kane, P.B., Considerations of the physiological variables that determine the blood carboxyhemoglobin concentration in man, J. Clin. Invest., 44, 1899, 1965.

Coburn, R.F. and Mayers, L.B., Myoglobin O_2 tension determined from measurements of carboxymyoglobin in skeletal muscle, *Am. J. Physiol.*, 220, 66, 1971.

Craig, T.V., Hunt, W., and Atkinson, R., Hypothermia — Its use in severe carbon monoxide poisoning, *N. Engl. J. Med.*, 261, 854, 1959.

Crocker, P.J. and Walker, J.S., Pediatric carbon monoxide toxicity, *J. Emerg. Med.*, 3, 443, 1985.

Davis, J.C., Dunn, J.M., and Heimbach, R.D., Hyperbaric medicine: Patient selection, treatment procedures, and side effects, in *Problem Wounds: Role of Oxygen*, Davis, J.C. and Hunt, T.K., Eds., Elsevier, New York, 1988, Chap. 11.

Elkharrat, D., Raphael, J.C., and Korach, J.M., Acute carbon monoxide intoxication and hyperbaric oxygen in pregnancy, *Intens. Care Med.*, 17, 289, 1991.

End, E. and Long, C.W., Oxygen under pressure in carbon monoxide poisoning, *J. Ind. Hyg. Toxicol.*, 24, 302, 1942.

Ferm, V.H., Teratogenic effects of hyperbaric oxygen, *Proc. Soc. Exp. Biol. Med.*, 116, 975, 1964.

Frimmer, M. and Hegner, D., Beeinflussung der Kohlenmonoxidvergiftung der Maus durch Persantin, *Klin. Wochenschr.*, 41, 1165, 1963.

Gadd, M.A., McClellan, D.S., Neuman, T.S., and Hansbrough, J.F., Effect of hyperbaric oxygen on murine neutrophil and T-lymphocyte functions, *Crit. Care Med.*, 18, 974, 1990.

Garland, H. and Pearce, J., Neurological complications of carbon monoxide poisoning, *Q. J. Med.*, 144, 445, 1967.

Gershon, S., Trethewie, E.R., and Crawford, M., The use of succinic acid in the treatment of acute carbon monoxide poisoning, *Arch. Int. Pharmacodynamic Ther.*, 134, 16, 1961.

Geyer, R.P. and Haggard, H.W., Review of perfluorochemical-type blood substitutes, in *Proc. 10th Int. Cong. Nutrition — Symposium on Perfluorochemical Artificial Blood*, Igakushobe, Osaka, Japan, 1976, 3.

Gilman, S.C., Bradley, M.E., Greene, K.M., and Fischer, G.J., Fetal development: Effects of decompression sickness and treatment, *Aviat. Space Environ. Med.*, 54, 1040, 1983.

Ginsberg, M.D. and Myers, R.E., Fetal brain injury after maternal carbon monoxide intoxication, *Neurology*, 26, 15, 1976.

Gorman, D.F., Clayton, D., Gilligan, J.E., and Webb, R.K., A longitudinal study of 100 consecutive admissions for carbon monoxide poisoning to the Royal Adelaide Hospital, *Anaesth. Intens. Care*, 20, 311, 1992.

Goulon, M., Barois, A., Rapin, M., Nouailhat, F., Grosbuis, S., and Labrousse, J., Intoxication oxycarbonee et anoxie aigue par inhalation de gaz de charbon et d'hydrocarbures, *Ann. Med. Interne (Paris)*, 120, 335, 1969.

Gros, J.F. and Leandri, P., Traitment de l'intoxication oxycarbonee par le cytochrome, *Presse Med.*, 64, 1356, 1956.

Haldane, J., The relation of the action of carbonic oxide to oxygen tension, *J. Physiol. (London)*, 18, 201, 1895.

Hardy, K.R. and Thom, S.R., Pathophysiology and treatment of carbon monoxide poisoning, *Clin. Toxicol.*, 32, 613, 1994.

Hemminki, K. and Vineis, P., Extrapolation of the evidence on teratogenicity of chemicals between humans and experimental animals: Chemicals other than drugs, *Teratog. Carcinog. Mutag.*, 5, 251, 1985.

Henderson, Y. and Haggard, H.W., The elimination of carbon monoxide from the blood after a dangerous degree of asphyxiation, and a therapy for accelerating the elimination, *J. Pharm. Exp. Therap.*, 16, 11, 1920.
Henderson, Y. and Haggard, H.W., The treatment of carbon monoxide asphyxia by means of oxygen CO_2 inhalation, *JAMA*, 79, 1137, 1922.
Hollander, D.I., Nagey, D.A., Welch, R., and Pupkin, M., Hyperbaric oxygen therapy for the treatment of acute carbon monoxide poisoning in pregnancy, *J. Reprod. Med.*, 32, 615, 1987.
Hyperbaric Oxygen Therapy: A Committee Report, Undersea and Hyperbaric Medical Society, Bethesda, MD, 1992.
Kaltwasser, K., Kleinau, H., and Pankow, D., Hypotherme und Kohlenmonoxid Toxitat, *Z. Exp. Chirug.*, 10, 45, 1977.
Koren, G., Sharav, T., Pastuszak, A., Garrettson, L.K., Hill, K., and Samson, I., A multicenter, prospective study of fetal outcome following accidental carbon monoxide poisoning in pregnancy, *Reprod. Toxicol.*, 5, 397, 1991.
Krantz, T., Thisted, B., Strom, J., and Sorensen, M.B., Acute carbon monoxide poisoning, *Acta Anaesthesiol. Scand.*, 32, 278, 1988.
Lamy, M. and Hauguet, M., Fifty patients with carbon monoxide intoxication treated with hyperbaric oxygen therapy, *Acta Anes. Belgica*, 1, 49, 1969.
Levine, E.M. and Tremper, K.K., Perfluorochemical emulsions: Potential clinical uses and new developments, *Int. Anesthesiol. Clin.*, 23, 211, 1985.
Linas, A.J. and Limousin, S., Asphyxie lente et graduelle par le charbon, traitment et gverison par les inspirations d'oxygene, *Bull. Mem. Soc. Therap.*, 2, 32, 1868.
Lowe, D.G. and Fendly, B.M., Human natriuretic peptide receptor-A guanylyl cyclase. Hormone cross-linking and antibody reactivity distinguish receptor glycoforms, *J. Biol. Chem.*, 267, 21691, 1992.
Mathieu, D., Nolf, M., and Durocher, A., Acute carbon monoxide poisoning: Risk of late sequelae and treatment by hyperbaric oxygen, *Clin. Toxicol.*, 23, 315, 1985.
Min, J.K., A brain syndrome associated with delayed neuropsychiatric sequelae following acute carbon monoxide intoxication, *Acta Psychiatr. Scand.*, 73, 80, 1986.
Myers, R.A.M., Carbon monoxide poisoning, *J. Emerg. Med.*, 1, 245, 1984.
Myers, R.A.M., Snyder, S.K., and Emhoff, T.A., Subacute sequelae of carbon monoxide poisoning, *Ann. Emerg. Med.*, 14, 1163, 1985.
Norkool, D.M. and Kirkpatrick, J.N., Treatment of acute carbon monoxide poisoning with hyperbaric oxygen: A review of 115 cases, *Ann. Emerg. Med.*, 14, 1168, 1985.
Norman, C.A. and Halton, D.M., Is carbon monoxide a workplace teratogen? A review and evaluation of the literature, *Ann. Occup. Hyg.*, 34, 335, 1990.
Olson, K.R., Carbon monoxide poisoning: Mechanisms, presentation, and controversies in management, *J. Emerg. Med.*, 1, 233, 1984.
Pace, N., Strajman, E., Walker, E. L., Acceleration of carbon monoxide elimination in man by high pressure oxygen, *Science*, 111, 652, 1950.
Pearce, E.C., Zacharias, A., Alday, J.M., Jr., Hoffman, B.A., and Jacobson, J.H., Carbon monoxide poisoning: Experimental hypothermic and hyperbaric studies, *Surgery*, 72, 229, 1972.
Peterson, J.E. and Stewart, R.D., Absorption and elimination of carbon monoxide by inactive young men, *Arch. Environ. Health*, 21, 165, 1970.
Prisco, L., Sull'azione dell'acido ascorbico nell'intossicazione ossicarbonica, *Bull. Soc. Ital. Biol. Sperim.*, 16, 641, 1941.
Radushevich, V.P. and Koroteeva, E.L., Parallel blood circulation with oxygenation of blood in severe poisoning with carbon monoxide fumes, *Vestn. Khir.*, 116, 131, 1976.
Raphael, J.C., Elkharrat, D., Guincestre, M.C.J., Chastang, C., Vercken, J.B., Chasles, V., and Gajdos, P., Trial of normobaric and hyperbaric oxygen for acute carbon monoxide intoxication, *Lancet*, 2, 414, 1989.
Roche, L., Bertoye, A., and Vincent, P., Comparison de deux groupes de vingt intoxications oxycarbonee traites par oxygene normobare et hyperbare, *Lyon Med.*, 49, 1483, 1968.

Schwerma, H., Wolman, W., Sidwell, A.E., and Ivy, A.C., Elimination of carbon monoxide from the blood of acutely poisoned dogs, *J. Appl. Physiol.,* 1, 350, 1948.

Shillito, F.H., Drinker, C.K., and Shaughnessy, T.J., The problem of nervous and mental sequelae in carbon monoxide poisoning, *JAMA,* 106, 669, 1936.

Simpson, C.A., Carbon monoxide poisoning in New Castle-upon-Tyne, *New Castle Med. J.,* 28, 67, 1963.

Sloan, E.P., Murphy, D.G., Hart, R., Cooper, M.A., Turnbull, T., Barreca, R.S., and Ellerson, B. Complications and protocol considerations in carbon monoxide-poisoned patients who require hyperbaric oxygen therapy: Report from a ten-year experience, *Ann. Emerg. Med.,* 18, 629, 1989.

Sluitjer, M.E., The treatment of carbon monoxide poisoning by administration of oxygen at high pressure, *Proc. R. Soc. Med.,* 56, 1002, 1963.

Smith, G. and Sharp, G.R., Treatment of carbon-monoxide poisoning with oxygen under pressure, *Lancet,* 905, 1960.

Smith, J.S. and Brandon, S., Acute carbon monoxide poisoning: Three years experience in a defined population, *Postgrad. Med. J.,* 46, 65, 1970.

Smith, J.S. and Brandon, S., Morbidity from acute carbon monoxide poisoning at three-year follow-up, *Br. Med. J.,* 1, 318, 1973.

Starkstein, S.E., Berthier, M.L., and Leiguarda, R., Psychic akinesia following bilateral pallidal lesions, *Int. J. Psychiatry Med.,* 19, 155, 1989.

Stelter, R., Zur Theraoie der akuten Kohlenmonoxidvergiftung mit Cytochrome C, *Die Med. WSchr.,* 4, 351, 1953.

Telford, I.R., Miller, P.D., and Hass, G.F., Hyperbaric oxygen causes fetal wastage in rats, *Lancet,* 2, 220, 1969.

Thom, S.R., Leukocytes in carbon monoxide-mediated brain oxidative injury, *Toxicol. Appl. Pharmacol.,* 123, 234, 1993a.

Thom, S.R., Functional inhibition of leukocyte B2 integrins by hyperbaric oxygen in carbon monoxide-mediated brain injury in rats, *Toxicol. Appl. Pharmacol.,* 123, 248, 1993b.

Thom, S.R., Lauermann, M.W., and Hart, G.B., Intermittent hyperbaric oxygen therapy for reduction of mortality in experimental polymicrobial sepsis, *J. Infect. Dis.,* 154, 504, 1986.

Thom, S.R., Mendiguren, I.I., Nebolon, M., Campbell, D., and Kilpatrick, L., Temporary inhibition of human neutrophil B2 integrin function by hyperbaric oxygen (HBO), *Clin. Res.,* 42, 130A, 1994.

Thom, S.R., Taber, R.L., Mendiguren, I.I., Clark, J.M., Hardy, K.R., and Fisher, A.B., Delayed neuropsychological sequelae following carbon monoxide poisoning and its prophylaxis by treatment with hyperbaric oxygen, *Ann. Emerg. Med.,* 25, 474, 1995.

Thom, S.R., Chen, Q., and Banick, P., Hyperbaric oxygen inhibits membrane-associated enzyme functions on polymorphonuclear leukocytes: implications regarding B_2 integrin function, *Undersea & Hyperbaric Med.,* 22, 49, 1995.

Tomaszewski, C., Rudy, J., Wathen, J., Brent, J., Rosenberg, N., and Kulig, K., Prevention of neurologic sequelae from carbon monoxide by hyperbaric oxygen in rats, *Ann. Emerg. Med.,* 21, 631, 1992.

VanHoesen, K.B., Camporesi, E.M., Moon, R.E., Hage, M.L., and Piantadosi, C.A., Should hyperbaric oxygen be used to treat the pregnant patient for acute carbon monoxide poisoning, *JAMA,* 261, 1039, 1989.

von Andrian, U.H., Chambers, J.D., McEvoy, L.M., Bargatze, R.F., Arfors, K.E., and Butcher, E.C., Two-step model of leukocyte-endothelial cell interaction in inflammation: Distinct roles for LECAM-1 and the leukocyte B2 integrins *in vivo, Proc. Natl. Acad. Sci. U.S.A.,* 88, 7538, 1991.

Walton, D.C., Eldridge, W.A., and Allen, M.S., Carbon monoxide poisoning, *Arch. Int. Med.,* 7, 398, 1926.

Warburg, O., Uber die Wirkung des Kohlenoxyds auf den Stoffwechsel der Hefe, *Biochem. Z.,* 177, 471, 1926.

Weaver, L.K., Larson-Lohr, V., Howe, S., Hein, S., Haberstock, D., and Weaver, M., Carboxyhemoglobin (COHb) half-life (t1/2) in carbon monoxide poisoned patients treated with normobaric oxygen (O_2), *Undersea Hyperbaric Med.*, 21, 13, 1994.

White, K.A. and Marletta, M.A., Nitric oxide synthase is a cytochrome P-450 type hemoprotein, *Biochemistry,* 31, 6627, 1992.

Winter, P.M. and Miller, J.N., Carbon monoxide poisoning, *JAMA,* 236, 1502, 1976.

Wong, S.K.F. and Garbers, D.L., Receptor guanylyl cyclases, *J. Clin. Invest.,* 90, 299, 1992.

Yee, L.M. and Brandon, G.K., Successful reversal of presumed carbon monoxide-induced semicoma, *Aviat. Space Environ. Med.,* 54, 641, 1983.

INDEX

D

H

M

N

O

P

Q

R

S

T

U

V

W

X

Z